Physiology of mammals
and other vertebrates

Physiology of mammals and other vertebrates

Second edition

P. T. MARSHALL
Head of Biology
The Leys School
Cambridge

G. M. HUGHES
Professor of Zoology and
Head of Research Unit
for Comparative Animal
Respiration
University of Bristol

CAMBRIDGE UNIVERSITY PRESS
CAMBRIDGE
LONDON · NEW YORK · NEW ROCHELLE
MELBOURNE · SYDNEY

Published by the Press Syndicate of the University of Cambridge
The Pitt Building, Trumpington Street, Cambridge CB2 1RP
32 East 57th Street, New York, NY 10022, USA
296 Beaconsfield Parade, Middle Park, Melbourne 3206, Australia

First published 1965
Reprinted with corrections 1967
First paperback edition 1967
Reprinted 1972
Second edition 1980

Phototypeset in V.I.P. Times by
Western Printing Services Ltd, Bristol
Printed and bound in Great Britain at The Pitman Press, Bath

British Library Cataloguing in Publication Data

Marshall, Peter Treharne
Physiology of mammals and other vertebrates. – 2nd ed.

1. Vertebrates – Physiology
I. Title   II. Hughes, George Morgan
596'.01   QP31.2   79–41436

ISBN 0 521 22633 3 hard covers
ISBN 0 521 29586 6 paperback
First edition ISBN 0 521 05678 0 hard covers
          ISBN 0 521 09451 8 paperback

Contents

Contents

Contents

Contents

Contents

Contents

Contents

Preface to the second edition

Since the first edition was published in 1965 there have been considerable advances in knowledge and understanding of physiology. This edition incorporates new findings, changes of emphasis and new directions in the comparative physiology of mammals and other vertebrates.

Thus while the general aims and organisation of the work remain largely the same as set out in the preface to the first edition, new knowledge and understanding have necessitated a thorough reassessment of the text.

The immediate changes will be seen in the depth of treatment of homeostatic mechanisms and of coordination and in the details of biochemistry and function at the level of the cell. The extensive use of the dogfish and the frog as 'set' types has been changed and much more use is made of comparative data from a wide range of non-mammalian vertebrates. The final chapter on reproduction has been greatly extended.

While the major rewriting of the text has been carried out by Peter Marshall, the co-author, Professor George Hughes, has read and commented on all the new material. For specialised sections we are grateful to Dr Robert Reid of the University of York for his comments on the cell biochemistry, to Dr David Aidley of the University of East Anglia and Dr Ian A. Johnston of the University of St Andrews for their help with the section on muscles, and to Dr Barry Roberts of the Plymouth Laboratory for his further help with the revision of the chapter on nervous coordination. Dr D. Brown of Addenbrooke's Hospital, Cambridge, was of great help in interpreting recent theories relating to immunity. Dr Peter Hogarth of the University of York has also read and made many helpful comments on the whole of the current text.

The checking and editing of this edition have been a formidable task and we are particularly indebted to Mrs Jane Farrell of the Cambridge University Press for her expert work in this respect. Many of the new drawings and diagrams, which form an important feature of the new edition, are the work of John Fuller and to him we also express our thanks.

August 1979 P.T.M.
G.M.H.

xiii

Preface to the first edition

Biology is a very large and varied subject which may be subdivided in many different ways. A common and usual one is to consider living organisms at a series of different levels, beginning with whole populations, then at the individual, organ system, tissue, cellular and molecular levels. Throughout the history of biology there have been changes in the particular level which has received most study and also shifting fashions in the approach to a given or to several levels which were in vogue at that particular time. Often these fashions can be related to developments of new techniques which require the repetition and interpretation of previous work. Some aspects of the biological approach remain constant despite these winds of change and one of these is the relation between structure and function. This relationship can be discussed at all levels of organisation and it is basic to the approach given in this book.

A great deal of this approach tends to be at the organ system level and as such continues to present problems to the biologist, but at the present time there is a great deal of emphasis at a molecular level so that no modern functional approach to the subject would be complete without some inclusion of the biochemistry of cellular activities. In this field we try to present a brief account of the rapidly expanding aspects in the context of more classical biology and to emphasise some of the principal biochemical processes rather than give a detailed account of metabolic pathways. Here, as well as elsewhere in the book, space has not been sufficient to allow a critical approach, and while much of the anatomical and physiological material is now well established the same is not necessarily true of the most recent biochemical work.

Despite the interest and importance of cellular function much of it is hardly suitable for teaching or demonstration to elementary classes and it is the physiological approach to the vertebrates that forms, and is likely to form, the bulk of first courses in animal biology. It is the experience of the authors and many others in teaching biology to sixth-formers and students at university that few recent textbooks have attempted to summarise in an elementary way the vast knowledge gained by mammalian physiologists. Although basically this is a textbook of physiology it differs from most standard texts in that it has not been written primarily for medical students. Because of this, much comparative material, both anatomical and physiological, has been included. Relatively large amounts of anatomical material are included in order to emphasise to the student the importance of considering form and function together and not in isolation from one another. Furthermore, comparative material has been included to show the need for further investigation in this sort of study, both for its own sake and also because of the light it may shed on the functioning of mammals. The value of close understanding between comparative physiologists, mammalian physiologists and clinical physiologists is apparent at the research level at the present time and perhaps, by emphasizing this in the early training of all three types of student, we may hope to encourage such co-operation further.

The presentation of such an integrated approach abounds with problems and we are aware that what is given here contains many faults both in detail and in its general attitudes. It is, however, because we believe there is a great need for integration at this level of teaching that we have thought such an attempt to be worth while. We also know that there are many others who are far more qualified to write a book of this sort than ourselves and hope that if any of them should read our attempt they will be good enough to let us know where they think we have made errors. Some of the information has been presented in a diagrammatic way which has inevitably required a great deal of simplification. We only hope that the simplifications that we have made and the selection of data presented will not give rise to any fundamental misconceptions at this elementary stage of teaching.

In summary then, we hope to have shown the relevance of the study of vertebrates in the A level syllabuses to the potential medical student or biologist. The major object of the book is to present data in a way which will prepare the sixth-former for the type of functional approach he will have at the university.

Because of our awareness of the great breadth of the field that is covered in this book we have sought advice from many people whom we should like to thank. First of all, we should particularly like to thank Dr George Salt for suggesting the cooperation between ourselves, and for his constant advice during the production of this book. We are grateful to Dr W. E. Balfour of the Physiological Laboratory, Cambridge, for reading through the whole typescript. Individual chapters have been read by several of our friends, including that on the endocrine system by the late Dr H. E. Tunnicliffe; that on disease by Dr F. E. Russell; on excretion by Dr A. P. M. Lockwood; and on the nervous system by Dr B. M. H. Bush. Much of the biochemical work was read critically by Dr R. Gregory of the Biochemistry Department. The diagrams of the cell and mitochondrion were devised by Dr A. V. Grimstone. We also wish to thank Mr B. Roberts for his helpful comments on the proof.

Throughout the many problems that have arisen during publication we have had much help from the editorial staff of the Cambridge University Press, to whom we would like to express our thanks.

We believe that an important feature of the book is the original drawings of histological and skeletal material made available by several laboratories, including Anatomy, Physiology and Zoology. The drawings were done by T. W. Armstrong, while still a pupil at The Leys School, and to him we would like to express our thanks.

August 1964   G.M.H.
                                                                P.T.M.

1 Nutrition

1.1 The basic biochemistry of mammalian metabolites

The mammals are heterotrophic, that is to say they derive their essential nutrients, via food chains, from the primary producers, or autotrophs, which alone can fix carbon into organic molecules. The diet that a mammal ingests must include both energy-yielding substances and the nutrients needed for synthesis of its own proteins and other complex molecules. Mammals have restricted biosynthetic capabilities and many compounds, such as certain amino acids (the 'essential' amino acids) and some vitamins, must be taken in the diet in the form in which they will actually be used by the body.

In order to understand the nature of the digestive process, as well as the synthesis of proteins, and the physiological roles of other metabolites it is necessary to have some elementary knowledge of the properties and biochemistry of the substances involved. The classes of chemicals to be discussed are, in order, carbohydrates, lipids, proteins and vitamins.

1.2 Carbohydrates

1.2.1 General features

Among the carbohydrates are the primary products of plant photosynthesis. They are by far the most important energy-providing substrates for mammals. They may also, like the ribose sugars of the nucleic acids, be of structural importance. Most carbohydrates contain only the three elements carbon, hydrogen and oxygen, the ratio of hydrogen to oxygen in the molecule being typically $2:1$. Compared with the lipids, there is a great deal more oxygen in the molecule for each atom of carbon. The general structural formula for carbohydrates is $C_x(H_2O)_y$.

Carbohydrates can be classified according to the number of basic sugar or saccharide units incorporated in the molecule. This saccharide unit is taken as a hexose or a pentose, that is a 6-carbon sugar or a 5-carbon sugar. However, trioses, or 3-carbon sugars, also exist and may be important intermediates in carbohydrate metabolism. A useful scheme of carbohydrate classification is shown in Fig. 1.1 (glycogen is not connected to a disaccharide as it goes directly to glucose on phosphorylation).

Foods with a high carbohydrate content, for example cereals, rice and potatoes, make up the bulk of human diet. As an adaptation to this diet

1

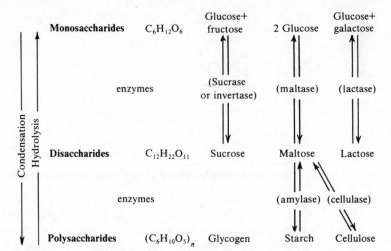

Fig. 1.1.
A scheme of
carbohydrate
classification. (The *n* in
the formula for
polysaccharides may be
some hundreds or
thousands of units.)

man has a collection of carbohydrases (or carbohydrate-splitting enzymes) capable of dealing with starch and the disaccharides maltose, sucrose and lactose. All these are hydrolysed in the alimentary canal to glucose or other monosaccharides which can be assimilated into the blood. Mammalian herbivores have symbiotic organisms in their gut for hydrolysing (the splitting of an organic molecule with the insertion of water) the polysaccharides such as cellulose with which they are unable to deal. (Cellulose in the walls of plant cells makes up the bulk of the herbivore's diet.)

Besides being able to hydrolyse carbohydrates, mammals are also able to synthesise them by condensation reactions involving energy-rich phosphates. (Condensation involves the removal of a water molecule and is the reverse of hydrolysis.) Many of the individual carbohydrates are interconvertible and fats may also be built up from monosaccharides.

On respiration carbohydrates yield 17.2 kilojoules per gram.

1.2.2 *Monosaccharides*

The most common monosaccharide is the hexose glucose. Its empirical formula is $C_6H_{12}O_6$ and the α form of the molecule may be represented as

CH₂OH

α-glucose

when two of the α forms of glucose condense together some form maltose and many of them form starch.

The β form of the molecule is represented by

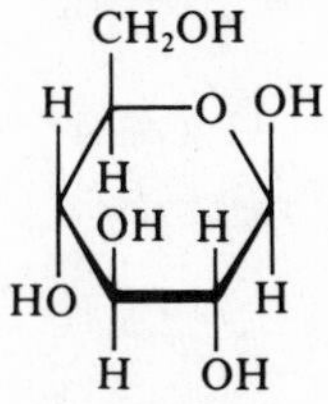

CH₂OH

β-glucose

Two of these molecules condense to the transition product cellobiose, which on further condensation forms cellulose.

A hexose isomer which is produced by the hydrolysis of lactose is galactose.

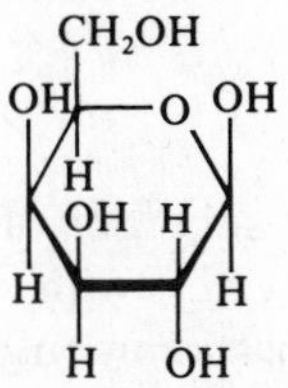

CH₂OH

Galactose

As all the above contain the aldehyde (–CHO) group they are called aldoses. Included in this group is the important structural sugar ribose. Ribose has five carbon atoms in its molecule and, bound to nitrogen bases, it is a constituent of nucleic acids, adenosine triphosphate (ATP) and hydrogen carriers such as nicotinamide adenine dinucleotide (NAD).

Fructose is a hexose that exists in a five-sided furan ring structure (as well as the six-sided pyran ring of glucose).

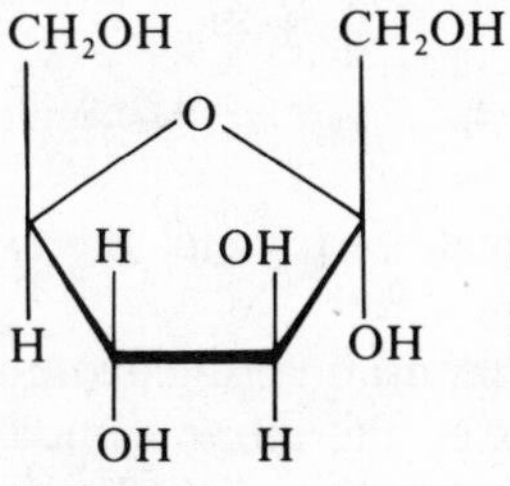

CH₂OH   CH₂OH

Fructose in a
furan ring structure

The empirical formula is $C_6H_{12}O_6$ and in the cyclic form the $>C=O$ group properties are not manifested.

All these monosaccharides can form condensation products by means of glycosidic bonds between particular atoms in adjacent sugars. The molecules thus produced are called glycosides. This bonding is similar to the formation of peptide links between amino acids and, as with these, polymers are built up. Glycosidic bonds may also join monosaccharides to sterols and other chemicals.

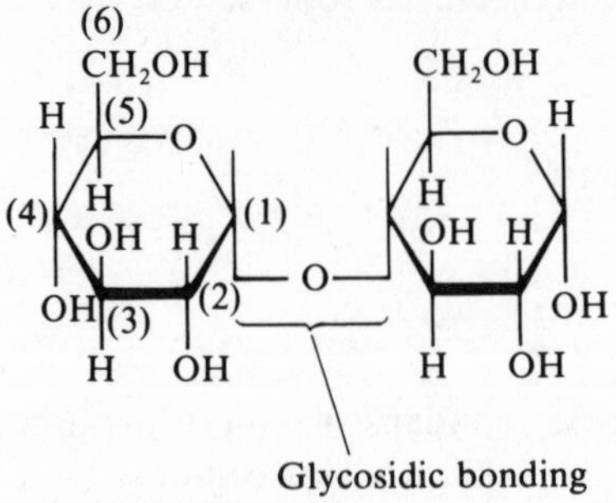

Glycosidic bonding

Monosaccharides may also bond to other molecules via a nitrogen bridge. In this manner ribose bonds to nitrogen bases.

1.2.3 *Disaccharides*

The disaccharides are the most important subgroup of a larger group, the oligosaccharides, which also includes polymers with up to 10 sugar units in their molecules. The disaccharides most commonly met with in mammalian physiology are maltose, sucrose and lactose. Maltose is formed by condensation of two glucose units thus:

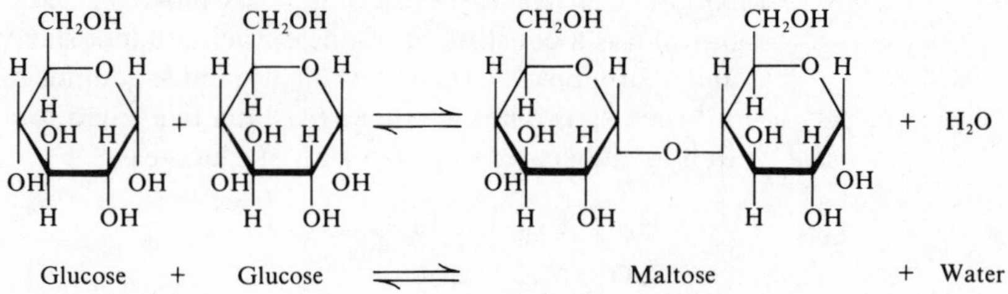

Glucose + Glucose ⇌ Maltose + Water

On hydrolysis by the enzyme maltase it yields the two monosaccharide units again.

Sucrose is formed by the condensation of glucose and fructose and is hydrolysed back to these units by the enzyme sucrase (sucrase is also called invertase because it inverts the optical rotation of the disaccharide).

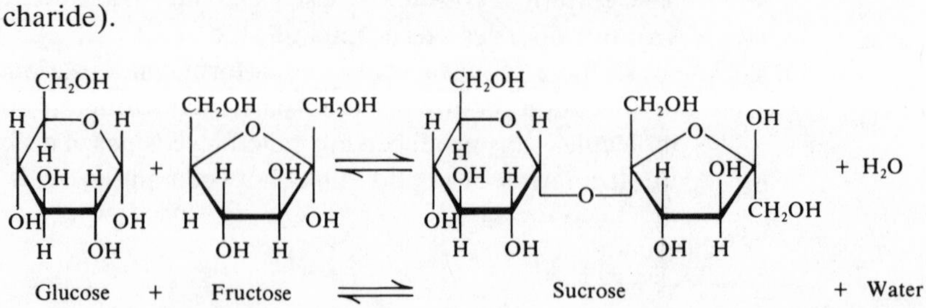

Glucose + Fructose ⇌ Sucrose + Water

4

1.2 Carbohydrates

Lactose, or milk sugar, is formed from glucose and galactose and in turn is hydrolysed by lactase. This enzyme is present in the young of mammals, including man. In some races of man it is not present later in life and ingestion of milk sugar can cause intestinal disturbance.

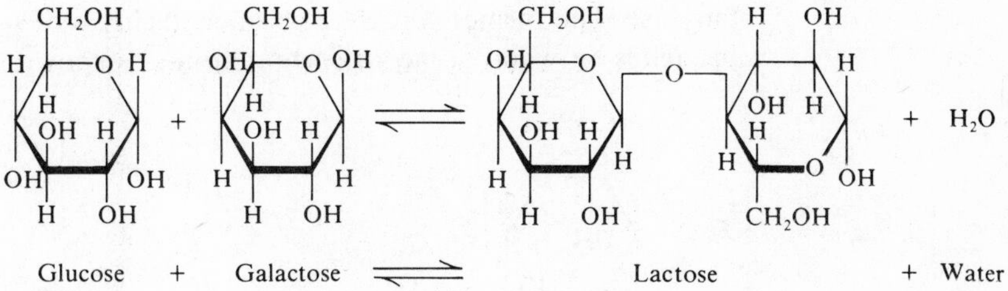

Glucose + Galactose ⇌ Lactose + Water

1.2.4 Polysaccharides

Where 10 or more sugar monomers are linked together the resulting molecule is termed a polysaccharide; when all the monomers are the same it is called a homoglycan polysaccharide and where they are different a heteroglycan polysaccharide. Cellulose is a structural polysaccharide which is synthesised from β-glucose units and is vitally important in plants. It is not found incorporated into animal tissues.

Mammals do not possess cellulase and cannot metabolise the polymer cellulose without the help of symbiotic micro-organisms which are able to hydrolyse the molecule via cellobiose to β-glucose. This latter can be utilised.

Chitin is another structural polymer. It is very important in the composition of the arthropod exoskeleton, though, like cellulose, it is not metabolised by mammals.

Hyaluronic acid is a heteroglycan with the following structure:

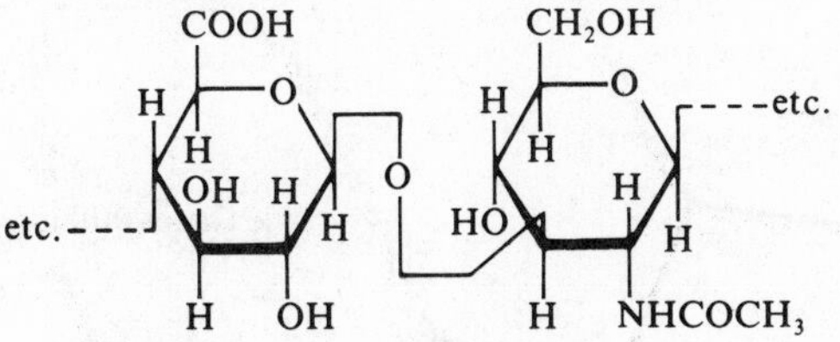

Repeating unit of hyaluronic acid

The chains are extremely long and the molecule has a molecular weight of many millions. It is an exceedingly important component of mammalian connective tissues, those tissues that hold all the other tissues together. It is also found in the vitreous humour of the eye and in the movable joints as part of the synovial fluid. Here it shows the remarkable property of changing its viscosity according to the strength of the shear forces that are acting on it at any given time. This makes it suitable for the lubrication of

5

joints over a wide range of working conditions. Hyaluronic acid is also associated with collagen, a protein which may make up as much as a third of all the protein of the mammalian body.

The food-storage polysaccharides important to mammals are starch and glycogen. Starch is a heteroglycan made up of amylopectin and amylose. The former consists of branched chains of α-glycosides with the branches between adjacent carbon atoms as shown:

CH₂OH

OH H

Amylopectin

H   OH

CH₂   CH₂OH

OH H

H   OH

Amyloses, the other components of starch, are long unbranched chains of α-glucoses held together by glycosidic bonds. Both amylopectins and amyloses are hydrolysed in the mammalian digestive tract.

Glycogen is the typical storage polysaccharide of mammals. It has an interesting structure which, while keeping the glucose in a polymeric form that prevents it from having undesirable osmotic effects, at the same time allows rapid mobilisation. Glycogen has a long 'spine' chain out of which, every fourth or fifth monomer, comes a side chain of some 10 glucose units (Fig. 1.2). The enzyme phosphorylase is able to hydrolyse these side

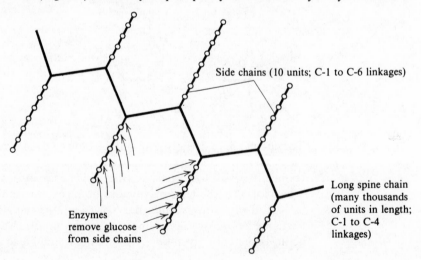

Side chains (10 units; C-1 to C-6 linkages)

Enzymes remove glucose from side chains

Long spine chain (many thousands of units in length; C-1 to C-4 linkages)

Fig. 1.2.
The glycogen molecule.
(After C. F. Phelps.)

6

chains rapidly and thus to free large amounts of glucose as required. It does not split the long spine chain, however, although this can be done by another enzyme. The glycogen can be likened to a tree covered in fruit, the latter being readily removed (and, in the case of the glycogen, also readily returned) while the tree itself remains intact.

1.3 Lipids

The major lipids of the mammalian body are the fats and these are esters of fatty acids with polyhydric alcohols, especially glycerol, and their formation and hydrolysis can be represented as:

$$
\begin{array}{ccccc}
R_1COOH & & HO.CH_2 & & R_1CO.O.CH_2 \\
 & & | & & | \\
R_2COOH & + & HO.CH & \rightleftharpoons & R_2CO.O.CH & + & 3H_2O \\
 & & | & & | \\
R_3COOH & & HO.CH_2 & & R_3CO.O.CH_2
\end{array}
$$

Fatty acids   Glycerol   A fat (a triglyceride)

Here R_1, R_2 and R_3 may be the same fatty acid or different ones. If not all the positions are occupied then this gives monoglycerides or diglycerides.

$$
\begin{array}{cc}
R_1CO.O.CH_2 & R_1CO.O.CH_2 \\
| & | \\
CH.OH & R_2CO.O.CH \\
| & | \\
CH_2OH & CH_2OH
\end{array}
$$

Monoglyceride   Diglyceride

Or if the third position on the glycerol is taken by a phosphate/nitrogen base complex this gives the important class of molecules known as phospholipids (see below).

The fatty acids most commonly involved in fat formation are stearic ($C_{17}H_{35}COOH$), oleic ($C_{17}H_{33}COOH$) and palmitic ($C_{15}H_{31}COOH$) acid. Of these stearic and palmitic are saturated, that is they have no double bonds, while oleic acid is unsaturated. Fats which are solid at room temperatures tend to contain more of the saturated fatty acids, whereas oils, which are liquid at the same temperatures, contain more unsaturated acids.

Weight for weight there is more energy in fats than in other foods. They yield 39 kilojoules per gram on oxidation, which is more than twice the energy yield from carbohydrate or protein. Fats are therefore good forms of energy store and they are found in adipose tissues under the skin and around various internal organs in mammals. In the liver fats are oxidised via the tricarboxylic acid (or Krebs) cycle and for their effective metabolism a supply of carbohydrate is also necessary.

Besides their role in the provision of energy, fats have other essential functions in mammals. They also function as phospholipids which, as mentioned above, are formed when glycerides combine with phosphate and nitrogenous bases. Lecithins are phospholipids where choline is the nitrogen base:

$$R_1CO.O.CH_2$$
$$|$$
A lecithin $R_2CO.O.CH$
$$|$$
$$CH_2O - Phosphate - Choline$$

The phosphate/base end of the molecule is hydrophilic while the fatty acid end is hydrophobic. Such lipids are essential constituents of cell membranes. Besides lecithins, molecules incorporating glycerol, fatty acids and either serine, inositol or ethanolamine are also important. A further class of lipids combines fatty acids with the nitrogenous base sphingosine; these sphingolipids may also be incorporated in cell membranes.

Lipid composition of rat liver cell membranes (as percentage total lipid)

Cholesterol	17
Free fatty acids ⎫ Triglycerides ⎭	10
Lecithins ⎫ Serine, inositol and ⎪ ethanolamine lipids ⎬ Sphingolipids ⎭	60
Others	13

Lipids are also found in mammals as waxes. In these the alcohol component of the lipid is more elaborate than glycerol and may be aromatic in construction.

The final group of lipids is the sterols. These have a ring structure and are often combined with fatty acids. Two important mammalian sterols are cholesterol and testosterone.

Cholesterol   Testosterone

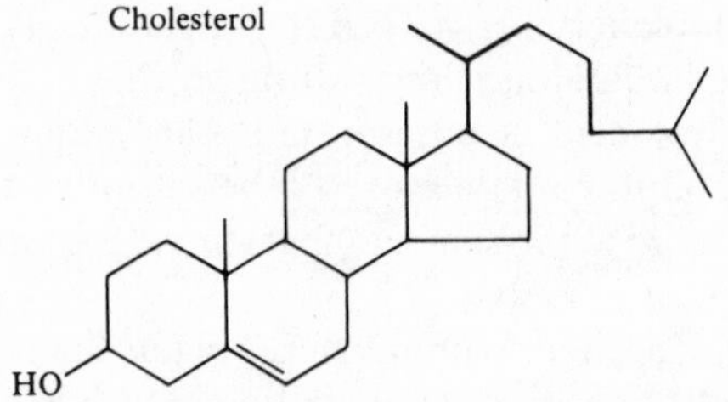

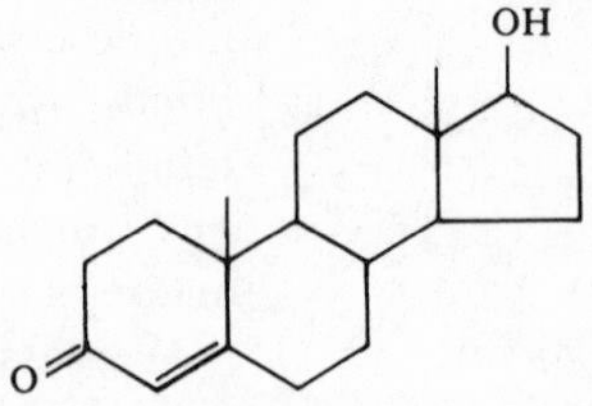

Cholesterol is a constituent of cell membranes. It is also the basic molecule from which various steroid hormones, such as the adrenal cortical hormones, and the gonadal hormones oestrogen and testosterone, are synthesised. Under certain conditions the body can manufacture vitamin D from a cholesterol-type precursor. The recently discovered and rapidly metabolised class of hormones known as prostaglandins also have a sterol-based structure. The role of one of these hormones is indicated on p. 178.

8

1.4 **Proteins**

Proteins are very complex organic molecules containing carbon, hydrogen and oxygen, and less commonly sulphur and phosphorus. All proteins are formed by the condensation of amino acids to give peptide chains. On hydrolysis the peptide linkages in this chain break as shown below:

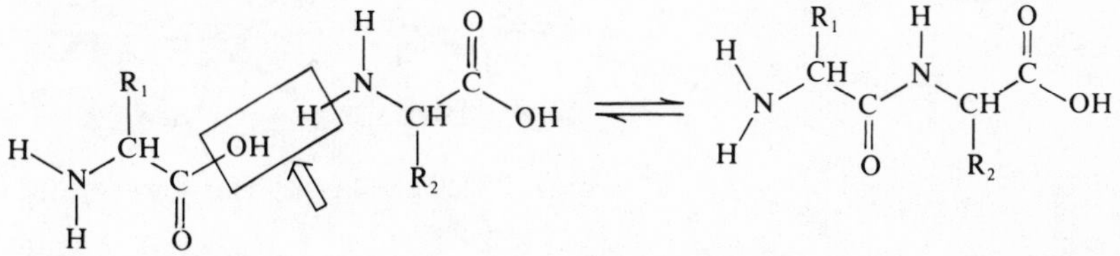

Although some 80 amino acids are known only 22 of these are commonly found in proteins, the great variety of proteins being due to the very large number of possible sequences of amino acids.

A possible classification of the common amino acids gives us six groups:

1. Monoamino (one $-NH_2$ group) and monocarboxylic (one $-COOH$ group)

e.g. glycine	leucine
alanine	isoleucine
valine	

2. Monoamino, monocarboxylic but also include a hydroxyl $(-OH)$ group

   e.g. serine
   threonine

3. Monoamino, dicarboxylic

e.g. aspartic acid	glutamic acid
asparagine	glutamine

4. Diamino, monocarboxylic

   e.g. lysine
   hydroxylysine
   arginine

5. Sulphur-containing

   e.g. cysteine
   cystine

6. Aromatic and cyclic

e.g. phenylalanine	histidine
tyrosine	proline
tryptophan	hydroxyproline

Proteins vary considerably in size. They range from small ones, such as insulin which has some 50 amino acids and a molecular weight of 6,000,

9

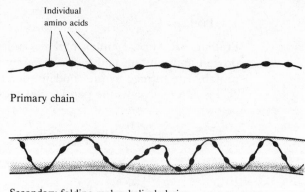

Individual amino acids

Primary chain

Secondary folding makes helical chains

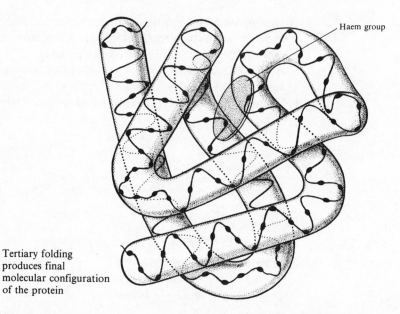

Haem group

Fig. 1.3.
Stages in the formation
of a tertiary protein
such as myoglobin.

Tertiary folding
produces final
molecular configuration
of the protein

through those of intermediate size, such as haemoglobin with a molecular weight of around 66,000, to the largest, such as fibrinogen with a molecular weight of nearly half a million. Their structures are further complicated by the fact that the primary chains of amino acids become folded into helices or lattices (secondary structure) and that these in turn take up tertiary arrangements. Fig. 1.3 shows these stages in the formation of a myoglobin molecule. Quaternary structure results from the joining of two, three or more molecules to form dimers, trimers or even larger conglomerates, respectively.

The various foldings of proteins are stabilised by the formation of cross-linkages between adjacent amino acid chains. These may be strong covalent links, as with the –S–S– bonds between cysteines, or may be weaker hydrogen bonding. Besides the amino acid sequences proteins also often include cofactors, which may be metals or vitamins. The whole complex arrangement of the final molecule may be designed to set the cofactor within a certain three-dimensional environment where it can be reactive in some metabolic function. This is the case with haemoglobin (see p. 77) where the Fe^{2+} ion has special properties *only within the precise protein configuration*.

On the whole, tertiary-structured proteins have the hydrophobic amino acids towards the centre and the hydrophilic ones facing outwards. The process of denaturing proteins involves the breaking of the cross-linkages between chains so that the functional tertiary structure is also broken up and the molecule no longer operates effectively.

1.4.1 *Major classes of proteins in the mammal*

Life processes depend essentially on the properties of proteins, and these have many separate roles in the physiology of mammals. One category of proteins is those that provide food for the young animal in the form of the casein of milk. This protein has a wide variety of amino acids strung together in such a way that the peptide linkages are freely exposed for ease of hydrolysis by the digestive enzymes.

Another important class of proteins is those involved in structure and locomotion. These have properties appropriate to their various roles. For example keratin of hair is structured in helices which allow flexibility with strength, while the collagen of connective tissue has units with covalent bonding and is thus strong and non-elastic. The soluble fibrinogen in the blood is denatured by thrombin at the site of a wound. This releases side links from neighbouring fibrinogen molecules so that large clots are formed, excessive blood loss prevented. The muscle proteins actin and myosin are fibrillar contractile molecules which also have the property of catalysing the breakdown of ATP to provide energy for this contraction and its reversal.

Yet other proteins are involved in the transport of oxygen and hydrogen, with which they form reversible bonds. For example haemoglobin bonds with oxygen, and cytochrome is a hydrogen carrier.

Immunoglobulins have a role in the body's defences. They remove circulating antigen molecules (foreign proteins, viruses, etc.) by bonding with them to form stable and inert complexes.

In a broad sense enzymes, and some hormones, are also proteins specialised in bonding, only here the bonds are such that they allow metabolic activities to occur. Enzymes bond temporarily with substrates and thus lower the activation energy for chemical reactions, while protein hormones are able to activate genes or prevent them from acting, or to affect other operations in the cell.

Many enzymes only work when a non-protein or prosthetic group is present and attached to the protein, or apoenzyme, part of the molecule (see also Chapter 2, p. 26).

1.4.2 *Nucleoproteins*

The nucleic acid DNA (deoxyribonucleic acid) is associated with proteins such as histones to form nucleoproteins. The histones make up a substantial part of the chromosomes and may play some part in determining which genes are operating at a given time. The nucleic acids themselves are not proteins.

1.4.3 *Proteins as part of the diet*

Proteins make up part of the mammalian diet, and their intake and digestion fulfils two separate purposes. Firstly an excess of proteins can be broken down and metabolised to yield energy or else changed into compounds that store energy, and secondly a certain amount of protein is essential for growth and repair.

As already described, the proteins of the mammalian body are composed of various combinations of some 22 amino acids. Some of these amino acids can be synthesised within the body from others or built up from simpler substances, but in man 10 have to be taken in with the diet. These are threonine, valine, leucine, isoleucine, phenylalanine, methionine, tryptophan, lysine, histidine and arginine, and are called the essential amino acids. Proteins which contain most of these essential amino acids are termed first class while those that do not are called second class. Animal proteins tend to come in the first category and plant proteins into the latter. Essential amino acids are used directly in the synthesis of the body's proteins. Fig. 1.4 shows the composition of insulin.

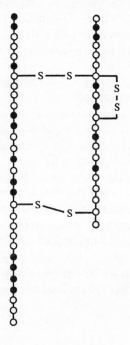

Fig. 1.4.
The amino acid composition of the insulin molecule, showing the proportion of essential amino acids (solid circles) and non-essential amino acids (open circles).

It is made from 52 amino acids of which rather less than half are essential.

In the normal diet much of the protein will necessarily be second class and a much greater quantity of amino acids will be taken in than is required for synthesis. The excess amino acids are deaminated in the liver, a process which involves removal of their nitrogen content. At the same time sulphur and phosphorus are removed and the resulting carbohydrate molecule can then be stored or oxidised to yield energy. On oxidation proteins have similar yields of energy to carbohydrates.

Probably in man, and certainly in ruminant mammals (an animal that can return its swallowed food from a rumen or pouch to its buccal cavity for further mastication by the teeth), the intestinal flora are able to synthesise both essential amino acids and vitamins. These powers may be of considerable importance in the nutrition of the animal; for example, the nitrogen requirement of a sheep can be satisfied by supplying it with urea $(CO(NH_2)_2)$. It is presumed that the micro-organisms in the gut are able to build up the necessary compounds from this simple source.

1.5 Vitamins

Vitamins are substances of which small quantities are required to maintain the health of the organism. It was originally thought that vitamins were amines, but it is now known that they include a number of other types of organic substance.

The role of vitamins in the cell is different from that of essential amino acids, for whereas the latter are concerned in synthesis, the vitamins mostly act as coenzymes directly facilitating metabolic processes. Vitamins are incorporated into the cell's machinery and there help to bring about a given step in respiration, growth or synthesis, but do not themselves become incorporated into protoplasmic substances.

Vitamins are named after letters of the alphabet, but it has been necessary to subdivide the groups and add further ones as new substances are discovered which operate as vitamins. As with the essential amino acids described, the abilities of individual mammal species to synthesise vitamins – and hence their vitamin requirements – vary.

1.5.1 *Vitamin A*

Vitamin A is related to β-carotene (one of the photosynthetic pigments of plants), but the latter is not so well assimilated from the gut as is the pure vitamin. Rich sources of β-carotene are certain green vegetables, tomatoes, carrots, etc. The vitamin itself is found in animal fats and oils, especially those of fish livers, and also in egg yolk, milk and butter fat. In mammals it can be stored in the liver, and in man some 95 % of the vitamin is found in this organ. Most mammals have limited powers of synthesis of the vitamin where a supply of β-carotene is available, and the amounts stored vary greatly from one species to another. Marine mammals store so much vitamin A in their livers that these organs may be toxic

13

to other animals. Man requires on average 1.7 milligrams per day of the vitamin and rather more during growth, pregnancy and lactation.

Vitamin A exists chemically in two forms, A_1 and A_2, which have slight differences at the cyclic end of the chain. It is an unsaturated monohydric alcohol and the A_1 form found in mammals and most vertebrates has the following structure:

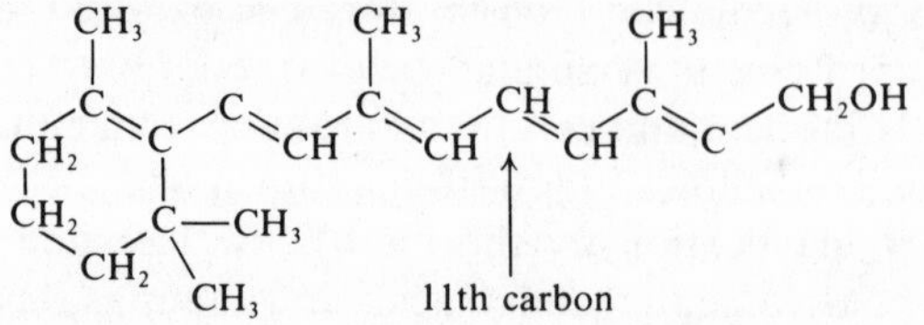

Vitamin A_1, *trans*-form

11th carbon

This vitamin plays an essential part in vision, combining with various opsins in both the rod cells and cone cells in the eye to produce a variety of visual pigments with different peaks of absorption of light. In the rods of mammals the long straight end of the vitamin A chain first becomes an aldehyde, by dehydrogenation of the hydroxyl group on the 15th carbon atom. This chemical is *trans*-retinene, which undergoes an isomeric change whereby the chain becomes bent at the 11th carbon atom giving *cis*-retinene. The *cis*-retinene combines with opsin to give the visual pigment rhodopsin, and this in turn is broken down in the presence of light to give *trans*-retinene and opsin once more.

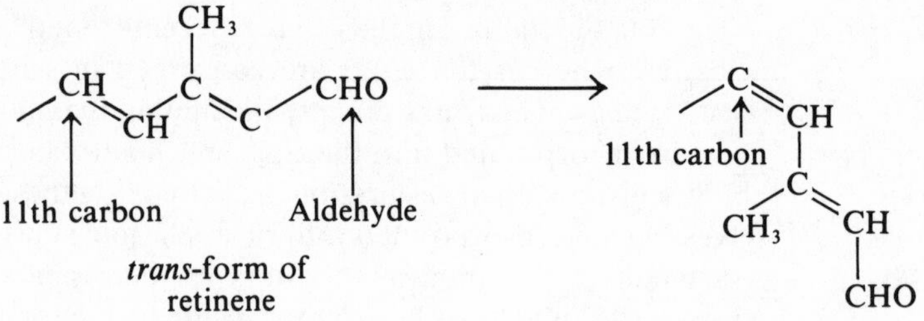

11th carbon   Aldehyde

trans-form of retinene

11th carbon

cis-form of retinene

The relations are thus:

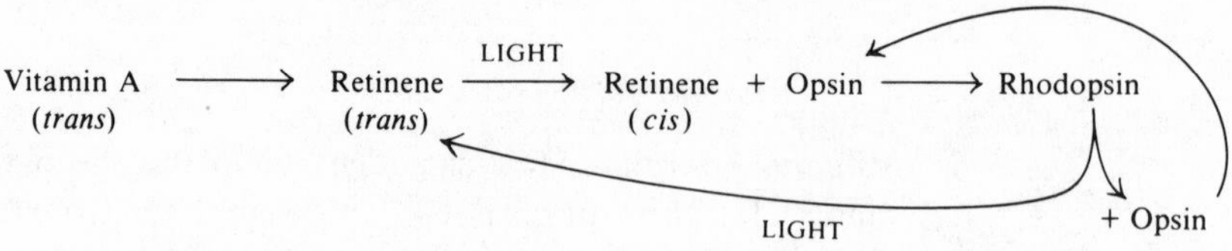

Vitamin A (*trans*) ⟶ Retinene (*trans*) ⟶ LIGHT ⟶ Retinene (*cis*) + Opsin ⟶ Rhodopsin ⟶ LIGHT + Opsin

A deficiency of the vitamin in the very early post-natal development of humans can lead to permanent blindness, and later in life a deficiency will produce poor vision in dim light, a condition described as night blindness.

Besides its role in vision, vitamin A plays a vital part in the normal laying down and functioning of epithelia. While its precise metabolic contribution is not certain, experiments suggest that the lipid–protein membrane complex of the epithelial cells is stabilised by the vitamin.

1.5 Vitamins

Where it is lacking, malformed and keratinised epithelia result, and should this occur in the cornea of the eye a disease called xerophthalmia is produced such that the cornea becomes cloudy and loses transparency. Generally, ill-formed epithelia in the skin and lining the tracts of the intestine and lung lead to susceptibility to infection by pathogens.

A final important (though again not fully understood) role of vitamin A is its function in normal growth. The vitamin promotes the build-up of proteins in young animals and without it they fail to develop. It was this particular function of the vitamin that was the one first noted by Hopkins in his pioneering experiments on vitamins using batches of young rats.

1.5.2 Vitamin B

The vitamin B group comprises a very heterogeneous collection of chemicals, which have little in common except solubility in water and perhaps similarity of sources. There are at least 12 substances classed as B vitamins and of these man needs at least 10 to maintain normal function in the body. Some of these chemicals are readily synthesised by the symbiotic bacterial flora of the mammalian rumen and intestine.

B1: thiamine
This consists of a pyrimidine linked to a sulphur-containing thiazole unit. It cannot be synthesised by man but is manufactured by his gut flora.

Vitamin B1

Thiamine acts as a vital coenzyme in the decarboxylation of (i.e. removal of carbon dioxide from) pyruvic acid:

$$CH_3CO.COOH \rightarrow CH_3CHO + CO_2$$

Pyruvic acid   acetaldehyde

The acetyl group passes via coenzyme A into the tricarboxylic acid cycle which is an essential stage in aerobic respiration. If the thiamine is not present then pyruvic acid and other toxic metabolites accumulate and upset membrane functioning so that paralysis may result. The body's fluid balance also becomes disturbed so that emaciation or oedema can occur, and commonly the heart and circulation are affected. All these symptoms add up to the deficiency disease beriberi which, untreated, is fatal.

Thiamine is also important in fat metabolism, where long-chain fatty acids are broken down and decarboxylated little by little before final respiration. Some of the unpleasant effects described above would be due to accumulation of ketoacids from partly metabolised fats.

Man requires some 0.5 milligrams thiamine for every 4,200 joules of

15

diet consumed, so it can be seen that this is the amount of the vitamin required during the respiration of this amount of substrates by the body.

B2: riboflavine

This is a yellow chemical with the following structural formula:

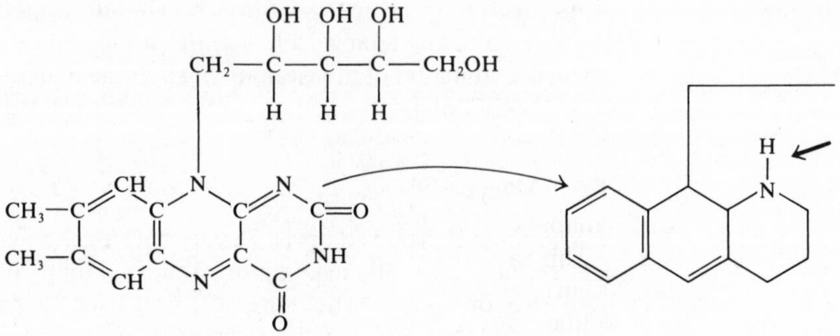

Oxidised flavine

Reduced flavine carrying hydrogen as indicated

In cells riboflavine acts as a coenzyme in the form of flavine adenine nucleotide (FAD). It is a hydrogen acceptor from the tricarboxylic acid cycle and passes the hydrogen on to the cytochromes.

Deficiency of riboflavine causes lesions of the skin and shedding of hairs in mammals, as well as an invasion of the cornea by blood capillaries, but as with some other B vitamins this molecule can be readily synthesised by micro-organisms in the gut. The human dietary requirement is about the same as for vitamin B1.

B3: nicotinic acid or nicotinamide

This acts in a similar way to riboflavine, forming the coenzyme nicotinamide adenine dinucleotide (NAD) or its phosphate (NADP). It is a hydrogen acceptor essential in the anaerobic stage of respiration, in the tricarboxylic acid cycle, and in the final chain of the aerobic pathways.

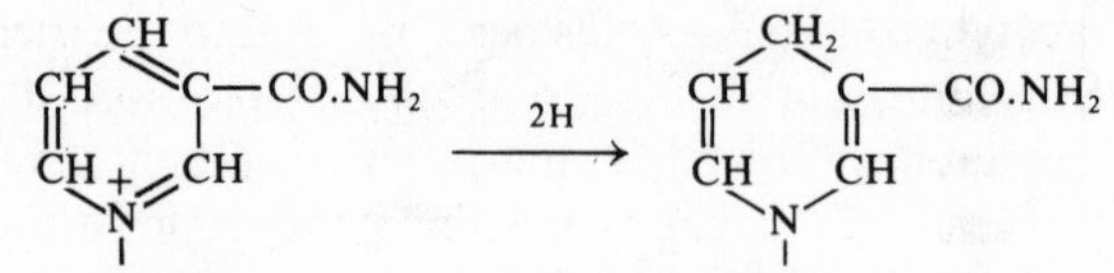

Nicotinamide

Reduced form of nicotinamide

(The other H attaches to one of the oxygen atoms of the phosphate group adjacent to the pyridine ring shown)

A lack of this vitamin causes the deficiency disease pellagra which is characterised by diarrhoea, dermatitis and dementia. As with beriberi the condition would be fatal in an extreme form. The symptoms indicate general breakdown in membrane functioning. The disease was first identified in maize-eating areas of the United States; maize is deficient in the essential amino acid tryptophan from which the body is able to synthesise its own nicotinamide.

Pantothenic acid
This is a part of coenzyme A whose molecule is made up of

Thiol ester – Pantothenic acid – 3 phosphates – Adenylic acid
  (HS–)

The enzyme combines with acetyl ($-CO.CH_3$) groups and transfers them from one substrate to another. It is the means whereby acetyl groups are fed into the tricarboxylic acid cycle from both anaerobic glycolysis and from fat respiration. It is also vital in the synthesis of acetylcholine, the synaptic transmitter made from a combination of acetic acid and choline.

    This member of the vitamin B group is seldom deficient, being of wide occurrence in the diet as well as being synthesised in the colon. If it is deliberately withheld from the diet of rats, upset in adrenal function takes place.

Biotin
This is a sulphur-containing coenzyme which is important in growth. It allows transfer of carboxyl (COO–) groups from one molecule to another and is involved in numerous biosyntheses within the mammal.

B6: pyridoxine
This has the chemical structure

Pyridoxine

It is a vital coenzyme in transamination of amino acids which are necessary in protein metabolism. For example:

| Glutamic acid | Oxaloacetic acid | α-ketoglutaric acid | Aspartic acid |

If the vitamin is absent then amino acids that would have been transaminated into more useful forms become, instead, deaminated, and much more urea is consequently lost from the body than normal. A deficiency produces poor growth, anaemia and dermatitis.

Nutrition

Folic acid

This is a complicated molecule based on glutamic acid. It is a coenzyme in the transfer of single-carbon fractions from one substance to another and is important in the synthesis of purines and pyrimidines, as well as some amino acids. Normally made in quantity by intestinal micro-organisms, this vitamin is seldom deficient, but in certain disorders of the lower gut deficiency can occur. A symptom of such deficiency in man is anaemia which, where due to this particular cause, can be treated by administration of the vitamin. Folic acid is found in green leaves and in kidney and liver.

B12: cobalamin (cyanocobalamin)

This is a cobalt-containing complex which is present in very minute amounts in the liver. The vitamin is among the most active known and seems to take part in a number of widely different transfer reactions, including transaminations and transmethylations. It is necessary in nucleoprotein and muscle metabolism.

In the disease pernicious anaemia a substance known as the intrinsic

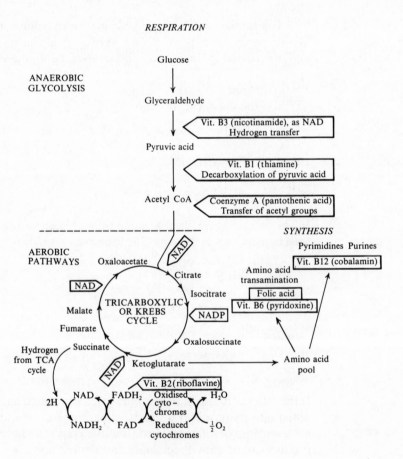

Fig. 1.5.
Summary of action in metabolism of the major constituents of the vitamin B group. (The skeleton pathways of metabolism used in this diagram will be discussed in more detail in the final part of Chapter 4 (p. 97).)

18

factor of Castle ceases to operate in the gastric mucosa. This intrinsic factor allows the take up of vitamin B12 (also called the extrinsic factor of Castle) from the diet. Without the vitamin, red corpuscles fail to mature properly and those that are made are large, fragile and low in haemoglobin, so that the patient shows acute anaemia.

1.5.3 *Vitamin C: ascorbic acid*

Ascorbic acid has the structure

```
    O=C
        |
HO — C
        ‖      O
HO — C
        |
 H — C
        |
HO — C — H
        |           Vitamin C
     CH₂OH
```

It can be very readily reduced and oxidised again, and in the cell operates joined to a copper-containing protein. Among other functions the vitamin is known to be necessary for the activation of folic acid (see above) and for the metabolism of aromatic amino acids.

A deficiency of vitamin C in man causes scurvy, which is characterised by haemorrhages. It has been suggested that the vitamin is necessary for synthesis of the 'cement' whereby individual cells are held together and that lack of it causes permeability of capillaries and other tissues. The aromatic amino acid tyrosine is involved in melanin pigment manufacture in the body and disturbances in normal pigmentation are another symptom of lack of vitamin C. It is also known to be present in high concentrations in the adrenal cortex and has been associated with production of the cortical hormones which are involved in the body's response to stress and infection.

Ascorbic acid is water-soluble and is destroyed on heating. Animal foods such as milk are low in the vitamin but it is present in fresh fruit and green vegetables. The normal human diet will contain an excess of the vitamin, which is commonly excreted unchanged in the urine.

1.5.4 *Vitamin D: calciferol*

Vitamin D is a class of substances all of which have similar activity. The ones found in man are D2 and D3 and both are sterol-like molecules which can be synthesised from the fat cholesterol. In the intestine cholesterol is dehydrogenated to dehydrocholesterol and in the skin this is changed to vitamin D by the action of the ultra-violet light from the sun. In addition to the body being able to manufacture the vitamin it can be obtained from meat, liver and milk in the diet.

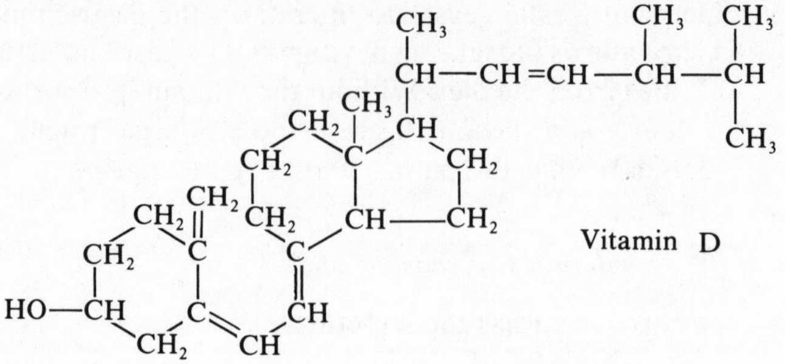

The chemical structure diagram shows:

$$CH_3$$
$$|$$
$$CH—CH=CH—CH—CH$$

with associated CH_3 groups and the label **Vitamin D**

The vitamin allows the absorption of calcium from the gut, that can then be used in blood, bone and muscle. The actual levels of calcium in the body are controlled by the hormone parathormone and if the gut absorption decreases then Ca^{2+} is released from the bones and muscles. This will eventually lead to demineralisation of the bones and to muscular paralysis. The former is a symptom of rickets, the disease due to a deficiency of vitamin D.

1.5.5 *Other vitamins*

A number of other vitamins have been described, some in particular vertebrates, some in mammals. Among those important to man are vitamin K which, by regulating the amount of prothrombin in the blood, affects its clotting, and vitamin E, concerned with reproductive processes. Lack of this vitamin may result in diverse effects in different species, ranging from sterility in both male and female rats to muscular dystrophy in rabbits.

Certain chemicals, related to the vitamins in structure, act as anti-vitamins, replacing the vitamin and disorganising the enzymic processes in which it is involved. These anti-vitamins may provide powerful anti-biotics (see p. 151).

In general, and as with the essential amino acids, it is supposed that during evolution the mammals, and, of course, other animals, have lost many of the powers of synthesis that their ancestral organisms had. Thus the essential amino acids and vitamins, which are indispensable to life processes, must be taken in with the diet. This intake may be greatly supplemented, however, by the activity of the micro-organisms of the alimentary tract.

1.6 **Mineral salts**

A large number of different minerals are required to fulfil structural and other functions in the body. These minerals may either be used as ions, as is the case with sodium and potassium, or they may be built up into complex organic substances as with iodine and cobalt.

The important minerals and some of their uses in the body are:

Calcium	bone, teeth, blood, muscle contraction
Chlorine	stomach, blood
Phosphorus	nerve, muscle, bone
Iron	haemoglobin (incorporated into molecule), cytochrome
Sulphur	hair, nails
Fluorine	teeth
Cobalt	bone, blood
Zinc	blood, enzymes

Many salts are present as ions in cytoplasm and extracellular fluids and their presence is essential for cell function, and for the operation of nervous and muscular tissues. Mineral salts are taken in with the diet, a deficiency of a certain salt sometimes leading to a craving for a food rich in the mineral. Under certain conditions it is possible to suffer from a deficiency disease due to lack of specific salts. Two examples of such diseases are the simple goitre resulting from lack of iodine, and dental caries, which can attack teeth deficient in traces of fluorine. However, it should be noted that a disease such as rickets, which involves shortage of Ca^{2+} in the body, is due to a vitamin deficiency and not to mineral shortage.

Though each mineral assimilated from the food may have a specific role in metabolism it is the concentration of salts that largely determines the osmotic pressure of the body fluids. For normal functioning of the body tissues the osmotic pressure must not vary by more than the equivalent of 0.1 gram NaCl in 100 millilitres of water. Body mechanisms dealing with correction of osmotic pressure are sensitive to much smaller changes than this. Ionic regulation, depending on both intake via the gut and excretion through the kidneys, ensures, as far as possible, that the salt concentration remains constantly within these limits.

Where excessive loss of salt takes place, as in heavy manual work, it is found beneficial to drink dilute sodium chloride solution. Certain salts, such as magnesium sulphate, which are very slowly assimilated, tend to hold back water in the gut and, by stimulating peristalsis, act as laxatives.

1.7 **Water**

Over and above the fuel and structural items of diet an intake of water is needed to complete the body's requirements. Though part of the body's water requirements is produced by metabolic processes, it must be regarded as a food, as without it the cell, and thus the whole organism, is soon unable to function.

Protoplasm is some 70–80 % water and all its reactions take place in solution. Water is the universal solvent in which the other constituents of living matter can remain either suspended or in solution. Only dormant seeds and spores of bacteria and other organisms contain less than 70 % water. In these the reactions within the dehydrated protoplasm are almost at a standstill.

Mammals are some 70 % water and to keep the degree of hydration necessary for the continuance of life they must replace the water lost by evaporation and excretion. A loss of more than 10 % body water is serious, and loss of more than 20 % leads to a thickening of the blood and stoppage of the circulation. The quantity of water passing through the body each day is some 3.5 litres, though this may be doubled where excessive sweating takes place. The way these 3.5 litres are made up depends largely on environmental factors, but, on average, some 500 millilitres water per day are lost via the lungs, some 2 litres via the kidney and in faeces and the remaining litre by evaporation from the skin.

Although the amount of sweat and urine lost may vary greatly the latter must not become too concentrated as the nitrogenous waste products it contains are toxic in high concentrations.

Of the quantity of water needed daily some half may be taken in as 'hidden' water in the food. Nevertheless, nearly all foods, except soft fruits, require additional water to balance the heat produced in their combustion or excretion.

1.8 Summary

The mammalian body is a living machine which requires the provision of fuel and also of the structural units which maintain and build up the machine itself. These two types of item comprise the balanced diet necessary for health.

The detail of a particular mammal's food will depend on the nature of its specific adaptations, that is, whether it is carnivorous, herbivorous or omnivorous. In their alimentary canals and feeding habits the mammals show great variation and all but a few are fully specialised for the obtaining and utilising of some major food supply – meat, grass, fruit, insects, etc. These adaptations, in so far as they concern the teeth and gut, will be discussed in Chapter 3.

Among the various types of mammals a good deal of difference is found in dietary requirements. On the whole carnivores, living on animal proto- plasm, are well supplied with the ingredients for their own synthesis, while herbivores depend on the micro-organisms which inhabit their alimentary canals for a supply of vitamins and amino acids. The synthetic powers of mammals are not very great compared with those of lower organisms and, by one means or another, it is necessary for them to select or provide themselves with the numerous constituents of their balanced diets. The nutrition of these animals illustrates the biological dictum that 'increase of morphological complexity is often accompanied by decreas- ing powers of biochemical synthesis'.

2 Enzymes

2.1 **General properties**

The functioning of the animal body at the tissue and organ level that we term physiology must obviously reflect chemical activities at the level of the cell. Thus major processes such as locomotion or digestion are seen ultimately to depend on the specialised chemistry of cells, and no depth of physiological understanding is possible without some knowledge of cellular physiology and its biochemical basis.

While the previous chapter was concerned with the basic chemicals that make up the animal body, the present one will examine those biological catalysts, the enzymes, whose special properties allow these chemicals to react together.

Enzymes are proteins, and this idea can be extended further in the notion that most proteins show enzymic properties. The essential work of enzymes is to lower the activation energy for a particular chemical process so that it will take place rapidly at the relatively low temperature of the living cell. To visualise how the enzyme works, the analogy of a boulder on a hillside is commonly taken and has much to commend it. In Fig. 2.1(*a*) the boulder represents a molecule which has a certain potential energy. If we wished to release this energy, in other words to get the molecule to decompose into smaller products, it would be necessary first to supply energy to push it over the hump or energy barrier. Once over the top of this it would run downhill to a lower energy state. What an enzyme does is effectively to tunnel through the energy barrier so that, with minimal supplies of energy to the system, reaction will occur. In Fig. 2.1(*b*) the boulder can be seen to be free to run down the hill via the pathway made by the enzyme.

Fig. 2.1.
An analogy to illustrate the method of functioning of enzymes.

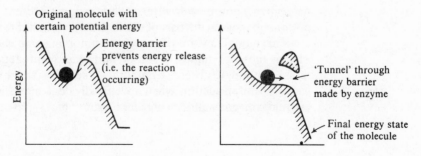

(*a*) No enzyme present   (*b*) Enzyme present

Original molecule with certain potential energy

Energy barrier prevents energy release (i.e. the reaction occurring)

'Tunnel' through energy barrier made by enzyme

Final energy state of the molecule

Energy

23

Enzymes are much more effective than inorganic catalysts; thus in the hydrolysis of sucrose to glucose and fructose, addition of molar hydrochloric acid and a reaction temperature of 100 °C has the same effect as 0.00001 molar sucrase at 15 °C!

Over 100 different enzymes have been obtained in pure crystalline form, but it is only for a few that actual details of primary and tertiary structures are known. It is estimated that something of the order of 1,000 different enzymes are present in a living cell at any one time.

A feature of enzyme reactions is the speed at which they occur. Where one substrate (substance on which enzymes act) will react to give decomposition products, or where two substrates react together, then the speeds of these reactions may be increased by a factor of 10 million times in the presence of the appropriate enzyme. The fastest-acting enzymes, such as pepsin, are able to turn over some 10^6 molecules of substrate in a minute and it has been pointed out that a normal meal would take around 50 years to digest if unassisted by enzyme activity.

Enzymes usually work only on one type of substrate – that is they exhibit specificity. They also require specific concentrations of hydrogen ions (i.e. specific pH) for optimum functioning and operate best within a narrow temperature range.

At the molecular level enzymes are seen to work by providing a particular chemical environment in which the bonds holding the substrate together become weakened so that it is able to react with other molecules. This very complex three-dimensional environment is produced by the very large size of the enzyme relative to the substrate upon which it acts. Ribonuclease has a molecular weight of 12,700, and this is one of the smallest enzymes known. Urease, for example, has a molecular weight of over 200,000.

Something of these ideas of spatial configurations can be gained by consideration of the lysozyme molecule shown in Fig. 2.2. Lysozyme, which is an enzyme found in tears and nasal mucus, hydrolyses the polysaccharide wall of some bacteria and is part of the body's defence mechanism. Lysozyme in its functional state is coiled up in typical protein form and provides a cleft into which the polysaccharide substrate fits exactly. When an enzyme is denatured by heat or by chemical means the essential tertiary structure breaks down and, as can be seen, the molecule no longer fits the substrate and the enzyme ceases to work.

As already mentioned in the previous chapter many enzymes work only when a non-protein or prosthetic group is present and linked to the protein or apoenzyme part of the molecule. Another term for these units is coenzymes. They are extremely common in some classes of enzymes, such as those mediating oxidation and reduction reactions (oxidoreductases), although absent from others such as the hydrolases. Other enzymes only function when specific divalent metal ions – commonly calcium, magnesium, manganese, zinc, etc. – are linked to their molecules.

2.2 How enzymes work

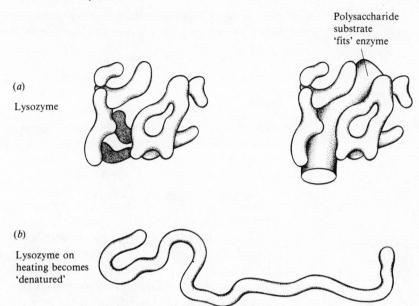

(a)

Lysozyme

Polysaccharide
substrate
'fits' enzyme

(b)

Lysozyme on
heating becomes
'denatured'

Fig. 2.2.
(a) Lysozyme in its
active configuration
showing how the
substrate fits into the
enzyme molecule.
(b) Lysozyme in its
denatured state
showing the loss of
tertiary structure.

Finally it should be noted that the particular enzymes a cell can produce must ultimately be determined by the cell's genetic instructions. These, however, may themselves be switched on or off by hormones, or may be induced by the presence of certain substrates. Such is the case with lactase, the lactose-digesting enzyme from the intestine. Produced in human infants it rapidly declines in certain races after weaning and the ingestion of lactose can cause serious intestinal disturbance. In contrast, the presence of maltose and sucrose in the diet actually induces production of the appropriate hydrolysing enzymes from the gut.

2.2 **How enzymes work**

As stated above, the configuration of the enzyme presents its substrate with a particular three-dimensional site into which it can fit. This is known as the enzyme's active site and it is usually made up of an arrangement of amino acids brought close together by the tertiary folding of the enzyme molecule. It is now definitely established that enzyme and substrate form a transition complex. In a very few instances this complex, which tends to be very unstable, has been isolated. Enzyme reactions thus take place as follows:

$$\text{Enzyme} + \text{Substrate} \rightarrow \text{Enzyme–Substrate complex} \rightarrow \text{Product} + \text{Enzyme}$$
$$\text{E} \qquad \text{S} \qquad\qquad \text{ES} \qquad\qquad\qquad \text{P} \qquad \text{E}$$

although there will usually be more than one substrate and product.

Two ideas have been put forward about the nature of the enzyme–substrate linkage. The earlier one was of a lock and key arrangement,

25

Enzymes

whereby the configuration of the enzyme represents the lock into which the particular substrate or key will fit. This can be represented as:

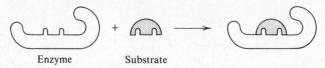

A modification of this hypothesis is that of induced fit, whereby the configuration of the enzyme actually changes as it takes up the substrate, i.e.

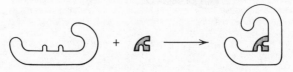

By acting as a template the enzyme greatly increases the chances of substrates reacting together, as it brings the relevant parts of the molecules into proximity. Also, binding of substrate(s) to the enzyme's active sites may cause differences in its molecular architecture which make the formation of new bonds possible.

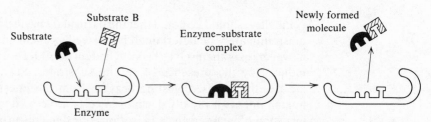

The role of coenzymes in enzyme functioning can also be shown by block diagrams. Thus:

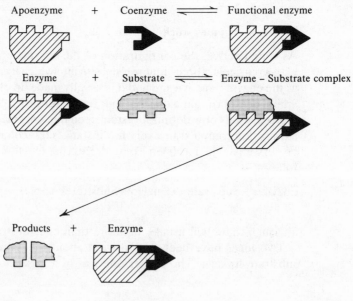

2.2 How enzymes work

In actual fact very few enzyme reactions have been worked out in their full chemical details but the general principles can be illustrated by one reaction that is known in detail. The enzyme is the protease from the pancreas called chymotrypsin, which hydrolyses peptide linkages adjacent to large or aromatic amino acids.

Chymotrypsin has 245 amino acids in three chains all coiled up in a three-dimensional complex. The active site of the enzyme consists essentially in the groupings between the 57th and the 195th amino acid, which are brought close together by the tertiary arrangement of the molecule. If this active site is examined chemically it is found that the following sequence of events take place during the hydrolysis of the peptide link in the substrate:

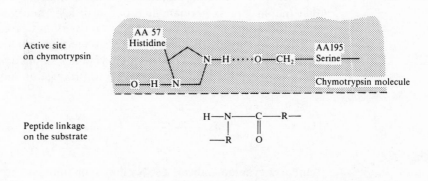

Active site
on chymotrypsin

Peptide linkage
on the substrate

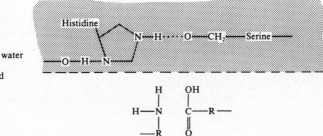

Enzyme–substrate
complex: new
linkages made

Addition of water
releases the
substrate and
peptide link
is cleared

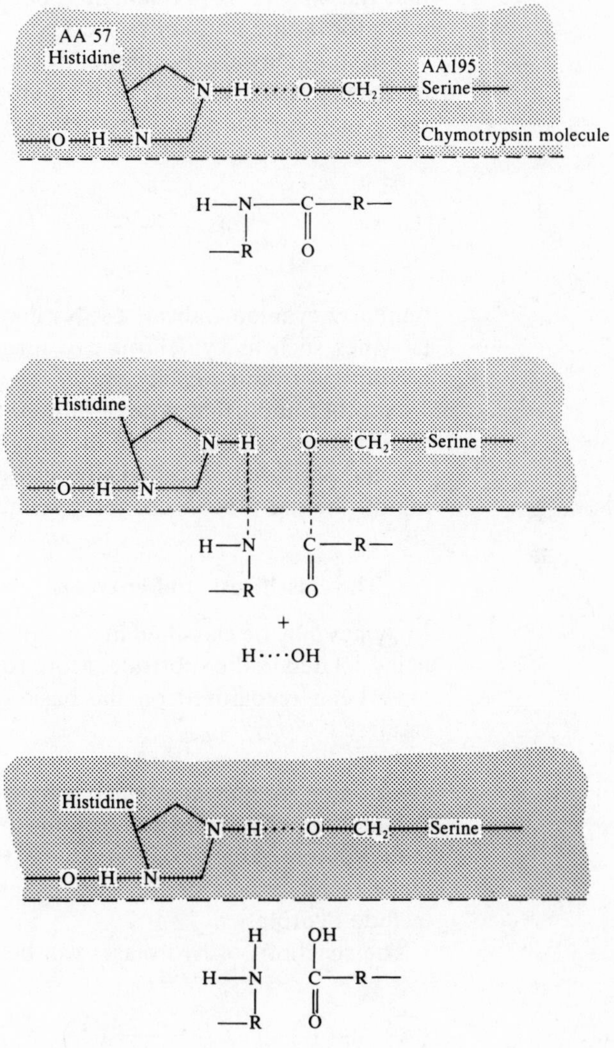

Appreciating that all these configurations exist in three dimensions the reason for the specificity of enzyme and substrate can be understood. Associated with specificity is the phenomenon of inhibition.

2.3 Enzyme inhibition

If a molecule has a shape which corresponds closely to that of the substrate of a given enzyme then it may become locked to the active site of the enzyme and prevent its normal functioning. Where such an inhibitor molecule exists side by side with the proper substrate then the two will 'compete' for the enzyme site, and if the inhibitor is present in excess then the enzyme will not work normally. This is called competitive inhibition. In addition, inhibitor–enzyme complexes may be stable, so that the enzyme is permanently 'out of order' after combining with its inhibitor. Some antibiotics are known to work as enzyme inhibitors.

Enzymes may also be inhibited by combination with heavy metals such as arsenic or mercury. Thus if an enzyme has sulphydryl (–SH) groups these may combine with arsenic (As) to produce an inactive complex:

Similarly cyanide radicals (–CN) may combine with iron or copper in enzymes such as cytochrome to form a non-functional complex. This accounts for their toxicity.

Another important type of inhibition is related to the action of modulators, which affect sites of the enzyme molecule other than the active site and produce a so-called allosteric effect. This may modify or actively inhibit the usual action of the enzyme.

2.4 The classification of enzymes

Enzymes may be classified in a number of different ways, a common one being related to the substrate. More recently, six broad classes of enzyme have been recognised on the basis of the type of reaction that they catalyse.

Hydrolases are enzymes which enable chemical bonds to be broken with the addition of water. The reactions they produce tend to be irreversible in the context of metabolic situations. Examples are the proteases, lipases and carbohydrases of the gut. Lysosome organelles inside cells also include hydrolases.

The reactions of hydrolases can be represented as:

$$\text{(A)}\!-\!\text{(B)} + H_2O \longrightarrow \text{(A)} + \text{(B)}$$

2.4 The classification of enzymes

Transferases catalyse the transfer of groups of atoms from one molecule to another. Groups commonly transferred are amino ($-NH_2$) groups in transaminations, and phosphoryl ($-PO_4$) groups in phosphorylations. The general change is therefore:

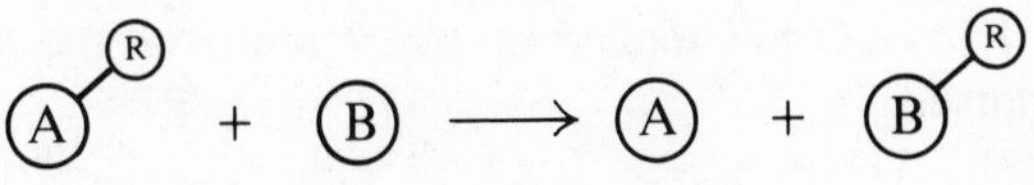

Ligases facilitate the junction of two molecules with the breakdown of a third. They are important in condensation reactions, as for example between two amino acids with the breakdown of ATP. Ligase enzymes work in the following way:

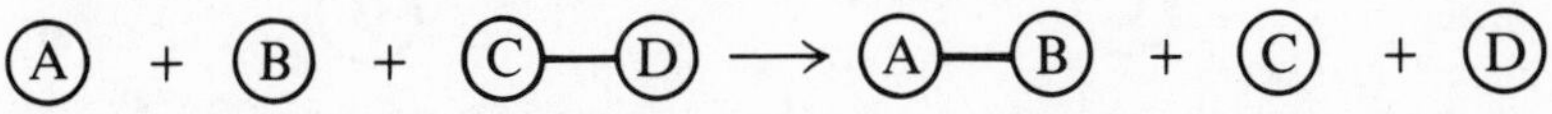

Lyases catalyse the linkage of two molecules by opening a double bond at one of them and linking the other where this bond is open. As a general reaction this can be shown as:

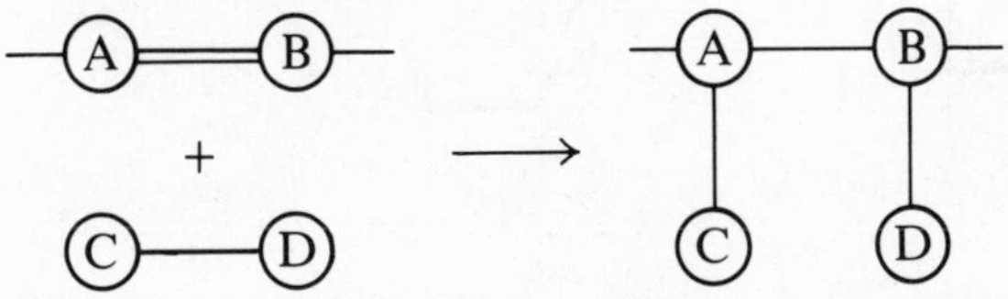

An example is the action of fumaric acid hydrolyase, where water is added across the double bond of the acid to form malic acid. In this case the additional molecule is water but other lyases work with different additions.

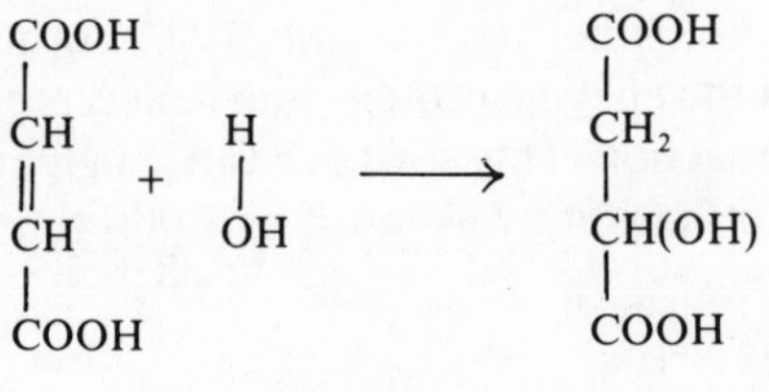

Fumaric acid   Malic acid

Oxidoreductases involve transfer of electrons from one molecule to another, often with associated hydrogen transfer. This group of enzymes is important in biological reductions and oxidations. The general reaction is:

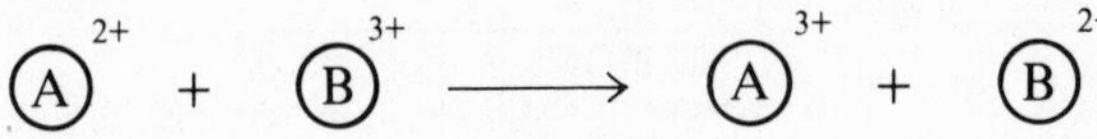

29

Enzymes

A specific example is:

CH₃CHOH.COOH + NAD ⟶ CH₃CO.COOH + NADH₂

$$CH_3CHOH.COOH + NAD \longrightarrow CH_3CO.COOH + NADH_2$$

Lactic acid   Pyruvic acid

Where the NAD has picked up hydrogen from the lactic acid, which is thus oxidised.

Isomerases facilitate internal changes of atoms within a molecule. The general change is:

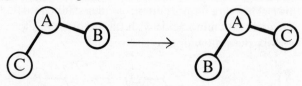

An example would be the conversion of glucose to fructose, i.e.

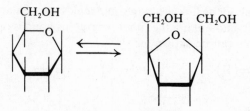

2.5 Factors controlling the rates of enzyme reactions

2.5.1 *Temperature*

As with other normal thermochemical reactions the speed of enzymic reactions approximately doubles with every 10 deg C increase in temperature. This holds between 0 and 50 °C but above this the rate of denaturation of the enzyme protein with heat is greater than the effective increase in the reaction. Thus a 10 deg C rise in temperature can increase the rate of denaturation by as much as 1,000 times; at 60 °C it may be

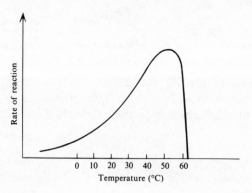

Fig. 2.3.
The effect of temperature on enzyme reaction rate.

virtually instantaneous (Fig. 2.3). However in the whole organism the effect of temperature may be minimised because in ectotherms some enzymes have a more or less constant rate of reaction over a wide temperature range. The action of other enzymes may appear unaffected by temperature because each enzyme consists of several isozymes (one of the many chemical forms in which an enzyme can exist in a single species) each of which has a different optimum temperature.

2.5.2 *Hydrogen ion concentration*

Hydrogen ion concentration (pH) affects the extent to which active groups at the surface of the enzyme are ionised, and thence the enzyme's degree of activity. Many enzymes work best at about neutrality (pH 7) but some, such as pepsin, require acid conditions (pH 2) while others, such as trypsin, need alkaline conditions (pH 9). The necessity for a stable pH and optimum temperature for the operation of enzymes indicates why mechanisms exist in the body for control of these factors.

2.5.3 *Particular property of the given enzyme*

With a fixed amount of substrate an increase in the initial amount of enzyme causes a proportional increase in the initial velocity of the re-action, until an optimum is reached when no further increase will occur (Fig. 2.4). Similarly with a fixed amount of enzyme, addition of substrate produces a proportional increase in the initial reaction rate until an optimum is reached. This is what we would expect from the law of mass action and the curve produced is as shown in Fig. 2.5.

For any given enzyme and its substrate the substrate concentration that produces half the maximum velocity is a fixed value and is called the Michaelis constant (symbol K_m) after its formulator. The lower the value of K_m the greater the affinity of the enzyme for its substrate, and the higher the value the less the affinity between the two. K_m is a fixed value for any single enzyme–substrate system and is a useful diagnostic feature of a reaction.

From what was said early in the chapter it may be appreciated that the value of K_m depends on both forward and backward reactions between enzyme and substrate and enzyme–substrate complex as well as on the

Fig. 2.4.
The relationship between initial rate of reaction and enzyme concentration at fixed substrate concentration.

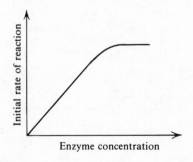

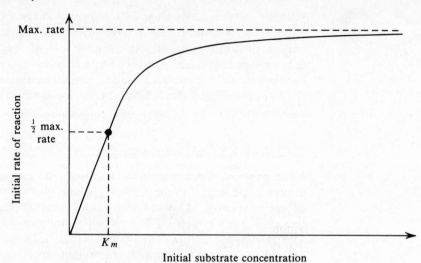

Fig. 2.5.
The relationship
between initial reaction
rate and substrate
concentration at fixed
enzyme concentration.

forward and backward reactions between the complex and the product of
the reaction. Enzyme kinetics in a multi-enzyme system such as that of a
cell is truly very complex!

3 Digestion

3.1 The nature of digestion

Digestion is the means whereby the various items of the diet become broken up into a form in which they can be assimilated into the blood or lymph. The breaking down of large molecules that takes place in digestion is based on the chemical reaction of hydrolysis, whereby water is inserted across the junctions of the initial molecule, making many smaller units. Typical examples of hydrolysis, as it occurs in digestion, are shown in Fig. 3.1.

Whereas in certain lower animals the food particles may be taken directly into the cells lining the gut (for intracellular digestion), in mammals and other vertebrates the digestive enzymes are liberated into the lumen of the gut and the products of their action absorbed (extracellular digestion). In digestive systems of the latter type two regions are usually differentiated, one specialised for digestion and the other for absorption. Within the first of these regions the various types of enzyme-producing cell have become grouped together so that concentrated solutions acting at an optimum pH can attack the food. In mammals digestion begins in the buccal cavity, where the teeth are differentiated according to the type of food the mammal eats and salivary enzymes start the chemical digestion of food.

Fig. 3.1.
Examples of hydrolysis in digestion. The enzymes that bring about these reactions are capable also of catalysing the reverse condensation reactions. In fact, although the enzymes control the rates of reactions, the final concentrations are at least partly determined by the fate of the products. Thus in the case of sucrose hydrolysis in the duodenum the products, glucose and fructose, are taken into the blood vessels of the ileum by a process of active absorption, so that, by the Law of Mass Action, the reaction tends towards the formation of more of these products. The reverse reaction for the condensation of monosaccharides into more stable sugars takes place in the liver and for this, since it is a building-up process, energy must be supplied. (Hydrolysis reactions yield energy.)

Carbohydrate

$$(C_6H_{10}O_5)_n + \frac{n}{2}H_2O \xrightarrow{\text{Amylase}} \frac{n}{2}C_{12}H_{22}O_{11}$$

Starch $\qquad\qquad\qquad\qquad\qquad$ Maltose

Fat

$C_{17}H_{35}COO.CH_2$
$\qquad |$
$C_{17}H_{35}COO.CH + 3H_2O \xrightarrow{\text{Lipase}} 3C_{17}H_{35}COOH + CH_2OH.CHOH.CH_2OH$
$\qquad | \qquad\qquad\qquad\qquad\qquad\qquad$ Stearic acid $\qquad\qquad$ Glycerol
$C_{17}H_{35}COO.CH_2$
Tristearin

Protein

$R.CH_2CO.NH.CH_2COOH + H_2O \xrightarrow{\text{Protease}} R.CH_2COOH + NH_2CH_2COOH$

33

3.2 **The organisation of the alimentary canal**

All vertebrate alimentary canals are organised along much the same lines; the sequence and secretions of the human alimentary tract are given in Table 3.1. Essentially the gut is a long tube the parts of which have been variously specialised for different functions. In general the upper regions have conditions favourable for the operation of specific hydrolysing enzymes while the lower parts are specialised for assimilation. The gut also has various sets of muscles which assist the mixing of food as well as its movement along the digestive tract. At some points the circular muscles form sphincters, which when closed prevent further movement and thus retain the food in a particular region of the gut.

The muscles and the cells and glands involved in secretion of enzymes are coordinated both by endocrines (hormones) and by the autonomic nervous system, so that gut movements and secretions to which the food is exposed are correctly timed.

The gut has a very large surface area, produced by its covering of finger-like projections known as villi. At the cellular level some parts of the surface are enormously increased in area by the presence of microvilli on the villi. Where assimilation of the products of digestion takes place the tissues concerned have a rich capillary blood supply involved in the removal of the food.

TABLE 3.1. *Secretions of the human alimentary tract*

Region	Name of secretion	Composition*	Daily amount (litres)	pH
Buccal cavity — Salivary glands	Saliva	Amylase, bicarbonate	1+	*c.* 6.5
Oesophagus				
Stomach	Gastric juice	Pepsinogen, HCl, rennin in infants, 'intrinsic factor'	1–3	*c.* 1.5
Pancreas	Pancreatic juice	Trypsinogen, chymotrypsinogen, carboxy- and aminopeptidase, lipase, amylase, maltase, nucleases, bicarbonate	1	7–8
Gall bladder	Bile	Fats and fatty acids, bile salts and pigments, cholesterol	*c.* 1	7–8
Duodenum	'Succus entericus' (see text for recent ideas on this)	Enterokinase, carboxy- and aminopeptidases, maltase, lactase, sucrase, lipase, nucleases	*c.* 1	7–8
Jejunum				
Ileum				
Caecum				
Colon				
Rectum				

*Excluding mucus and water, which together make up some 95% of the actual secretion.

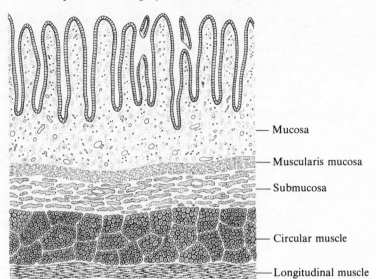

— Mucosa

— Muscularis mucosa

— Submucosa

— Circular muscle

— Longitudinal muscle

Fig. 3.2.
A generalised
transverse section of
the alimentary canal.

The alimentary canal is more like a surface structure than an internal system, and the cavity of the gut is quite distinct from the true body cavity or coelom, which is surrounded by mesoderm. Within the gut flourishes a vast bacterial flora. Some 40% of human faecal matter is bacteria, and this flora is normally commensal or symbiotic, that is, non-harmful or actually useful, provided it remains in the food canal. On the other hand, entry of micro-organisms into the true body cavity leads to immediate infection and disease. There is no one typical mammalian alimentary canal, but on the whole those of herbivores are more complex than others. A mammalian herbivore depends on its internal flora to carry out the bulk of its digestive processes and its gut is specially adapted to house micro-organisms and encourage their activity.

A generalised transverse section of the alimentary canal is shown in Fig. 3.2. Once this is clearly understood it becomes much easier to follow its modifications as found in the different regions along its length.

3.3 The epithelial lining of the alimentary canal

The epithelial cells lining the alimentary canal are derived from embryonic endoderm. All epithelia have certain structural and other properties in common. Structurally they consist of a single layer (in simple epithelia) or several layers (in stratified epithelia) of cells joined to a fibrous basement membrane which they themselves secrete and which anchors them firmly to the underlying tissues. Adjacent cells are held together by an intercellular connective 'cement' and may also have pro-toplasmic strands between them. Each cell has a round or oval central nucleus and on the whole tends to be more differentiated on the side facing the external surface.

Every metabolic exchange with the external medium must take place

35

Digestion

via the epithelia and they may be specialised in a number of ways according to the region of the body in which they are found. Most of such cells do not live very long and a characteristic of the tissue is its rapid rate of replacement. The common forms of epithelia in the alimentary canal are squamous, cubical and columnar, the names being related to the shapes of the cells. These three types, together with the organisation of a stratified epithelium, are illustrated in Fig. 3.3.

An interesting adaptation of epithelial cells lining the assimilatory regions of the small intestine is the presence of a brush border, formed by a folding of the cell's surface into microvilli. Although this can just be

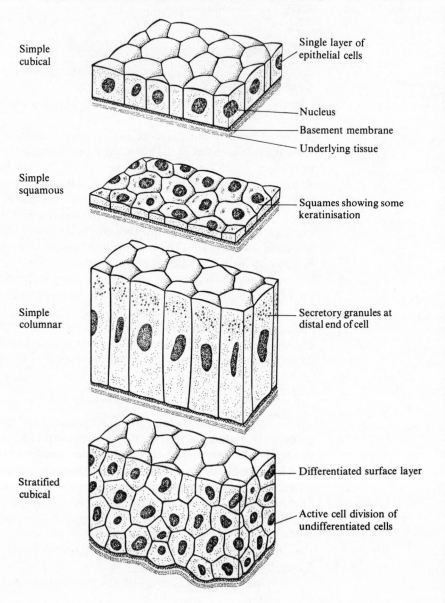

Simple cubical

Single layer of epithelial cells

Nucleus

Basement membrane

Underlying tissue

Simple squamous

Squames showing some keratinisation

Simple columnar

Secretory granules at distal end of cell

Stratified cubical

Differentiated surface layer

Active cell division of undifferentiated cells

Fig. 3.3.
Types of epithelium.

36

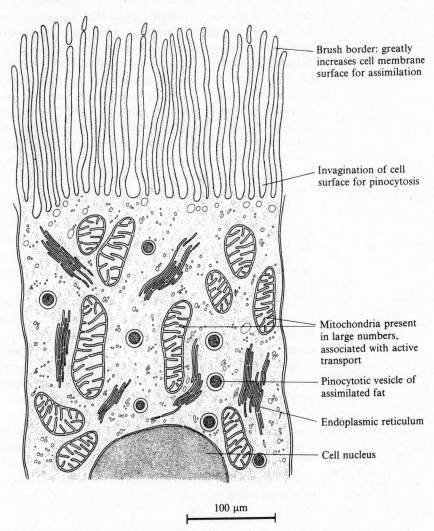

Brush border: greatly increases cell membrane surface for assimilation

Invagination of cell surface for pinocytosis

Mitochondria present in large numbers, associated with active transport

Pinocytotic vesicle of assimilated fat

Endoplasmic reticulum

Cell nucleus

100 μm

Fig. 3.4.
Drawing based on an
electron micrograph of
a single brush-bordered
epithelial cell.

made out with a light microscope it is revealed in much greater detail at the very high magnifications possible in the electron microscope. A drawing of such a cell based on an electron micrograph is shown in Fig. 3.4. It is calculated that the brush borders of the human intestine give it a total surface area of some 2×10^6 square centimetres, and must thus considerably increase the rate at which assimilation can occur. Besides the microvilli, the cell shows other modifications for its assimilatory function.

3.4 The principles of coordination of secretion in the alimentary canal

Both nervous and endocrine mechanisms are used in the coordination of the digestive system. That part of the nervous system concerned in the operation of the viscera is called the autonomic nervous system and itself

37

consists of two systems, the sympathetic and parasympathetic. The former is an extension of the segmental motor paths of the visceral spinal arcs and is stimulated by the hormone adrenaline (it is dealt with more fully in Chapter 10).* Its basic function is to increase the efficiency of the body for immediate action, and besides the positive actions it produces (increase in respiration, blood pressure, release of glycogen, etc.) it also tends to inhibit digestive activity. Thus the connections from the sympathetic system which supply the parts of the alimentary canal or its associated glands depress their activity, or, under extreme stimulus, cause a reversal of the normal gut movements that is called antiperistalsis. Sympathetic stimulation also causes a reflex emptying of the bladder and bowel.

The parasympathetic system consists of fibres in the vagus cranial nerve (Xth) together with fibres of the trigeminal nerve (Vth) and the facial nerve (VIIth), as well as part of the pelvic plexus. The general action of the parasympathetic system is to stimulate digestion and secretion. The system of nerves that innervates the intestine is shown in Fig. 3.5.

Besides these two parts of the nervous system and their associated endocrines a number of localised hormones are used in the coordination of digestion. These usually cause a particular gland to produce secretions under the stimulus of food, but may also set a gland in action on the approach of food, as with the effects of duodenal secretion on the pancreas. Local nerve reflexes are also used to bring about peristalsis (see p. 43), villi movements and all the activities concerned in the passage of food and elimination of waste through the gut.

The general picture is thus of a complex system continuously breaking down and assimilating the food that is eaten by the animal. This system works largely under its own local control measures and yet in time of stress its activity ceases, becoming subordinated to the survival of the whole organism.

3.5 **The human alimentary canal**

3.5.1 *The teeth and the mechanism of chewing*

The teeth are contained in the buccal cavity, which lies between the mouth and the oesophagus, and are differentiated according to the diet, which, in the case of man, is a variety of plant and animal food. Human teeth occur as laterally flattened incisors at the front, rather small canines (the eye teeth) outside the incisors and then premolars and molars at the back of the jaw (see Fig. 3.7). In the case of omnivorous dentition, the differentiation of the various types of teeth is not well marked. Before the adult condition is reached a milk dentition exists and may be summarised

---

* This section will be more readily understood after the autonomic nervous system has been studied (p. 260).

Lumen of gut

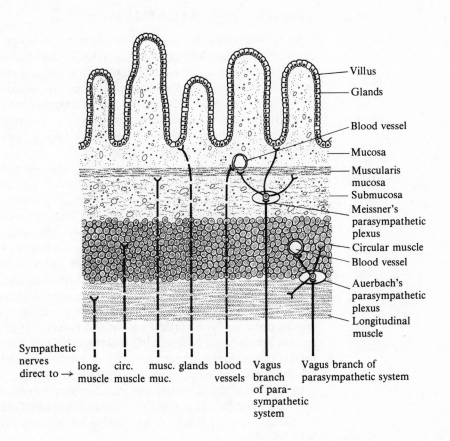

ig. 3.5.
he double motor
nnervation of the
ntestine.

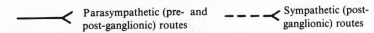

by the following dental formula. (The dental formula is an expression for the teeth on one side of the mouth.)

Milk dentition: $I\frac{2}{2} C\frac{1}{1} PM\frac{2}{2} M\frac{0}{0} = 20$

Each tooth develops from an odontoblast or tooth bud. This secretes a layer of prismatic calcium phosphate or enamel. Below this, forming the bulk of the tooth, is dentine, a substance related to bone and permeated with minute canals in which run nerve endings and blood capillaries originating from the pulp cavity. Fixing the tooth into the socket of the jaw is a further calcareous substance, cement. These features are illustrated in Fig. 3.6.

Human teeth have small canals leading up through the roots to the pulp cavity. These control the amount of food entering the tooth and thus its rate of growth, which in the case of man is very slow.

39

Man has 32 teeth, midway between the herbivorous and carnivorous condition. His dental formula is as follows:

Adult dentition: $I \frac{2}{2} C \frac{1}{1} PM \frac{2}{2} M \frac{3}{3} = 32$

In all mammals the shape of the dentary or lower jaw reflects the chewing action performed which in turn is related to the nature of the food. The lower jaw is lowered by gravity and the action of a small muscle called the digastric. The variations that are found exist in the muscles closing the jaws, that is the masseter and temporal, and in the nature of their attachment to the dentary. There is also considerable variation in the sort of joint employed between upper and lower jaws according to the movements these are required to make.

Compared with those of other mammals the jaw muscles of man are somewhat reduced. The dentary is composed of a tooth-bearing ramus and a vertical coronoid process which bears an anterior angular process, where the temporal muscle is attached, and a posterior condyle (Fig. 3.7). This latter makes a loose roller joint with the glenoid fossa of the squamosal of the skull. In man the temporal muscle leads upwards to the side of the head and exerts a direct pull upon the lower jaw, thus providing the biting action of closing the teeth. The second large muscle of mastication, the masseter, runs from the posterior flange of the dentary to the jugal and squamosal of the zygomatic arch or cheek bones. This muscle provides both vertical and lateral movement to the dentary and powers the chewing action of the molar teeth. Both these muscles are of similar size and importance and work at a similar mechanical advantage in the distance of their points of attachment from the articulation.

In the size of the coronoid, the muscle arrangement, the nature of the articulation and pattern of the teeth, man is intermediate between herbivore and carnivore. These more specialised mammals will be described later (p. 56).

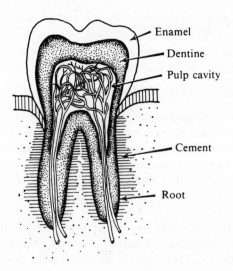

Enamel

Dentine

Pulp cavity

Cement

Root

Fig. 3.6.
Longitudinal section of
a human molar.

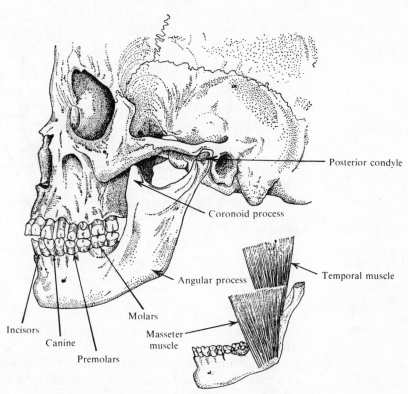

Fig. 3.7.
Human skull showing
dentition.

3.5.2 *The buccal cavity*

The top of the buccal cavity is bounded by the palate, the base by the tongue, and the sides by the muscles of the cheeks (buccinators). The cells lining the cavity form a stratified epithelium, as distinct from the columnar epithelium found further down the gut.

Into the buccal cavity empty three pairs of salivary glands – the sublingual, submandibular and parotid – which together secrete the digestive juice, saliva. Posteriorly the soft palate acts as a valve preventing food getting into the back of the nasal cavity, while the entrance of the windpipe, or trachea, is provided with another valve, the epiglottis, which stops food passing into it. Reflex arcs involving the facial (VIIth), glossopharyngeal (IXth) and vagus (Xth) nerves ensure that the bolus of food formed by the tongue is passed smoothly down into the oesophagus during swallowing.

The buccal cavity is well provided with sensory apparatus to test the substances passed into the mouth. Both distant and immediate perception of olfactory (concerned with smell) stimuli cause the secretion of saliva, as does stimulus of the taste organs of the tongue. These stimuli are transmitted as sensory impulses from the olfactory (Ist), the trigeminal (Vth) and facial (VIIth) nerves to the medulla of the hind brain, and thence via motor nerves to the salivary glands. The presence of food in the mouth also stimulates secretion and from all these sources of stimulus

41

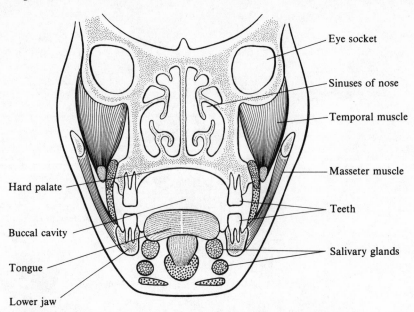

Eye socket

Sinuses of nose

Temporal muscle

Masseter muscle

Hard palate

Teeth

Buccal cavity

Tongue

Salivary glands

Lower jaw

Fig. 3.8.
A frontal section
through the lower
portion of the human
head showing the
relationship of the
buccal cavity to jaws
and glands.

pathways leave via the vagus (Xth) nerve, which leads to the stomach. The latter is thus prepared in advance to receive the food. It should be noted that stimulation of the sympathetic system under conditions of fear or shock causes small quantities of viscid saliva to be produced and inhibits the normal copious flow.

Saliva itself consists of sodium bicarbonate (which provides an alkaline medium in which the salivary enzyme ptyalin or amylase can work), the ptyalin itself, water and mucus. The ptyalin hydrolyses carbohydrate starch to maltose according to the reaction:

$$(C_6H_{10}O_5)_n + \frac{n}{2} H_2O \rightarrow \frac{n}{2} C_{12}H_{22}O_{11}$$

This change takes place in a number of steps including the production of soluble starch, erythrodextrin, and achroodextrin. This reaction can be followed in detail as the original starch gives a purple colour with iodine, while erythrodextrin gives a red colour and the achroodextrin produces no colour. Besides having a digestive action saliva also lubricates the food and assists in swallowing. Meanwhile the teeth and tongue help to mix the enzyme with the food and although carbohydrates are the only substances actually hydrolysed in the buccal cavity the mastication does greatly assist the subsequent digestion of all foodstuffs by rendering them into small particles upon which the enzymes can act.

3.5.3 *The oesophagus*

This is a tube some 25 centimetres in length connecting the buccal cavity to the stomach. The swallowing reflex relaxes the upper part of the oesophagus and a wave of relaxation followed by one of contraction

42

3.5 The human alimentary canal

pushes the food bolus down to the stomach. This sort of movement is common to all regions of the gut and is called peristalsis. It depends, normally, on local nerve reflexes triggered by the presence of food relaxing and contracting the longitudinal and circular muscles alternately, the contraction following the food and the relaxation preceding it. The innervation of the oesophagus, however, is derived from those parts of the nervous system concerned in swallowing and is not a local reflex.

3.5.4 *The stomach*

The stomach is a muscular bag, capable of great distension, which is divided into cardiac and pyloric parts and leads into the duodenum via the

Gastric pit, containing mucous cells
near the lumen. Further from
the lumen are pepsin-secreting cells
and oxyntic cells which secrete
hydrochloric acid

Lumen of stomach

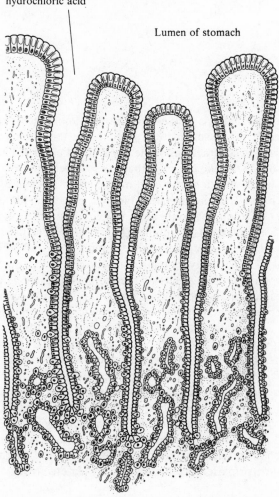

Fig. 3.9.
Transverse section of
the stomach wall.

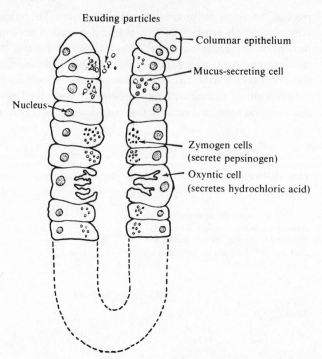

Fig. 3.10.
Transverse section of a
gastric pit.

pyloric sphincter. The active churning movements of the stomach take place in the lower, pyloric portion, the food being passed into this by peristaltic contractions of the upper, cardiac part. These movements mix the food with the enzymes that are secreted into the stomach, rendering the whole into a semi-liquid mass known as chyme.

The walls of the stomach (Fig. 3.9) are much folded and covered by a layer of columnar epithelium. Indented into the mucosa are some three million gastric pits or tubules (Fig. 3.10). Specialised cells in the tubules secrete the gastric juice whose composition and action is as follows:

(*a*) A copious layer of mucus is secreted by the upper cells of the gastric pits and the lining cells of the stomach. Mucus is a constituent of all intestinal juices and protects the tissues from the action of enzymes. It is a glycoprotein and a type of mucopolysaccharide. It is amphoteric (that is it can act as both acid and base according to conditions) in nature and can thus neutralise both acids and alkalis, as well as providing a mechanical buffer between hard food substances and the living tissues.

(*b*) Hydrochloric acid is secreted to give a final concentration in the stomach of approximately 0.1 molar which gives a pH of 2. The effects of this acid are to bring about direct hydrolysis of food substances and to provide a favourable medium for the activity of the enzyme pepsin. The acid also kills the majority of the bacteria taken in with the food. The mechanism whereby the acid-producing cells of the gastric pits (known as oxyntic or parietal cells) are able to secrete strong acid depends on the dissociation of water present in the cell to give hydrogen (H^+) and hydroxyl (OH^-) ions. The latter combine with carbon dioxide in the presence of carbonic anhydrase enzyme to form bicarbonate (HCO_3^-) which then pass out of the cell into the bloodstream. This is found to become more alkaline shortly after the intake of food. Meanwhile the H^+ ions diffuse out of the cell into the lumen of the tubule or pit and at the same time, to maintain the neutrality of charges across its membrane, Cl^- is secreted. Thus:

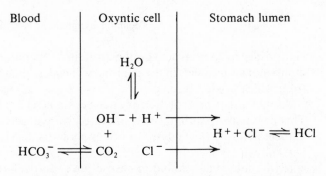

(*c*) Pepsin is an enzyme secreted in most vertebrates and comes from the zymogen cells (also called peptic cells or chief cells) of the gastric pits and is a proteolytic enzyme (that is it breaks down protein) acting in the acid medium provided by the hydrochloric acid. It is not found in other phyla. The nature of protein digestion is very complex because proteins are such large molecules, and in order to understand the differences between the various proteolytic enzymes of the digestive system some outline of the way in which the protein molecule is hydrolysed must be given.

In any given protein there may be very many peptide linkages, these being the –CO.NH– groups which join the constituent amino acids. The group on either side of these linkages will vary with different amino acids and each peptide linkage may present a different surface to the hydrolysing enzymes, which are very specific in their action. For this reason we find a number of proteolytic enzymes, each specific for peptide linkages in a particular molecular environment.

It is possible to differentiate two main classes of protein-splitting enzyme: the exopeptidases which attack the terminal peptide linkages and the endopeptidases which attack linkages within the molecule. Both classes act in the following general way:

$$R_1CO.NH.R_2 + H_2O \rightarrow R_1COOH + NH_2R_2$$

Pepsin attacks peptide linkages within the protein molecule and is therefore an endopeptidase. It is specific for peptide links where the –NH– part of the link is provided by the aromatic amino acids tyrosine and phenylalanine, and as these commonly occur inside protein molecules most of the latter will be hydrolysed by pepsin. The action of pepsin may be represented as follows:

(further amino acids) R.CO ⋮ NH.CH.CO.NH.R (further amino acids)

Pepsin cleavage here

Hydrolysis by pepsin leaves the protein as peptone molecules. The number of amino acids in the peptone will be less than that in the protein from which it is derived.

As with other proteolytic enzymes pepsin is secreted as its precursor. This precursor (pepsinogen) is activated by hydrochloric acid and then autocatalytically (it catalyses or brings about its own hydrolysis) to form pepsin.

(*d*) Rennin is also found in gastric juice, particularly in young mammals, and its function is to convert, in the presence of calcium salts, the soluble protein casein (previously called caseinogen) that is found in milk into paracasein. The solubility of paracasein depends on pH and a further change, 'clotting', takes place under the action of rennin and calcium ions in acid conditions.

(*e*) Haemopoietic factor, or intrinsic factor of Castle, is secreted by the gastric mucosa and allows the uptake of vitamin B12 from the diet (see p. 19).

Coordination of the activities in the stomach takes place initially by impulses passing down the vagus nerve (Xth) into the nerve net at the base of the mucosa. These are caused by the presence of food in the buccal cavity or by the smell or sight of food. Once the food actually reaches the stomach it causes the release of a hormone gastrin from the stomach epithelia into the blood. Gastrin is one of a number of hormones produced in the digestive system, and is secreted by groups of mucosa cells in the pyloric region of the stomach. It passes rapidly round the bloodstream and, circulating in the tissues of the gastric pits, increases their activity. It has recently been shown to be a relatively simple polypeptide with 17 constituent amino acids, of which the sequence of the last five is critical to its effect of stimulating secretion of gastric juice and hydrochloric acid. Another hormone, called enterogastrone, has an antagonistic effect to gastrin. It originates from the duodenal mucosa and its effect is to 'switch off' the stomach by decreasing its secretions and movements. The action of the sympathetic connections to the stomach also tends to depress enzyme secretion although not necessarily acid secretion.

After some 3–4½ hours in the stomach chyme (semi-digested matter) begins to pass via the pyloric sphincter at the lower end of the stomach into the duodenum. The actual opening of the sphincter muscle may be because of a change in the acidity or in osmotic strength of the stomach contents at the end of digestion. This chemical change relaxes the sphincter so that the peristaltic contractions of the stomach can push food through it into the duodenum.

By the time this happens pepsin and rennin have done their work and some hydrolysis of food substances has taken place in the presence of hydrochloric acid. The acid has also broken down connective tissue in the food, releasing its digestible contents. The warmth of the stomach has melted any fat present and the continuous churning action of the stomach wall has done much to reduce the food to very small particles.

Substances of small molecular weight, such as monosaccharide sugars and ethyl alcohol, may also start to be absorbed through the mucosa of the stomach.

3.5.5 *The duodenum*

The duodenum is the region of the gut after the stomach and is the first part of the small intestine. The same general tissues can be recognised in the wall of the duodenum as are found in the stomach (Fig. 3.11). The outer layers consist, as before, of mesentery, of longitudinal and circular

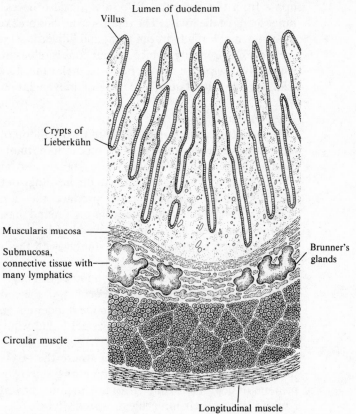

Fig. 3.11.
Transverse section of
the duodenal wall.

47

muscles followed by a nerve plexus, and then the submucosa and the mucosa. The submucosa is full of the characteristic Brunner's glands, whose presence distinguishes the duodenum from other regions of the small intestine. In the mucosa there are many secretory crypts of Lieber-kühn while the surface of the mucosa consists of many villi which effectively increase the surface area in contact with the food. Into the duodenum opens the hepato-pancreatic duct which brings important digestive juices from the gall bladder and the pancreas. As these secretions act upon the food at the upper end of the duodenum and before the digestive juices of the duodenum itself, their composition and function will be described first.

The bile

This originates from the liver where it is passed across from the canals of the lobules (Fig. 6.31) into the small intracellular bile canaliculi. These merge into larger vessels which drain into the gall bladder, embedded within a lobe of the liver, and from the bladder the bile duct runs down to meet the pancreatic duct. Bile consists of certain salts such as sodium glycocholate and sodium tauroglycocholate which reduce surface tension and are important in emulsifying fats. Emulsification breaks up the fats into small globules and thus makes them more readily digested by the lipase from the pancreas. Bile salts are also necessary for the normal functioning of the lipase. The bile also contains breakdown products from blood haemoglobin bilirubin (red) and biliverdin (green) – which colour the faeces. A quantity of bicarbonate ions is released with the bile fluids, which gives a pH of 8 and helps to make the duodenum alkaline. In addition bile stimulates the peristaltic movements of the gut.

The pancreas

This is a diffuse pink organ found in the first loop of the small intestine after it leaves the stomach. It produces important digestive enzymes, termed collectively pancreatic juice, and is also an endocrine organ. These two functions are reflected in the histology of the pancreas, where the islets of Langerhans, which produce the hormones insulin and glucagon, can be distinguished from the ground mass of secretory tissue where the enzymes are formed. Collecting ducts in the pancreas join up into the pancreatic duct and this, joining with the bile duct, enters the upper end of the duodenum.

The pancreatic juice contains a number of enzymes:

(*a*) *Trypsinogen* is the inactive precursor of the endopeptidase trypsin. It is activated by enterokinase from the duodenum and autocatalytically. In its active form it operates best at pH 8 and will show less activity in more acid conditions.

Trypsin attacks peptide linkages where the –CO– part of the link is provided by the amino acids arginine or lysine (Fig. 3.12). It converts proteins or peptones to di-, tri-, or polypeptides having two, three or more amino acids in their molecules respectively.

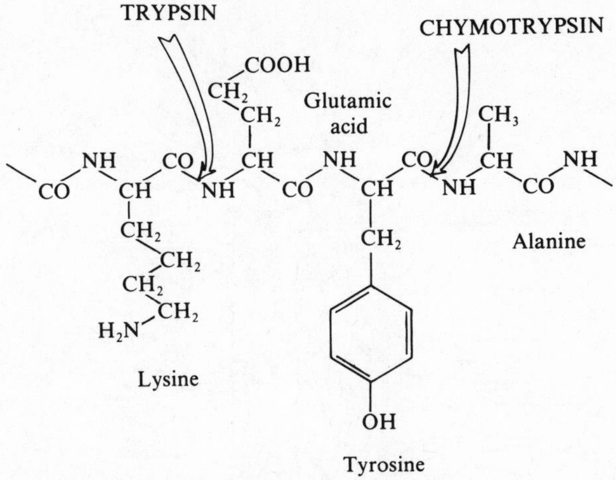

Fig. 3.12.
Method of action of trypsin and chymotrypsin.

(b) *Chymotrypsinogen* is activated by trypsin and like the latter is an endopeptidase. It cleaves peptide links wherever the amino acids phenylalanine or tyrosine provide the –CO– part of the link (Fig. 3.12), and like trypsin results in the hydrolysis of proteins and peptones to smaller units.

(c) A *carboxypeptidase* is an exopeptidase which acts on peptide linkages next to a carboxyl group. It works on small protein units (peptones or polypeptides) after they have been released by endopeptidases (see Fig. 3.14, p. 51).

(d) A *lipase*, or a fat-splitting enzyme, hydrolyses the links between glycerol and fatty acids to release these substances in the duodenum. A fairly typical hydrolysis would be:

$$C_{57}H_{110}O_6 + 3H_2O \rightarrow 3C_{17}H_{35}COOH + CH_2OH.CHOH.CH_2OH$$
Tristearin   Stearic acid   Glycerol

Some of the fat present may not be hydrolysed, however, and will be assimilated directly, while much may be converted to glycerides (molecules that contain glycerol and small hydrocarbon side chains).

(e) An *amylase* is similar in action to the salivary ptyalin already described and it completes the hydrolysis of starch started by the latter. In the duodenum amylase converts starch to the disaccharide maltose which is later broken down to monosaccharides.

Amylases exist in α and β forms and these attack starch molecules in different ways – but both bring about hydrolysis.

All the enzymes of the pancreas work best at a pH of 8–9, and to create a suitably alkaline environment alkaline fluids are secreted in the bile and from the pancreas and duodenum (see, however, p. 52 (f)).

The enzymes of the duodenum
Previously it was customary to think of the secretions of the glands of the duodenum as making up the 'succus entericus', and that this collection of enzymes acted in the lumen of the gut.

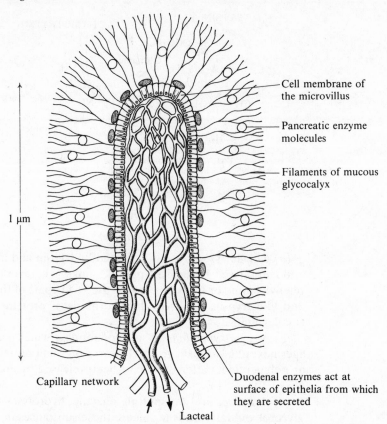

Cell membrane of
the microvillus

Pancreatic enzyme
molecules

Filaments of mucous
glycocalyx

1 μm

Fig. 3.13.
Schematic diagram of a
single microvillus from
a duodenal epithelial
cell.

Capillary network

Duodenal enzymes act at
surface of epithelia from which
they are secreted

Lacteal

Recently, however, these ideas have been considerably revised (Fig. 3.13). It is now thought that the enzymes of the duodenum are secreted from the surfaces, including the microvilli, of the epithelial cells. The enzymes remain attached to the surface of the epithelia and here they act, their products being released into the lumen of the gut or being immediately assimilated into the epithelial cells. Investigations using colloidal iron and other special histological techniques indicate that a matrix of mucopolysaccharide sticks out permanently from the surface of the cells and it is within this matrix that the duodenal enzymes are found. This filamentous network is sometimes called the glycocalyx and is also found in the stomach. The enzymes from the pancreatic juice will act also within the glycocalyx but not attached to the epithelial surface. The glycocalyx greatly increases the surface area of the gut, and promotes juxtaposition and interaction between substrate and hydrolytic enzyme in the gut.

Alkaline secretions are released from the pancreas into active trypsin by hydrolysis of a terminal peptide link. This means that trypsin is only in its active form at the site of its activity and in the presence of food in the duodenum.

Duodenal enzymes include:

(*a*) *Enterokinase*. This converts the trypsinogen from the pancreas into

50

active trypsin by hydrolysis of a terminal peptide link. This means that trypsin is only in its active form at the site of its activity and in the presence of food in the duodenum.

(*b*) *Erepsin* is really a number of different enzymes which complete the digestion of the protein molecule. After the activity of trypsin and chymotrypsin described above the proteins are in the form of oligopeptides which may consist of small numbers of amino acids linked together, say two to six. The various enzymes in erepsin are each specific for a particular sort of oligopeptide. Thus one that acts on a peptide linkage next to a –COOH group is called a carboxypeptidase, while a type that acts on the peptide linkage next to an –NH group is called an aminopeptidase. Finally the oligopeptides are reduced to groups of only two or three amino acids and these are finally hydrolysed into single amino acids by dipeptidases and tripeptidases respectively. All these enzymes are included in the term erepsin. A summary of protein digestion is given in Fig. 3.14.

A theoretical protein molecule

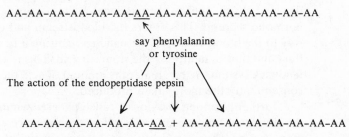

AA–AA–AA–AA–AA–AA–AA–AA–AA–AA–AA–AA–AA–AA–AA

say phenylalanine
or tyrosine

The action of the endopeptidase pepsin

AA–AA–AA–AA–AA–AA–AA + AA–AA–AA–AA–AA–AA–AA–AA

These substances are peptones and are acted on by endopeptidases
such as trypsin

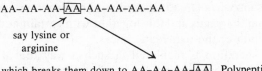

AA–AA–AA–AA–AA–AA–AA

say lysine or
arginine

which breaks them down to AA–AA–AA–AA   Polypeptides   AA–AA–AA–AA

or AA–AA–AA   Tripeptides or AA–AA–AA

or AA–AA   Dipeptides or AA–AA

These in turn are hydrolysed by the exopeptidases, which consist of aminopeptidases and carboxypeptidases

AA–AA–AA–COOH by carboxypeptidase
AA–AA–AA–NH₂ by aminopeptidase

and finally rendered to individual amino acids·

AA

AA

in which form they are absorbed   AA

Fig. 3.14.
Pathways of protein
digestion in the
mammal.

51

(*c*) *Lipase* acts on the hydrolysis of fat, as already described for pancreatic lipase.

(*d*) *Sucrase* acts on the disaccharide sucrose, converting it to glucose and fructose. This enzyme is also called invertase.

(*e*) *Maltase* acts on the disaccharide maltose, turning it into two molecules of glucose.

(*f*) *Lactase* converts the disaccharide lactose, to glucose and galactose.

These duodenal enzymes act best in a medium of pH 8.3, but owing to the very acid state of the stomach the general pH of the duodenum is about 5.5. It may reach 6 in the ileum but alkalinity is never obtained.

All classes of food substance have now been broken down into small molecules which can be assimilated into the bloodstream or lymph. The carbohydrates are in the form of glucose or other monosaccharides; the fats remain as fat, or have been hydrolysed to fatty acids, glycerol or glycerides; and proteins are in the form of amino acids.

Coordination of the duodenum and pancreas

The acid which passes through the pyloric sphincter with the semi-digested chyme acts on the duodenal mucosa causing the release of the hormone secretin. This enters the bloodstream and eventually finds its way to the blood vessels of the pancreas causing it to secrete an alkaline fluid and thus to neutralise the stomach acid. This is a good example of a feedback system, as the greater the amount of acid the more secretin and consequently the more alkali produced.

A further hormone, previously called pancreozymin, is also formed by the duodenal mucosa and released in the presence of food. (This hormone has now been recognised as the same chemical as cholecystokinin which stimulates contraction of the gall bladder. It is therefore called CCK–PZ.) This travels to the pancreas and causes the secretion of the various enzymes already described. The pancreas is under nervous control from the vagus and this together with certain hormones controls the production of the pancreatic juices.

Bile secretion is regulated in such a way that it is released at the appropriate time into the duodenum by the contraction of the gall bladder. When food is eaten the bile starts to be released after about 1 hour and reaches its maximum flow some 2–5 hours later. Fats entering the duodenum cause production of CCK–PZ which acts as a hormone on the gall bladder stimulating it to empty. This should not be confused with secretin which besides its effects on the pancreas also controls the rate at which the liver produces bile, but not the actual emptying of the gall bladder.

Finally the duodenum itself is stimulated to secrete its enzymes by the usual double mechanism, that is, nervous and hormonal. The vagus innervation stimulates secretion and sympathetic fibres from the solar plexus depress it, while local stimulation of the mucosa of the jejunum and ileum causes liberation of the hormone enterocrinin which in turn initiates secretion by the ileal glands. The movements of the small intestine whereby the food is pushed along towards its lower end will be described at the end of the next section (p. 54).

3.5.6 *The ileum*

This is the second half of the small intestine and is where the bulk of assimilation takes place. The lining epithelium secretes mucus but no enzymes and has some 20–40 villi to the square millimetre. Each villus is lined by epithelial cells and each of these cells may have as many as 2,000–3,000 microvilli, each of which is approximately 1 nanometre in length and 0.1 nanometre in diameter. The total surface area of the intestine is in excess of 2,000,000 square centimetres. These villi are larger than those of the duodenum (Fig. 3.15) and are provided with a complex network of capillaries arising from the mesenteric artery and returning into branches of the hepatic portal vein. In the centre of each villus runs a lacteal, derived from the lymph vessels which permeate the submucosa.

Some of the smaller molecules in the food, in particular glucose and galactose as well as smaller quantities of fats and amino acids, will already have been assimilated before reaching the ileum, but, as this is the main

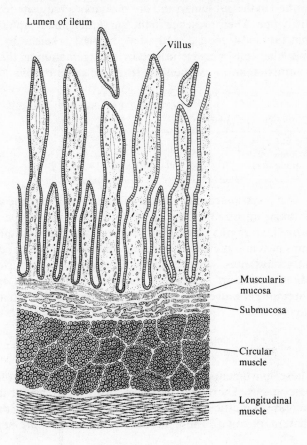

Lumen of ileum

Villus

Muscularis mucosa

Submucosa

Circular muscle

Longitudinal muscle

Fig. 3.15.
Transverse section of the ileum wall.

region of uptake, the various systems whereby the food actually enters the tissues of the body will be described at this point.

The mucosa of the small intestine is rich in the enzyme phosphatase and it seems that sugars are phosphorylated before being absorbed. Like other classes of food substance they can be assimilated against a concentration gradient and it may be that the phosphorylated sugar is able to pass across the cell membrane while the sugar alone is not. Once inside the cells the sugar would be released and as it would not be able to escape back into the gut it could be used or transported away. Such an active assimilation requires the expenditure of energy.

Amino acids are also actively assimilated as can be shown by the fact that metabolic poisons prevent their uptake. They pass into the capillaries of the villi for transport to the liver in the hepatic portal vein.

Fats may be assimilated as fatty acids, glycerol, glycerides, or as very small molecules of emulsified fat. The fatty acids become water-soluble in the presence of bile salts with which they combine. Some two-thirds of the fatty substances present are passed into the lacteals of the villi while the rest enters the blood capillaries. Fat that gets into the lacteal is probably unchanged fat; that which enters the blood capillaries does so as short-chain fatty acids and glycerol. Chylomicrons (i.e. small fat droplets) are formed in the gut lumen and are probably a requisite for absorption into the lymph. Their presence in the gut lumen and lymph supports the view that fat is absorbed directly. Fat droplets taken in by pinocytosis into epithelial cells increase in size by amalgamation as they move into the lymph system, as can be seen from the table below:

Position:	Lumen	Epithelial cell	Lymphatic vessel
Size of chylomicrons (nanometres)	50	100–200	500–1000

Other substances assimilated into the small intestine are water – which is taken up osmotically, as the body fluids are hypertonic (of higher osmotic strength) to the gut contents – vitamins and mineral salts. Vitamins pass into the gut by diffusion, the fat-soluble A and D entering with fatty substances, which is one of the reasons for requiring fat in the diet. Mineral salts are selectively absorbed: thus sodium and chloride ions enter the gut wall easily, whereas magnesium and sulphate do so only very slowly.

3.5.7 *Movements of the small intestine*

These movements are mainly peristaltic, as already described for the oesophagus. There are, however, two other types of movement which increase the efficiency of the digestive processes in this region of the alimentary canal. These subsidiary movements are segmenting movements, which nip the food into segments and occur at a rate of about 12 per minute, and pendular movements which cause swaying of the intestine from side to side. The three systems cause a very thorough churning of the food as they pass it to and fro across the villi.

At the junction of the small intestine and the colon is the ileo-colic sphincter. This is normally shut and serves to prevent the forward passage of food and, more importantly, its backwards return from the colon along with harmful micro-organisms. After several hours the sphincter is reflexly opened and the contents of the gut move on, by peristalsis, into the large intestine.

3.5.8 *The large intestine*

A tube 8 or more centimetres in diameter which, in man, is seen to be made up of three portions: an ascending limb, a transverse limb and a descending limb. Its functions in different mammals are very variable, but in man its importance is less than that in herbivores, where it plays a vital part in digestion, and more than that in carnivores, where it is very reduced. The structure of the large intestine is similar to that of the rest of the gut, but it has no villi and no secretions other than mucus.

In man the functions of the large intestine, or colon, are to receive excretory substances removed from the blood, in particular calcium, magnesium, iron and phosphate ions, and to assimilate water and other substances. Some 500 millilitres of water pass along the large intestine in 24 hours but in the same period only 100 millilitres are eliminated in the faeces.

The large intestine has a rich bacterial flora and these live on undigested and indigestible residues entering from the ileum. Prominent species include varieties of staphylococci, including the well-known *Escherichia coli* (which can be used as an index of sewage pollution of fresh water), *Bacillus pyrocyaneus*, etc. One estimate of the number of bacteria in the human large intestine is 1.28×10^{11}, and about 40 % of the weight of the faeces is made up of bacteria, mostly dead. Recently it has been observed that prolonged treatment by antibiotics causes vitamin, especially vitamin B, deficiency disease, so that it would appear that in our own case a proportion of our vitamin requirements is provided by our intestinal flora.

The descending portion of the large intestine leads to the rectum, and this opens to the exterior by means of the anus.

3.5.9 *Movements of the large intestine and elimination of faeces*

The large intestine receives parasympathetic stimulation from the vagus nerve and sympathetic innervation from the ganglia of the lumbar and sacral regions of the spinal cord. The movement of the contents of the large intestine takes place by mass peristalsis, which is intermittent and caused by the taking in of food, or movement higher up in the gut. These movements automatically fill the rectum and when the pressure of the contents increases beyond some 40 mm Hg the anal sphincter will send stimuli to the central nervous system which leads to the elimination of the faeces.

Other than bacteria, the waste matter voided consists of indigestible matter such as cellulose which has passed unchanged through the gut and is called roughage. It is beneficial to include a certain amount of such matter in the diet as it gives the intestinal muscles work to do and thus maintains their tone. Other materials include a part of the gut lining and bile pigments.

3.6 Modifications of the alimentary canal found in mammals other than man

3.6.1 *The teeth and jaws*

In a mammal these indicate quite clearly the feeding habits of their owner. As already described for man the teeth of mammals fall into four categories. In the front of the mouth are the chisel-shaped incisors, which are followed by a single sharp canine. Behind this are the premolars which may be modified for grinding but do not usually show the complex cusp patterns of the back teeth or molars.

Herbivores

In herbivores the incisors are large (Fig. 3.16*a*), and are continuously self-sharpening by having a leading edge of enamel working against softer dentine in the opposing tooth. The canines are usually absent and in their place a large gap, or diastema, exists which allows the food to be freely circulated in the mouth. The premolars tend to become molariform and work in conjunction with the molars, the two types of teeth making up a broad grinding surface. Like the incisors, they show continuous growth and also the most specialised of mammalian cusp patterns. In order to achieve a close-fitting and indented surface the cusps are made up of an

Enamel

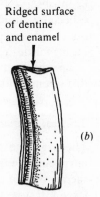

Ridged surface
of dentine
and enamel

(a)

(b)

Fig. 3.16.
Teeth of a rabbit:
(*a*) an incisor; (*b*) a
molar.

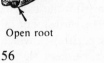

Open root

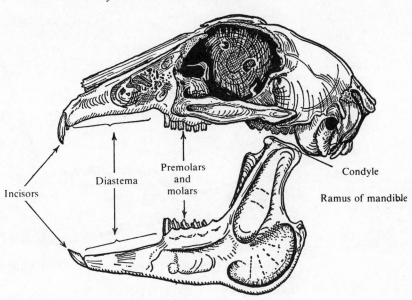

Incisors

Diastema

Premolars
and
molars

Condyle

Ramus of mandible

Fig. 3.17.
The dentition of a
rabbit (herbivore).

arrangement of enamel hills and dentine valleys placed in such a way that those of the one jaw fit into those of the other (Fig. 3.16*b*). Owing to the different hardnesses of these two substances an excellent grinding surface is maintained throughout the life of the animal. Besides the cusps of the individual teeth the whole system falls into a series of ridges running either transversely or along the line of the jaw. The direction of these ridges will conform with the direction of movement of the jaw. Fig. 3.17 illustrates the dentition of a rabbit.

As a herbivorous adaptation the importance of close grinding, which reduces the food to a cellular level, cannot be overemphasised. In the first place the food is rendered into a form in which it is suitable for rapid bacterial action in the rumen or intestine, and in the second place destruction of the cell structure leads to processes of autolysis or self-digestion. By this process the complex proteins and other chemicals of the protoplasm are destroyed by the activity of their own enzymes.

The mandible of a herbivore also shows several adaptations towards an efficient chewing action. For example the coronoid process is much elongated and this gives the masseters a large moment about the articulation. The temporal on the other hand has little leverage and is much reduced (Fig. 3.18). The chewing action is performed by the masseters rolling the narrow lower jaw from side to side against the upper. There is little vertical force on the mandible and it is weakly constructed compared with those of carnivores.

The joint that the posterior condyle makes with the glenoid is loose and allows the lower jaw to move freely from side to side. This is only possible in a system which uses small vertical forces.

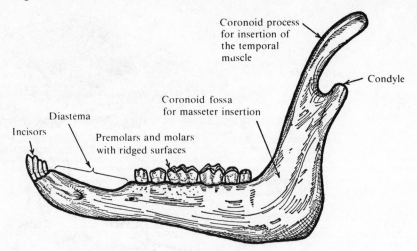

Fig. 3.18.
The lower jaw of a
sheep, showing
adaptations to a
herbivorous diet.

The dental formula of three typical herbivores is as follows:

Rabbit   I $\frac{2}{1}$ C $\frac{0}{0}$ PM $\frac{3}{2}$ M $\frac{3}{3}$ = 28
Horse   I $\frac{3}{3}$ C $\frac{1}{1}$ ♂ PM $\frac{3}{3}$ M $\frac{3}{3}$ = 40
Sheep   I $\frac{0}{3}$ C $\frac{0}{1}$ PM $\frac{3}{3}$ M $\frac{3}{3}$ = 32

Carnivores

In carnivores all the teeth find some use in feeding and, on the whole, there tend to be more of them than in the herbivores. In the front of the mouth, the incisors are not particularly well developed though they find some use in grooming and nibbling meat from bones. Behind the incisors are a pair of very large canines in each jaw and these stabbing teeth are used to catch the prey. The premolars and molars are covered in enamel and rise to sharp edges along the line of the jaw, the joint action of upper and lower teeth providing a powerful shearing action like scissors. Further adaptation of the teeth at the back of the jaw includes an enlarged 4th premolar in the top jaw working against a similarly enlarged 1st molar in the bottom jaw. These teeth, adapted for cutting and bone crunching, are called the carnassials and are found in the cat family and other families of carnivores.

Thus the dentition of carnivores (Fig. 3.19) is more adapted to tearing than to grinding, as flesh does not require the same degree of mastication as plant tissues. For this reason, though the teeth are large and very strong, they do not have the same wear as those of herbivores and are not open-rooted.

The articulating condyle of the dentary is placed low down and consequently the forces exerted by the temporal and masseter are more evenly distributed about the joint. Part of the strain on the jaw is provided by the forward and downward struggles of the prey and these are balanced by the opposite force of the large temporal. The other main strain involved in bone crunching and tearing meat is taken by the masseter, which acts perpendicularly to the line of the teeth.

58

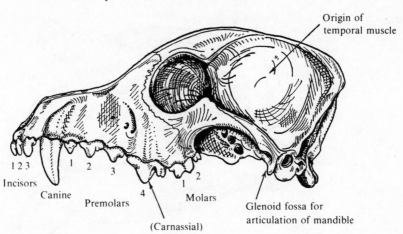

Origin of
temporal muscle

1 2 3   1   2
              3   1   2

Incisors
     Canine   4   Molars
         Premolars   Glenoid fossa for
                                  articulation of mandible
            (Carnassial)

Fig. 3.19.
The dentition of a dog
(carnivore).

The joint that the lower jaw makes with the glenoid (Fig. 3.19) is a tight roller joint and this is as important to the efficient working of the jaw as is a tight rivet to scissors. Held by this joint the jaw moves only in the vertical plane. A further point is that the positioning of the temporal and masseter at equal distances from the point of articulation enables them to take up the forces developed during the crunching of bones, and prevents these being carried to the joint itself which could cause dislocation. The dentary is very strong, to transmit these powerful forces to the teeth.

In the mammal the tongue assists in the mastication of food, presenting material to the teeth and circulating it in the buccal cavity. It also assists in swallowing.

3.6.2 *The alimentary canal of herbivores*

There are two main modifications of the alimentary canal of herbivores and these can best be illustrated by taking the horse and cow as examples. The former is less advanced and will be described first.

The stomach of the horse is large, but simple in construction, and resembles the type described for man. There are few micro-organisms present and little fermentation of cellulose takes place. The large intestine is extremely long and thrown into many bulges, or sacculated, to provide a stagnant environment for the activity of micro-organisms. These are able to break down cellulose and release simple food substances into the lumen of the gut whence they can be assimilated. The breakdown of bacterial protoplasm also provides vitamins. Thus the micro-organisms have come to have a symbiotic relationship with their host that is essential to it, and increase its range of potential foodstuffs to include things which would not be digestible without their aid. On the other hand their activity below the region of assimilation (that is, the ileum) is not very efficient as the large intestine has only limited powers of absorption. In some herbivores of this type, such as the rabbit, the products of bacterial digestion in the large intestine or the appendix are

59

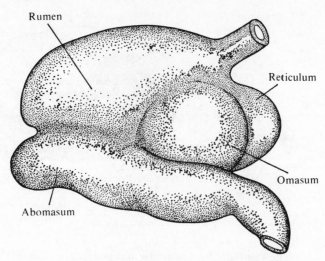

Rumen

Reticulum

Omasum

Abomasum

Fig. 3.20.
A cow's stomach.

passed out of the anus and eaten again so that nutritive matter can be assimilated. (This habit is called coprophagy.)

A more specialised herbivorous adaptation is found in certain ruminants, such as the cow. These have very elaborate stomachs divided into four regions (Fig. 3.20). The first of these regions is a large rumen, and into this food passes from the oesophagus. In the rumen it is churned and returned to the buccal cavity for further chewing; this corresponds to the 'chewing the cud' phase of digestion. If the food is fine, it may pass directly into the reticulum or second compartment of the stomach, but the major fermentation takes place in the rumen and the food may remain there for as long as 7 days. After passage through the rumen and reticulum the food enters the omasum where surplus moisture is absorbed or squeezed out, this part of the stomach being particularly muscular. Finally the semi-digested food passes to the abomasum, where the normal complement of enzymes operate. It is interesting to note that the epithelia of the first three regions of the ruminant stomach are cornified and tough like the lining of the oesophagus and provide a rough surface for mechanical breakdown. They are not impermeable, however, and many of the products of micro-organism digestion pass into the blood circulating the rumen. This is particularly true of fatty acids.

In contrast to the stomach, the large intestine of ruminants is simple, though long, and little fermentation takes place in it. Despite the complex anatomy of the stomach an oesophageal groove exists whereby certain types of food can be passed rapidly along the wall of the stomach from oesophagus to duodenum.

3.6.3 *The chemistry of ruminant digestion*

A typical ruminant such as a cow will have some 10^{11} bacteria and 10^6 protozoa per millilitre of rumen contents and will secrete up to 100 litres of saliva each day. In all, this mass makes up as much as one-seventh of the whole weight of the animal.

60

3.6 The alimentary canal in other mammals

Carbohydrates in the food are broken down and fermented by the micro-organisms to give fatty acids and carbonic acid and methane. The sodium bicarbonate of the copious saliva reacts with the carbonic acid to produce carbon dioxide, which the cow releases by belching. Meanwhile the fatty acids are absorbed directly into the blood where some, such as acetates and butyrates, are directly oxidised while others, such as proponates, are used in hexose sugar production. In fact all ruminants are noted for their extremely low levels of blood sugars, levels which in any other mammals would constitute serious hypoglycaemia.

Carbohydrate digestion and subsequent metabolism is thus:

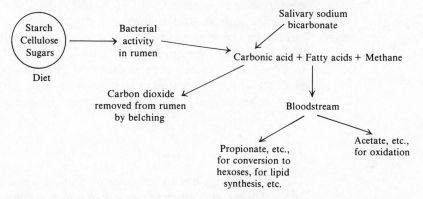

Protein intake is also somewhat unusual. The dietary protein is taken up by the flora of the rumen and converted to bacterial protein or else is deaminated and its nitrogen released as ammonia. While the ruminant eventually digests the bacterial protein, the ammonia passes direct to the bloodstream and thence to the liver. There it is changed to urea and the urea then passes to the salivary glands where it is released. This salivary urea enters the rumen and in its turn is taken up by micro-organisms and made into protein. Thus there is a cycle of nitrogenous substances from rumen to blood to saliva and back to rumen:

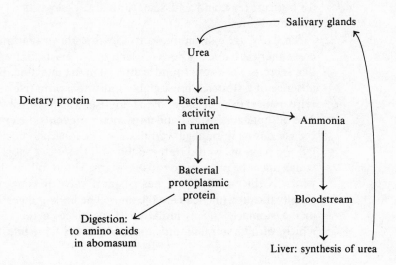

61

Phosphate goes through a comparable cycle, being also circulated via the saliva.

One further interesting characteristic of ruminants is the very high amounts of nuclease enzymes that they secrete in their alimentary canals. This is thought to be an adaptation to the vast amounts of micro-organism nucleic acids that they need to hydrolyse.

3.7 **The digestive system in non-mammalian vertebrates**

In general the digestive tract and its secretions are among the most conservative organ systems of the vertebrates. Even the neuronal and endocrine coordination mechanisms of the alimentary canal, where they have been investigated, are virtually identical throughout the group. It must be assumed that from the earliest vertebrates a satisfactory alimentary system had already been evolved, and that this was to need very little modification subsequently. Indeed by far the most elaborate divergences from the archetypal gut are found in the mammalian ruminants described above.

3.7.1 *Fishes*

The majority of fishes are carnivorous and seize and swallow their prey intact. Their teeth, which are modified scales, are homodont (i.e. all the same) and they have an infinite capacity to replace missing ones. Teeth are not confined to the jaws but may occur on the palate or even the pharynx.

Very primitive fishes such as cyclostomes, and a few of the more advanced teleosts (e.g. carp), do not have a stomach or any pepsin-like acid-requiring enzyme, so that food passes directly to an alkaline intestine. Other proteinases, lipases and amylases, however, are present in primitive fishes without a stomach. There are also a number of fishes which, although having a stomach, possess only one type of secretory cell for both protease and acid, secretions always separated in other classes of the vertebrates.

Similarly, though the majority of fishes have a normal discrete pancreas there are a few where no such organ exists, but where pancreatic-like secretory cells are found scattered in the intestine. Both this and the absence of a stomach presumably indicate a primitive state of affairs. A fairly typical fish gut is that of the dogfish, which is illustrated in Fig. 3.21. The oesophagus is short and a sphincter prevents the entry of water into the gut during normal gill ventilation. The lobed stomach has cardiac and pyloric portions well differentiated and there is a short duodenum into which open the pancreatic and bile ducts. This is followed by the intestine, which is thick-walled and has a spiral valve passing down its centre. Finally the intestine leads to the anus. The bulk of digestion is completed in the stomach and assimilation takes place in the valvular intestine, which, with its spiral modification, has a very large internal surface. The

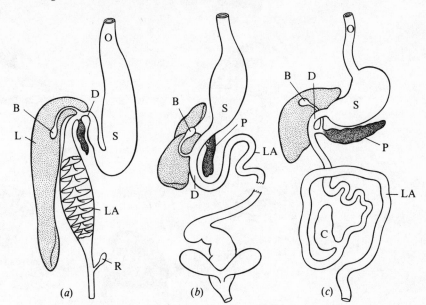

Fig. 3.21.
Comparative
alimentary canals of
(a) dogfish, (b) frog
and (c) mammal
(generalised).
O, oesophagus;
S, stomach;
D, duodenum;
P, pancreas; B, bile
duct; L, liver; R, rectal
gland; C, caecum;
LA, large intestine.

rectal gland has a function in the control of ionic equilibrium, particularly in the excretion of sodium chloride.

The typical teleost alimentary canal is much the same except that no spiral valve is found.

3.7.2 *Amphibians*

All amphibians are carnivorous as adults and the intestine of those that have herbivorous tadpoles shortens and changes as the diet alters. While the early Amphibia had large teeth the modern forms use a horny pad with which to seize their food; it is swallowed whole with the aid of the tongue and retracted eyeballs.

The frog is atypical of Amphibia in producing salivary amylase, but is typical in having a glandular oesophagus. From this a pepsin-like protease is secreted and passes into the stomach (as the pH of the oesophagus is not low enough) with the food. The stomach contains the acid-secreting cells so that the enzyme, as in other vertebrates, is active in this organ.

After the stomach is the duodenum, with secretory and mucous cells, and into this the bile duct and pancreatic duct open. The frog does not possess the crypts of Lieberkühn found in the mammalian duodenum. Finally the ileum has special absorptive cells but no villi, and leads into the short and thick-walled large intestine, which terminates in a cloaca.

3.7.3 *Reptiles*

This class is also mainly carnivorous (though some turtles and tortoises are herbivorous) and here the teeth, which are more or less homodont in most reptiles, have been retained. Snakes and lizards swallow their food

63

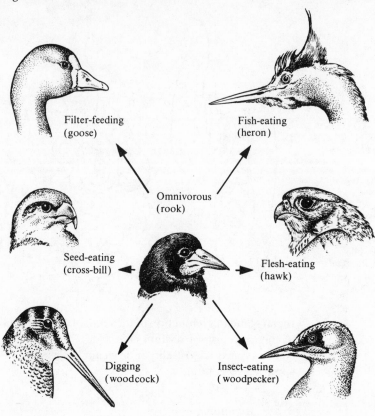

Fig. 3.22.
Adaptive radiation of
beaks of birds.

whole, while crocodiles tear portions off and also swallow stones into their stomach to assist mechanical breakdown of food.

Snakes have elaborate jaws which allow a gape larger than the snake's own head and also provide a means whereby the body of a relatively large prey can be pushed down into the stomach. Many snakes also have modified venom glands which besides producing poisons also secrete a number of hydrolases that help in digestion of their food. Their venoms are proteins which act in a variety of ways, some paralysing while others are haemolytic in action.

3.7.4 *Birds*

These vertebrates have no teeth and the jaws have a horny covering, the beak. This shows great diversity according to the type of diet and its adaptive radiation is comparable to that of the teeth of mammals (Fig. 3.22). As with the mammals, such adaptive radiation allows efficient exploitation of a wide range of biological niches and undoubtedly contributes to the success of the group as a whole.

Like the beak, the crop which ends the oesophagus may be variable, but generally its function is to sort, and if necessary reject, substances taken in to the mouth. An interesting adaptation of the crop occurs in the pigeon, where a milk-like fluid is produced by the sloughing off of the lining epithelium and this is used to initiate feeding in the young.

Below the crop is a muscular gizzard which may contain stones and serves to grind up the food. This adaptation fulfils the function of the teeth in other vertebrates and allows the inclusion of a wide range of foodstuffs in the diet. The remainder of the gut is as already described for other vertebrates except that no gall bladder is present.

Like Amphibia and Reptilia the birds have a cloaca at the end of the intestine through which not only waste from the gut but also urine and genital products must pass. In the birds the cloaca is an elaborate structure divided into three distinct regions. Dorsal to the final region, the proctodaeum, is situated the bursa of Fabricius, an organ now realised to be important in the initiation of the immune reactions of the animal.

4 Respiration, gas exchange and transport systems

Respiration is vital to all living organisms because it provides the energy required not only to maintain their body structure but also to carry out their varied activities. The energy is liberated by oxidative processes in the cells and for this purpose the oxygen is taken in at a respiratory surface (where carbon dioxide is usually liberated). It is convenient, therefore, to separate external respiration from tissue or cellular respiration, but in animals the size of vertebrates a third transport process is required. In very small organisms, external respiration is mainly by diffusion across the external membranes to the sites of oxidation. Knowing the rates of oxygen consumption and diffusion it has been estimated that diffusion alone can supply sufficient oxygen only if the organism is less than 1 millimetre in diameter. This is due partly to the relative decrease in the size of the surface absorbing the oxygen to the volume it has to supply (surface increases by the square of the linear dimensions, volume by their cube); it is also due to the length of the diffusion path involved from the surface to the internal tissue.

The volume of oxygen required by an animal is often expressed as the number of millilitres per kilogram per hour. Animals vary considerably in their oxygen requirements, mammals and birds using larger amounts than the lower vertebrates. This generalisation only holds for animals of the same size, because in all groups there is an inverse relationship between body size and metabolic rate. For example, a small mammal such as a shrew may require 10–20 times more oxygen per unit weight than a man. Mammals and birds require a continuous supply of oxygen if they are to maintain their metabolic rate and high body temperature. In some cases this high level of metabolism is interrupted and the animal passes into a state of hibernation for a long period. Some small mammals and birds undergo similar reductions in temperature and metabolism which may occur in a 24-hour rhythm. For example, humming birds and bats enter a so-called 'torpid state' during the night and day respectively.

In both aquatic and terrestrial vertebrates the final path of oxygen from the external medium to the blood is by diffusion across an aqueous phase that covers the respiratory epithelium. Diffusion in air is rapid; in water it is 300,000 times slower. There is a danger, greater in water than air, that stagnant layers may form at the respiratory surface with consequent reduction in the rate of gaseous exchange. Animals partly overcome this danger by rhythmic ventilation of the respiratory organ. In aquatic verte-

4.1 Respiration in mammals

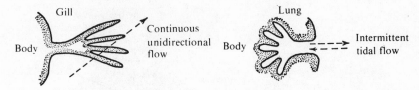

brates the respiratory surface is the gill, a finely divided external expansion of the surface epithelium which is readily ventilated and is supported by the aqueous medium (Fig. 4.1). The danger of water loss is absent here, but on land this is a problem and such an organ would be useless. The respiratory organs of land animals are usually in-tuckings of the surface that have only a small proportion of their moist epithelium directly in contact with the external air and thus enable gaseous exchange to take place with the minimum loss of water. Ventilation of such an organ is more difficult than of a gill where a continuous flow of water is readily achieved, but in most lungs the ventilation of the respiratory epithelium is tidal (Fig. 4.1). The lungs of lower vertebrates are simple sacs and as we go up the vertebrate series an increase in the folding of the respiratory epithelium is found. In all vertebrates there is a relationship between the size of the respiratory surface and the activity of the organism (Table 4.1, p. 78). The thickness of the barrier to gas transfer between the respiratory medium and the blood tends to be lower in the most active vertebrates.

4.1 Respiration in mammals

4.1.1 The respiratory tract

The lungs are contained within the thorax (Fig. 4.2) where they are surrounded by the inner and external pleural membranes, mesodermal in origin, which are normally closely apposed to one another. In certain diseased conditions, for example pleurisy, they may become inflamed, or the pleural space may be filled with liquid or air.

Air enters the respiratory tract through the nostrils, which act as filters and warm the air before it passes above the soft palate to the back of the throat on its way to the lung. Posteriorly the nasal passage enters the pharynx into which also opens the cavity of the mouth. Leading off from the pharyngeal cavity are several passages. The most important of these are the oesophagus, down which the food passes, and, ventral to this, the opening to the respiratory tract proper. This opening is protected by the epiglottis, which prevents food entering the larynx during swallowing. The epiglottis projects upwards from the floor of the pharynx and is stiffened by an elastic cartilage. The larynx is supported by several cartilages and internally contains the vocal cords, which may be vibrated by the air stream. From the larynx the trachea leads posteriorly and bifurcates into the two primary bronchi. All these tubes are supported by incomplete rings of cartilage which prevent their collapse when the pressure inside is reduced.

67

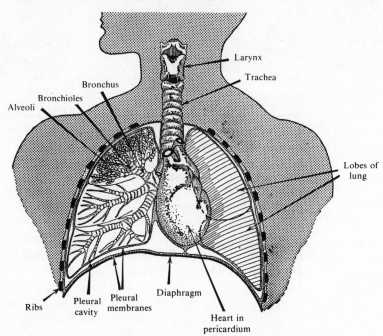

Fig. 4.2.
The human lung and
its associated
structures.

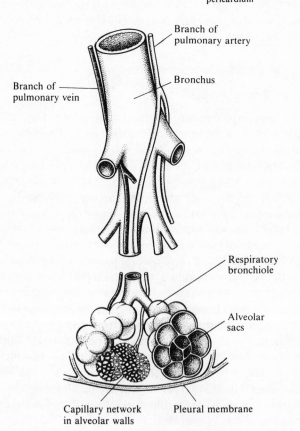

Fig. 4.3.
The alveolar region of
the lung.

4.1 Respiration in mammals

The bronchi branch three more times and give rise to smaller tubes called bronchioles. These latter also exhibit a pattern of dichotomy, although not a regular dichotomy, rather like plant stems, branching with up to 20 sets of bifurcations from each original tube. As they approach the alveoli, or air sacs, the individual bronchioles become very small (approximately 0.5 millimetres in diameter) and lose their cartilage plates. They are lined by a cuboidal ciliated epithelium and near the alveoli themselves even their connective tissue layer is absent, but a smooth muscle layer is still present even in the wall of the alveolar ducts. For the last six generations of bifurcation the respiratory bronchioles show other changes. At first individual alveolar sacs jut out from their walls, and these become steadily more frequent until the complete alveolar sac system is reached.

The alveolar sac system and its associated capillaries (Fig. 4.3) makes up 60 % or so of the whole lung, and it is in this part of the lung that most of the gaseous exchange in the body occurs.

4.1.2 The alveoli

The average adult human male has some 300 million alveoli in his lung, which give a total surface area of some 80 square metres (equivalent to a full-sized tennis court). The alveoli are thin, moist and supplied with a rich capillary network; they are thus highly adapted for efficient gaseous exchange.

The walls of adjacent alveoli have capillaries between them, separated by elastic connective tissue which allows expansion and also gives the wall support (Fig. 4.4). Between the blood and the lumen of the air sac are squamous epithelial cells which are very thin and readily allow diffusion of gases. The thickness of the air-to-blood barrier is only $c.0.5$ micrometres in most parts of the alveolar sac and it is calculated that each sac is surrounded by some 50×10^{-6} millilitres of blood or by some 2,300 red blood corpuscles. The total length of the lung capillary system is of the order of 19,000 kilometres, so it can be seen that the connections between blood and air in the lung are extremely intimate.

Oxygen and carbon dioxide pass to and from the air spaces into and out of the blood capillaries from both sides (see Fig. 4.4). Many alveoli also have pores through to adjacent chambers which probably assist the circulation of air.

Besides the epithelial lining cells and the endothelial cells around the capillaries there are other important cell types found in the alveoli. Cubical secretory cells in the walls are associated with the release of surfactant, a fluid which decreases surface tension on the alveolar walls. This allows them to open more readily, not only at birth but also during normal respiratory exchanges. Respiratory distress syndrome in newly born infants can be due to a failure of the lungs to expand normally because of a deficiency in lung surfactant.

As might have been predicted, large numbers of phagocytes (alveolar

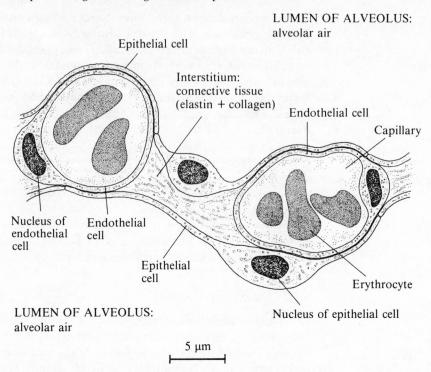

LUMEN OF ALVEOLUS:
alveolar air

Epithelial cell

Interstitium:
connective tissue
(elastin + collagen)

Endothelial cell

Capillary

Nucleus of
endothelial
cell

Endothelial
cell

Epithelial
cell

Erythrocyte

LUMEN OF ALVEOLUS:
alveolar air

Nucleus of epithelial cell

5 μm

Fig. 4.4.
The wall of the
alveolus. Note that
there are three distinct
layers between the
lumen and the blood:
the epithelium, the
interstitium and the
endothelium.

macrophages) that have by chance reached the gas exchange surface are present on the surface of the alveolar tissues and these function in the removal of dust and other foreign bodies that have inadvertently entered the lung.

4.1.3 *Ventilation*

The elastic lungs are contained within the thoracic cavity, which is a closed box supported externally by the ribs and separated posteriorly from the abdominal cavity by the dome-shaped diaphragm (see Fig. 4.2, p. 68). Changes in volume of the thorax immediately cause slight rarefaction or compression of the air in the lung. These changes in pressure result in the flow of air into and out of the lungs (Fig. 4.5). The pressure within the thoracic cavity (intrathoracic pressure) is normally about 4 mm Hg less than atmospheric, which is the pressure in the lung when at rest. This difference in pressure is probably due to changes which occur during development. In the early stages the lungs completely fill the thoracic cavity but later the thorax increases in volume to a greater extent than the lungs. This results in the lungs becoming stretched away from the thoracic cage. Consequently a reduced pressure within the thorax is produced. During inspiration the intrathoracic pressure is reduced by some 5 mm Hg and this reduction is transmitted to the lung so that air flows in to restore the intrapulmonic pressure to atmospheric. During expiration

70

the intrathoracic pressure rises again to its resting level and the increase in pressure compresses the air in the lung causing its pressure to rise above atmospheric. Air is forced out until once more the pressure within the lung is restored at atmospheric before the start of the next inspiration.

Changes in volume of the thoracic cavity are brought about by two main types of muscular action. The diaphragm may contract and effectively increase the antero-posterior dimensions of the thoracic cavity, or the ribs may be raised, so increasing the cross-sectional area of the thorax. These two movements are not entirely separate, for contraction of the diaphragm also has the effect of raising the costal margin. Normally a combination of diaphragmatic and costal pumping is used, but under certain circumstances sufficient ventilation can be achieved by only one of these actions. The diaphragm is dome-shaped anteriorly and is attached to the lumbar vertebrae and the posterior ribs. The central tendon is not domed upwards as much as on the two sides so that during relaxation the right and left sides of the diaphragm are elevated above the central tendon by the pressure of the abdominal contents. These arches are flattened when the diaphragm contracts and the central tendon moves down about 1.5 centimetres. The diaphragm is innervated by paired phrenic nerves which contain the motor fibres. Cutting one phrenic nerve abolishes the contraction of the diaphragm on that side.

The ribs articulate with the thoracic vertebrae by two heads, a tuberculum to the transverse process and a capitulum to the centrum. Movements of the ribs are about the line joining these two articulations. At rest, the ribs are inclined backwards and ventrally with respect to the vertebral column but during inspiration they move anteriorly with a consequent increase in the transverse dimensions of the chest. In human infants, the ribs are nearly at right angles to the vertebral column and hence any movement, whether forward or backward, decreases the chest volume; consequently ventilation is largely diaphragmatic. The upward and outward movements of the ribs has been likened to the action of a

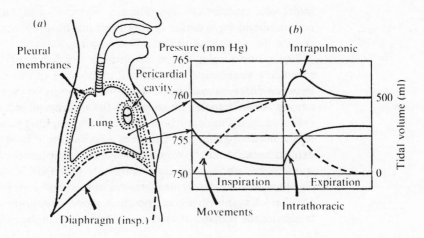

Fig. 4.5.
Pressure and volume changes during the respiratory cycle in man.

71

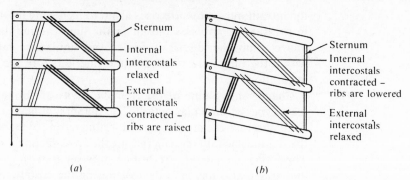

Fig. 4.6.
The use of the antagonistic intercostal muscles in (*a*) raising and (*b*) lowering the rib cage.

bucket-handle about the articulations with the vertebral column and sternum. The intercostal muscles which produce these movements are divisible into external and internal groups. During quiet breathing no distinction can be made between the action of these two groups, but in more active ventilation the external group is clearly involved in the inspiratory movement (Fig. 4.6). This action is due to their attachment to the anterior rib being nearer to the vertebral column than the insertion on the more posterior rib. The internal intercostals have the relative positions of their insertions reversed and thus tend to lower the rib cage when they contract. The first rib and its attachment to the sternum remains fixed during quiet breathing and forms a point relative to which the other ribs move.

Expiration is mainly produced by passive relaxation of the inspiratory musculature and a return of the ribs and diaphragm to their resting position. Even in quiet breathing, however, there is evidence that expiration is not entirely passive because some of the inspiratory muscles remain slightly active even when the volume of the thoracic cavity is increasing. This leads to a more controlled outflow of air from the lung. During vigorous breathing active expiratory muscles are brought into action, mainly those of the abdomen, which increase the pressure within the abdominal cavity and help to force air out of the thorax. The internal intercostal muscles also assist in this. The size of the nostrils and glottis are increased by muscles during inspiration, movements which clearly reduce the resistance to the inflow of air.

The relative roles of the diaphragm and intercostal musculature in producing ventilation vary among individuals of the same species and between different mammals. Human infants rely on their diaphragm, but among adults females make use of the intercostal muscles to a greater extent than do males, who breathe more diaphragmatically. Mammals which normally stand on four legs also use the diaphragm to a greater extent, because the rib cage is involved in a supporting function. Correspondingly, amongst aquatic mammals the body is supported by the medium and the intercostal musculature is more developed. The diaphragm of many diving mammals is more horizontal in the body and facilitates the transmission of pressure changes in the external medium to

72

4.1 Respiration in mammals

the pleural cavities. Whales have relatively small lung volumes (2.5 millilitres per 100 grams weight) and a large proportion of dead space. In seals the lungs are about 5 millilitres per 100 grams body weight which is similar to the figure for a man. The tidal air of diving mammals makes up a large proportion of the total lung volume (80 % in a porpoise) and assists in the rapid exchange of gases when they surface.

During normal breathing the respiratory movements occur about 32 times per minute in a new-born child, 26 times in a child of 5, 16 times in a man of 25, and 18 times in a man of 50.

In man, the total capacity of the lungs is about $5\frac{1}{2}$ litres (Fig. 4.7). But even following the most forced expiration $1\frac{1}{2}$ litres or more of this cannot be forced out of the lung. The maximum volume which can be ventilated during forced breathing is referred to as the vital capacity of the lungs. This is normally between 3 and 4 litres and may be as high as 5 or 6 litres among athletes. The volume of air moving in and out of the lung during normal breathing makes up the tidal volume. During quiet breathing a man takes in about 500 millilitres of air. Of this volume only a certain part reaches the alveoli as the rest fills what is known as the *dead space* – the air tubes, etc., leading to the respiratory epithelium. The dead space normally amounts to 140 millilitres of the tidal volume. Consequently the volume actually reaching the alveoli (the alveolar air) is little more than 360 millilitres. Because ventilation is tidal, during expiration the dead space is filled with air of the same composition as alveolar air and by forced expiration it may be sampled.

The composition of the alveolar air at rest and under normal atmospheric conditions remains very constant; in fact most of the regulatory mechanisms of respiration are designed to maintain the constancy of this air. The compositions of the inspired and expired air differ mainly in the proportions of oxygen and carbon dioxide. Inspired air contains 20.95 % oxygen and 0.04 % of carbon dioxide, whereas the expired air has only

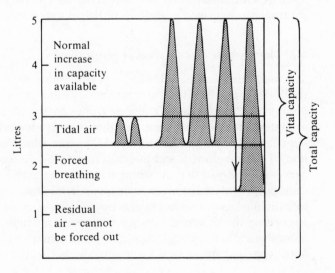

Fig. 4.7.
The functional divisions of the capacity of the lungs.

Fig. 4.8.
Changes in the
concentrations and
partial pressures of
oxygen and carbon
dioxide as they pass
through the lungs,
circulation and tissues.
(Data are for man,
with the composition of
the atmosphere as it is
at sea level.)

Atmosphere		Oxygen		Carbon dioxide	
		Concentration	Partial pressure (mm Hg)	Concentration	Partial pressure (mm Hg)
Inspired air	Gaseous	(%) 20.95	159	(%) 0.04	0.30
Alveolar air	Gaseous	13.80	98	5.5	40
Pulmonary vein and arteries of the body	In solution	(ml/100ml blood) 19	96	(ml/100ml blood) 50	40
Capillaries of the body	In solution	19–12.5	96–40	50–56	40
Pulmonary artery and veins of the body	In solution	12.5	40	56	46
Expired air	Gaseous	(%) 16.4	116	(%) 4.1	28.5
Atmosphere	Gaseous	20.95	116	0.04	0.30

16.4 % oxygen but 4.1 % carbon dioxide (Fig. 4.8). The percentage of carbon dioxide increases when the tidal volume is increased because the dilution effect of the dead-space air diminishes. The marked difference in composition of the inspired and expired air contrasts with the relative constancy of the alveolar air. This is partly because the tidal air normally forms a small proportion of the total air in the alveoli. Thus at the end of normal expiration the alveoli still contain 2½ litres of gases. In inspiration, 360 millilitres of atmospheric air are taken into this space and mixed with the 2½ litres already present. Consequently the effects of this small ventilation on the total lung air are very slight and will only amount to less than 0.5 %. Haldane determined the alveolar carbon dioxide at the end of inspiration and expiration respectively; his figures were 5.54 % and 5.70 %. Thus the alveolar air can function as a sort of buffer between the external air and the gas tensions of the blood.

4.2 The nervous coordination of respiration

The neurones concerned with regulating the rhythm and depth of respiratory movements are contained within the pons and medulla oblongata of vertebrates. These neurones have intrinsic properties and are interconnected in such a way that they produce rhythmic bursts of impulses in the phrenic and intercostal nerves, which lead to the contraction of the diaphragm and intercostal muscles. Recordings from these nerves have shown that the strength of the contraction of the respiratory muscles is regulated partly by an increase in the frequency of the impulses to individual motor units but also by bringing more of them into action. Recording the electrical activity within the medulla shows that some neurones are active during inspiration and others during expiration, but both types are intermingled with one another so that a clear separation

4.2 *The nervous coordination of respiration*

into inspiratory and expiratory centres cannot be made. The neurones continue to show their rhythmic discharges when the medulla is isolated from any sensory input either by cutting all the cranial nerves or by abolishing the respiratory movements by the injection of curare (South American Indian arrowhead poison) which paralyses the muscles. The basic respiratory rhythm is therefore intrinsic to these neuronal networks; but it can be modified by inputs from proprioceptive sense organs in the bronchial tree. During normal respiration discharges from these sense organs can be recorded from the vagal nerves, and they initiate by reflex the next phase (inspiration) of the cycle (Fig. 4.9).

The rate and depth of ventilation are precisely regulated according to the composition of the alveolar air. Small increases in carbon dioxide (0.25 %) may double the ventilation volume. The carbon dioxide changes the pH of the cerebrospinal fluid and this change is detected by special cells in the medulla. Cells in the medulla are extremely sensitive to carbon dioxide and immediately alter the respiratory rhythm. Injection of carbon dioxide in bicarbonate buffer produces effects similar to those produced by heightened carbon dioxide tensions of the alveolar air. The carbon dioxide tension is also detected by receptors in the carotid body and aortic arch. These regions are most sensitive, however, to changes in the oxygen tension, which again has some controlling effect, though not

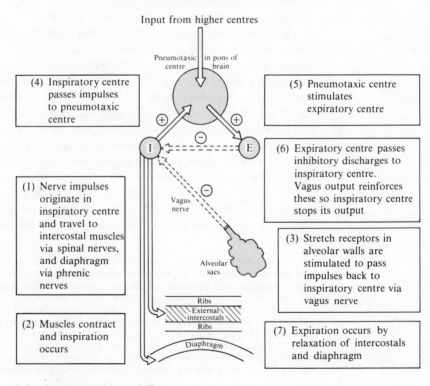

Fig. 4.9.
A possible scheme of some major interactions in normal breathing. The scheme is based on brain section preparations and represents only an incomplete picture of all the events that occur.

I, inspiratory centre in medulla.
E, expiratory centre in medulla.

75

so great as the carbon dioxide tension. The respiratory centres of the medulla are very insensitive to lowering of the oxygen tension.

In addition to these factors, which are clearly respiratory in nature, the respiratory rhythm may also be influenced by effects produced in other parts of the central nervous system. For instance, stimulation of certain areas within the cerebral hemispheres or of the cerebellum may affect the respiratory rhythm. It is an everyday experience that the respiratory movements can be controlled voluntarily.

4.3 **Transport of the respiratory gases**

The source of oxygen is one factor which limits the rate at which the gas can be obtained by an animal. Air is a much richer source than water. One litre of air contains approximately 200 millilitres of oxygen, whereas natural waters contain between 0.05 and 9 millilitres of oxygen per litre depending on the type of water and the temperature.*

The oxygen-carrying capacity of the blood is increased by the presence of the respiratory pigment haemoglobin (although there are a very few vertebrates, such as the leptocephalus larvae of eels and some antarctic fish, that do not possess haemoglobin). The haemoglobin is contained within corpuscles – the red blood corpuscles – and this arrangement enables the blood viscosity and osmotic pressure to be kept smaller than if the same amount were dissolved in the plasma.

The study of the functioning of haemoglobin as a respiratory pigment has been greatly aided by its very characteristic absorption spectrum under different conditions. The oxygenated form of the molecule, oxyhaemoglobin, has two chief absorption bands, whereas reduced haemoglobin has a single broad band in the yellow-green part of the spectrum. Haemoglobins readily combine with carbon monoxide to produce carboxyhaemoglobin which also has two well-defined bands. Unlike haemoglobin, carboxyhaemoglobin does not combine reversibly with oxygen, which is the reason for its poisonous effects. The importance of haemoglobin to an animal can therefore be studied by following the effects of carbon monoxide poisoning, which in most cases is lethal. Some fishes, however, are able to survive when all their haemoglobin is rendered useless, so long as they are kept at low temperatures and inactive.

* The amount of oxygen which will dissolve in a solution is determined by its solubility and partial pressure in the gas mixture. The value of the partial pressure of the gas with which the solution is in equilibrium is referred to as the tension of the gas in the solution. For example, water saturated with oxygen has an oxygen tension of 152 mm Hg, which is the partial pressure of oxygen in air. This tension will be reduced if the partial pressure of oxygen in the air falls, as for example at high altitudes. It is customary to refer to the oxygen in terms of its tension when considering diffusion and uptake at the respiratory epithelium because the equations relating to these phenomena are expressed as partial pressures. On the other hand, when considering the amount of oxygen available to an aerial or aquatic animal the volume of gas per litre is a more useful way of expressing the concentration.

4.3.1 *The nature of haemoglobin*

The haemoglobin molecule is made up of two α chains, each of which has 141 amino acid residues, and two β chains with 146 residues each. To each of the four chains is linked a haem unit, with the amino acid histidine acting as the linkage site; the total molecular weight is 34,000. Haem is the respiratory carrier part of the haemoglobin molecule. Each haem unit contains a central iron atom which is in the ferrous (Fe^{2+}) state, and it is this iron atom that combines with oxygen, although its valency remains unchanged. This is illustrated below.

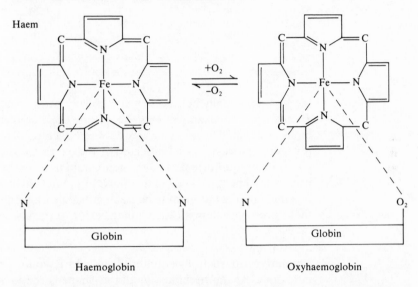

Haemoglobin   Oxyhaemoglobin

The detailed configuration of the molecule, as well as some suggestions as to its mode of functioning, was elucidated by M. F. Perutz in the early 1960s.

Both α and β chains are very similar in the coiling of their tertiary structures. In the quaternary structure of the molecule, formed by the linking of the α and β chains, α chains connect mainly with β and vice versa.

Differences in the properties of various haemoglobins are due to small changes in the polypeptide chains. Foetal haemoglobin differs from adult haemoglobin only in its polypeptide chains. The species-specific properties of oxygenation are probably due to small changes in the chains at the regions where they coil in the vicinity of the iron atom. The very marked effects of small changes in these chains have been well worked out for the defective haemoglobin responsible for the disease sickle-cell anaemia. The hundredfold reduction in solubility of this haemoglobin is due to the place of one glutamic acid residue, out of a total of 290 residues, being taken by a valine residue.

4.3.2 *Combination with oxygen*

An important property for any blood pigment is that it must increase the volume of oxygen carried by a given volume of blood when fully saturated (Table 4.1). In most birds and mammals the oxygen-carrying capacity of

TABLE 4.1. *Oxygen consumption of various vertebrates*

Animal	Oxygen requirements (ml/kg/hour)	Lung or gill surface area (cm²/g)	Oxygen carrying capacity of blood (volume %)
Man	330	7	20
Mouse	2,500–20,000	54	*c.* 17
Sparrow	6,700	—	*c.* 16
Frog	150	2.5	10
Dogfish	54.5	1.86	5

the blood is usually between 15 and 20 volume %, being higher in some diving forms, for example seals (29 volume %), but lower in most amphibians and fish. The combination between haemoglobin and oxygen is a reversible process which is affected by the oxygen tension with which the blood is in equilibrium. The oxygen combines loosely with the ferrous iron in the haem part of the haemoglobin molecule in the proportion of one molecule per atom of iron. Each molecule of haemoglobin (Hb) can therefore combine with four molecules of oxygen:

$$Hb + O_2 \leftrightharpoons HbO_8$$

The curve for such an equilibrium would have an extremely marked inflexion, but the peculiar sigmoid (S-shaped) shape of the oxyhaemoglobin dissociation curve is explained by the intermediate compound theory:

$$Hb + O_2 \leftrightharpoons HbO_2$$
$$HbO_2 + O_2 \leftrightharpoons HbO_4$$
$$HbO_4 + O_2 \leftrightharpoons HbO_6$$
$$HbO_6 + O_2 \leftrightharpoons HbO_8$$

The actual mechanism of oxygen transport by the haemoglobin molecule is still somewhat controversial but Perutz and his co-workers have shown that considerable changes occur in the quaternary configuration of the molecule as it becomes oxygenated. Entry of the first two molecules of oxygen causes the chains to rotate apart, making the entry of subsequent oxygen molecules easier. It is suggested that haemoglobin exists as a high-affinity form for oxygen and as a low-affinity form, and that the first combination with oxygen takes place with the low-affinity form. The first two oxygen molecules occupy haem sites on the α chains, and by their presence cause a separation of the α and β chains. This in turn exposes the high-affinity form of the haemoglobin and subsequent oxygen molecules are taken up easily, to sites on the β chains. In fact it is found that the haemoglobin of a given individual is made up of varying amounts of the high and low affinity forms – thus the sum affinity of the blood for oxygen can show variations in different environmental situations. The ability to modify the affinity of the blood for oxygen, albeit over a period of time, must confer survival advantage.

On deoxygenation the reverse process occurs, the β chains giving up their oxygen first and then the α chains, so that the chains once again rotate closer to each other. Throughout the whole process of oxygenation and deoxygenation the iron atoms of the haem units maintain their ferrous state.

Haemoglobin functions by giving up oxygen at low oxygen tensions (in the tissues) and combining with it at high oxygen tensions (in the lungs). The relationship between tension and percentage saturation of the blood is not linear, however, as can be seen from the dissociation curve in Fig. 4.10. Most haemoglobins become fully saturated at oxygen tensions below that at their respiratory surfaces. This provides a 'safety factor' whereby the ambient oxygen tension can be lowered quite considerably before the blood is no longer saturated with oxygen. The steep part of the sigmoid curve usually coincides with the normal range of tensions over which the haemoglobin functions in the animal. Some indication of the position of the curve is given by the unloading tension (T_u), sometimes called P_{50} because it is the tension at which the haemoglobin is 50 % saturated. The higher the value of P_{50} maintained in the animal the better, as it gives some indication of the level of oxygen tension in the tissues. Birds have high P_{50} values (duck, 50 mm Hg), in mammals it is about 30 mm Hg and for fishes the value is in the range 10–20 mm Hg. The transfer of oxygen from the maternal to foetal circulation is aided by differences in the dissociation curves of the two haemoglobins. The foetal curve is to the left of the maternal curve, so that it becomes oxygenated at lower partial pressures (Fig. 4.10). There is also a counter-flow between foetal and maternal bloods in the placenta of some mammals.

The dissociation curve of bloods varies according to the carbon dioxide tension. At higher carbon dioxide concentrations (with consequent lowering of the pH) the curve is moved to the right – the so-called 'Bohr shift'.

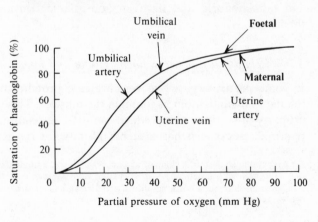

ig. 4.10.
'omparison of the
xygen dissociation
urves for human
aaternal and foetal
lood. The foetal blood
vith its curve to the
ft of the maternal)
ill load up with
xygen from the
aother's blood at a
iven partial pressure.

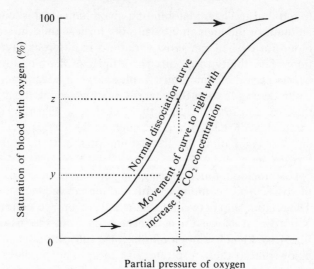

Fig. 4.11.
The Bohr shift. At a given partial pressure of oxygen, say x mm Hg, the haemoglobin normally carries y % oxygen. But if the carbon dioxide concentration rises and the pH falls it is able to carry more oxygen (z %).

This has the important effect of increasing the amount of oxygen liberated from the haemoglobin in tissues where the carbon dioxide content is high (Fig. 4.11). The actual curve along which the haemoglobin functions in the body lies between the curves determined experimentally at different carbon dioxide tensions. This effective curve is steeper than those at a single carbon dioxide tension and consequently leads to a greater liberation of oxygen for a given difference in oxygen tension between the lungs and tissues. The haemoglobins of tetrapods and fishes which live under conditions of high carbon dioxide and low oxygen tension tend to be relatively insensitive to carbon dioxide, whereas fish living in well-aerated conditions have blood which is markedly affected by the carbon dioxide tension.

There are other factors which influence the carriage of oxygen by red corpuscles. Substances such as DPG (diphosphoglycerate) can be important in some vertebrates, while ATP has been found to play such a role in fishes. The investigation of the affinity of non-mammalian erythrocytes for oxygen is complicated by the fact that they contain nuclei and mitochondria and have an appreciable consumption of oxygen themselves.

4.3.3 *Relationship between haemoglobin and myoglobin*

In some particularly active vertebrates a protein myoglobin,* closely related to haemoglobin, is found in the muscles. This myoglobin has the property of storing oxygen which can then be released to the muscle as required. Because it has a higher affinity for oxygen than the blood

* Myoglobin consists of a chain of 153 amino acids and a single haem unit, and it was the second protein whose structure was elucidated. This was achieved by J. Kendrew in 1961 using the same techniques of X-ray diffraction that were to be successful with haemoglobin a few years later.

haemoglobin it automatically loads up from this latter source. Diving mammals such as whales have very large amounts of myoglobin in their 'red' muscles and before they dive this myoglobin will have become fully charged with oxygen obtained from the circulatory blood. This oxygen store is used up during the dive and is one of the adaptations involved in increasing the length of time that these animals can remain below the surface.

Myoglobin is also found in the flight muscles of birds, the most active of all vertebrates, and in powerful swimmers such as salmon and tuna fish.

If a section is taken through a fish, such as a herring, strips of red muscle are seen along its sides, distinct from the white myotomal muscles. The fish uses these aerobic red muscles for 'cruising' and the oxygen-debt-producing white muscles for rapid accelerations and escape movements. In fact the storage function of myoglobin may not be as important as its ability to increase the rate of diffusion of oxygen between the capillaries and the mitochondria.

4.3.4 *Transport of carbon dioxide*

Haemoglobin in vertebrate blood also aids the transport of carbon dioxide (Fig. 4.12). It does this partly by combining directly with the carbon dioxide to form a carbamino compound. About one third of the carbon dioxide that is transported and liberated is in this form:

$$HHbNH_2 + CO_2 \rightleftharpoons HHbNHCOOH$$

Most, however, is carried in the plasma as bicarbonate. In the tissues carbon dioxide liberated by tissue respiration diffuses into the plasma where it dissolves by a slow process. Some enters the corpuscles and forms carbonic acid by a rapid process catalysed by the enzyme carbonic anhydrase. This carbonic acid subsequently dissociates into bicarbonate and hydrogen ions:

$$\overset{\text{Carbonic}}{\underset{}{\text{anhydrase}}}$$
$$CO_2 + H_2O \rightarrow H_2CO_3 \rightarrow HCO_3 + H^+$$

Oxyhaemoglobin is a stronger acid than reduced haemoglobin and this increase in hydrogen ions encourages the liberation of oxygen by dissociation of the oxyhaemoglobin:

$$H^+ + HbO_2 \rightarrow HHb + O_2$$

The bicarbonate ions diffuse out of the red corpuscles into the plasma and their place is taken by an inward movement of chloride ions which restore electrochemical neutrality. The membrane of the red blood cell is relatively impermeable to positive ions and this 'chloride shift' is the only way in which neutrality can be established. By this mechanism a large amount of bicarbonate is carried in the plasma where it forms the so-called 'alkali reserve' of the body. In the lungs, the proportion of oxyhaemoglobin is

81

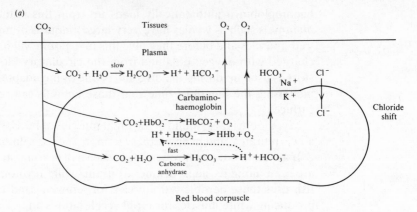

Fig. 4.12.
The exchanges between
(a) tissue, plasma and
red blood corpuscles,
and (b) lung, plasma
and red blood
corpuscles, in the
transport of respiratory
gases.

increased, and being a stronger acid than the reduced form the equilibria of the reactions are displaced in the opposite direction so that carbon dioxide is liberated from the carbonic acid (Fig. 4.12b). Transport of oxygen and carbon dioxide are therefore complementary to one another.

4.4 Respiration in other vertebrates

4.4.1 Dogfish

As we have seen, the respiratory current in a mammal enters and leaves by the same opening and the flow is tidal. In fishes water passes from the mouth to the gill slits and although it enters and leaves these openings intermittently it has been shown that the flow across the gill filaments is almost continuous. In cartilaginous fishes the gill slit in front of the hyoid arch persists as a spiracle and water enters through this opening as well as through the mouth. Water which enters the spiracle leaves through gill slits 1–3 on the same side. That entering the mouth generally emerges through the three posterior gill slits (3–5) and again the flow tends to be unilateral. The respiratory movements take place about every 1 or 2 seconds. Each cycle includes a rapid expulsion phase (Fig. 4.13a) followed by a slower period of expansion of the branchial region (Fig.

4.13*b*). The openings between the respiratory chambers and the external medium are protected by valves. Those which cover each of the five gill slits on each side are easily seen and form passive flaps which prevent the entry of water when the gill pouches expand and their pressure is lower than that in the external water. The spiracular valve is an active valve (i.e. moved by muscular action) which shuts when the mouth cavity decreases in volume. Inside the upper and lower jaws there are small flaps of skin which project backwards; these are not obvious in the dogfish, but in some species may be very clear, especially among bony fishes.

The wide separation of the gill slits on the outside of the dogfish is because the septum which separates neighbouring gill slits is very well developed. In transverse section this septum can be seen to extend from the branchial arch skeleton to the outside where it continues as the gill flap. The septum contains gill rays and muscles which can compress the gill region. On either side of the septum are the gill filaments. These are flat plates which are stacked one above the other round each of the five branchial arches. The two rows of gill filaments attached to a given branchial arch form a holobranch. The filaments on one side form a hemibranch and these come into contact with those of the neighbouring branchial arch but only at their tips (Fig. 4.14). This contact between the filament tips leads to the functional separation of the respiratory cavities into two parts. Water drawn in through the mouth and spiracle first enters the large oro-branchial cavity which extends between the gill arches as far as the tips of the filaments. Outside the filaments are found the parabranchial cavities. There are therefore five parabranchial cavities on each side which when compressed result in water being forced out through the gill slits. In the common dogfish the slits communicate with only the ventral region of the parabranchial cavities but in other elasmobranchs, for example the basking shark, they extend along the whole length of the

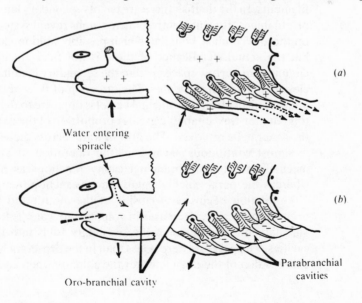

(*a*)

Water entering
spiracle

(*b*)

Parabranchial
cavities

Oro-branchial cavity

Fig. 4.13.
Ventilation in the
dogfish.

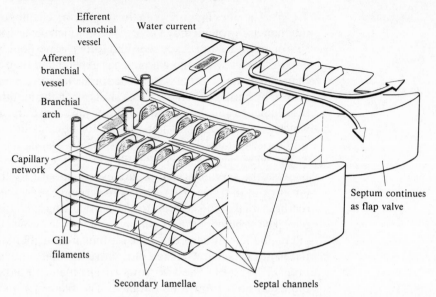

Efferent branchial vessel

Water current

Afferent branchial vessel

Branchial arch

Capillary network

Septum continues as flap valve

Fig. 4.14.
Two branchial arches
and attached gill
filaments of a dogfish.

Gill filaments

Secondary lamellae

Septal channels

parabranchial cavities. When compressed by hand in an anaesthetised dogfish, water is similarly ejected and the branchial region expands again because of the elasticity of the skeleton. Recordings of the electrical activity in muscles have confirmed that the hypobranchial muscles, which run from the pectoral girdle to the jaw and other visceral arches, play little part in the expansion phase of a dogfish during quiet respiration. Activity in the superficial constrictor and most other muscles mainly takes place when the jaws and branchial region are constricted.

Gaseous exchange takes place within the secondary lamellae. These fine plates extend upwards and downwards on both sides of the many gill filaments. In the dogfish there are twenty secondary lamellae on each side of a filament. Recent electron micrographs reveal systems of microridges on these secondary lamellae which possibly hold mucus on the surface, but the actual significance of these is not fully elucidated. The blood circulates in narrow spaces within the secondary lamellae and comes into close contact with the water. The direction of flow of the blood from an afferent to an efferent branchial vessel is opposite to that of the water and this counter-flow enables a greater proportion of the oxygen contained in the water to be removed. The flow of water across the secondary lamellae is almost continuous because of the operation of a double pumping mechanism. The oro-branchial cavity functions as a pressure pump, whereas the parabranchial cavities act as suction pumps.

Ventilation begins with closing of the mouth and the spiracle. The orobranchial cavity decreases in volume and water is forced through the gill filaments and between the secondary folds into the parabranchial cavities. The pressure here is less than in the oro-branchial cavity but as it exceeds that of the outside water the gill flaps open and water is ejected.

During this phase of the cycle therefore, the water flow across the gills is mainly due to the pressure-pump action of the oro-branchial cavity. Following this phase the constrictor muscles relax and the whole of the branchial region expands, chiefly because of the elasticity of the components of the visceral arch skeleton and their ligamentous connections with one another. During this stage the mouth and spiracle open and water enters because of the decreasing pressure within the oro-branchial cavity produced by the lowering of the floor of the mouth and pharynx, together with lateral expansion of the branchial arches. Almost simultaneously the parabranchial cavities enlarge because of the elasticity of the skeleton. The pressure within both oro-branchial and parabranchial cavities is less than that of the surrounding water and hence the flaps over the gill slits are closed. The pressure within the parabranchial cavities is lowered more than that within the oro-branchial cavity and hence the flow across the gills is mainly due to the suction exerted by these cavities. When the branchial region has reached its resting size there is a brief respiratory pause before the cycle begins once again with a compression of the respiratory cavities. Water passes almost continuously over the gills although it enters and leaves the system intermittently.

This double pumping mechanism is perhaps more readily understood by reference to the bony fishes. Here the five parabranchial cavities on each side are replaced functionally by a single opercular cavity as a result of a reduction in the septum between the branchial arches and the development of an operculum attached to the hyoid arch. In these fishes there is no spiracle and all the water enters the mouth, which is guarded by maxillary and mandibular valves. About one-fifth of a cycle after the mouth has opened, the operculum expands laterally. Expansion of the opercular cavity lowers its pressure below that of the buccal cavity and water flows across the gill filaments. These filaments are splayed out and are only in contact with one another at their tips. Because of the reduced septum, the water flow between the secondary lamellae is more clearly counter to that of the blood. The efficiency of this mechanism in bony fish is indicated by the utilisation of up to 80 % (compared with 50 % in dogfish) of the oxygen contained in the water entering the mouth. If the current is reversed experimentally the utilisation falls by more than half.

The ventilation of the gills in some sharks is achieved by them swimming with their mouths open, and there are no pumping movements except when the fish is at rest. In bottom-living cartilaginous fishes such as skates and rays nearly all the water enters through the dorsally situated and very large spiracles. Among bony fishes the tuna is an example of a form which respires entirely by the current entering the open mouth during swimming. Some bottom-living species tend to have a more developed suction pump mechanism; flatfishes are a notable example which also have adaptations preventing the flow of any water from the outside (which might include sand) into the opercular chambers.

Active fishes have a much larger gill surface than less active species, as can be seen from the following figures:

Fish	Surface area (mm²) per gram of fish
Toadfish	160
Mackerel	1,040
Tuna	2,000

More active fishes also have thinner tissue barriers separating the blood and water (e.g. less than 1 micrometre in tuna).

Where water is poor in oxygen some fishes such as catfish and climbing perch may show an intermittent respiratory exchange pattern taking some air directly from the surface in addition to obtaining oxygen from the water. This enables them to maintain a relatively constant oxygen consumption over a fairly wide range of external environmental conditions. However, these fishes that can use both air and water as a source of oxygen release most of their carbon dioxide into the water rather than as a gas into the atmosphere.

For the lungfish increasing concentration of carbon dioxide in the surrounding water cause the cessation of gill breathing and the bringing in of the ventilation of the lungs, which are inflated by a buccal force pump.

4.4.2 *The lungs and ventilation mechanisms in amphibians and reptiles*

The structure of the lung in amphibians and reptiles is essentially similar to that of mammals. The respiratory surface is much smaller, however, and in the frog the surface area per millilitre of the lung's volume is only one-fifteenth of that in man. The lungs of some newts are simple sacs and have scarcely any folding of their internal surfaces. The least folding occurs in perennibranchiate forms which utilise their external gills for

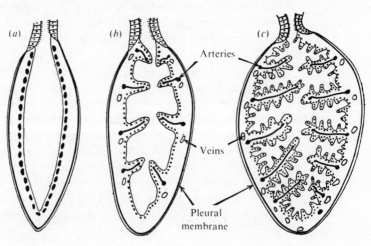

Fig. 4.15.
Stages in the increase in surface area by folding, in the lungs of newts (*a*, *b*) and frogs (*c*).

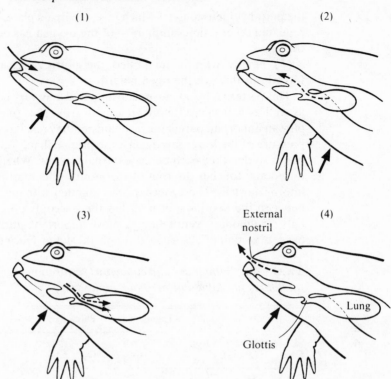

Fig. 4.16.
Pulmonary ventilation
in a frog. For
explanation see text.

respiration (Fig. 4.15). All amphibians make great use of their skin for respiration but this is not so important in reptiles. Reptile lungs have greater surface areas than those of the frog and true bronchi supported by cartilaginous rings are present. The single trachea leads off from the glottis, unlike the amphibians where both lungs originate directly from the glottis. In both groups, buccopharyngeal movements of the throat are present. In frogs ventilation of the buccopharynx by these movements results in a certain amount of gaseous exchange, but in reptiles they are more likely to be concerned with olfaction. The buccopharyngeal movements occur at frequencies between 80 and 120 per minute in frogs but are periodically interrupted by pulmonary ventilation. The mechanism is essentially a buccal force-pump operated by changes in volume of the buccal cavity produced by movements of the hyoid. It is illustrated diagrammatically in Fig. 4.16, where four main phases are recognised.

(1) The glottis is closed and the nostrils open. The floor of the mouth is lowered by the sterno-hyoideus muscles and air enters the buccopharynx because of the reduction in pressure.

(2) The nostrils are closed, the glottis opens and air is forced from the lungs into the buccopharynx partly because of the elasticity of the lungs and also by contraction of the flanks. Consequently, the floor of the mouth is lowered still further.

(3) The mixed air now contained in the buccopharynx is forced into the lungs through the open glottis. This is brought about by contraction of

87

the peto-hyoideus muscles which raise the hyoid plate. This phase may be repeated several times until most of the oxygen has been removed from the air.

(4) Finally, with the lung filled, the glottis closes and the extra air is forced out through the open nostril.

This pattern is by no means constant and may vary between individuals as well as between different species of frog. Movements of the nostrils play an important part in the mechanism; they result partly from the extra pressure of the lower jaw against the premaxillae, which move the nasal bones so that they occlude the external nostrils. When the hyoid plate is raised and touches the roof of the mouth, the anterior horns block the internal nostrils. These mechanisms, together with the closely fitting joint between the two jaws, ensure that the buccopharynx can be made airtight. Pulmonary ventilation is most important during summer when oxygen is mainly absorbed through the lung. During the winter most

TABLE 4.2. *Pulmonary and cutaneous respiration (millilitres per kilogram per hour in the frog)*

Date	Weight of frog (grams)	Cutaneous		Pulmonary	
		O_2	CO_2	O_2	CO_2
Oct.	54	54	92	51	15
Apr.	46	51	145	160	70

oxygen enters through the skin (cutaneous respiration), and this is the chief route of carbon dioxide loss throughout the year (Table 4.2). On first inspection the lung surface area appears to be proportional to the metabolic rate in reptiles but not in amphibians. If, however, the respiratory surface of the skin is taken into account for the latter and added to the lung surface then the total respiratory surface is found to be proportional to the metabolic rate.

Ventilation of the lungs in reptiles is aided by the action of the ribs, which are absent in modern amphibians. These are moved by the intercostal muscles which produce a reduction in pressure within the thoracic cavity resulting in air being drawn into the lung. Relaxation of the inspiratory muscles and activity of abdominal muscles takes place during expiration. In most lizards respiration begins with an initial expiratory phase when air is forced out of the lungs and is then followed by rapid inspiration. The lung is now inflated and remains so during a pause before the next respiratory act. At low temperatures these pauses may be very long but under hot conditions the frequency is quite rapid (30 per minute.) In the chameleon the lungs have extensions posteriorly into non-vascularised air sacs which may be inflated by the buccal force-pump. This ability to increase its size is sometimes used by the chameleon as a defence mechanism.

4.4.3 Birds

Air sacs, which are extensions of the bronchial tree, form an important part of the respiratory system of birds, and occupy as much as 80 % of the total body cavity. Fig. 4.17 shows the position of the air sacs and their connections with the lungs. The lungs themselves are relatively small and are permanently attached to the dorsal thoracic wall where they are embedded in the ribs. In fact the bird lung occupies only about one-tenth of the volume occupied by the lung of a mammal of comparable size, but this is compensated for by the bird lung having some 10 times more surface area for each millilitre of air it contains.

The detailed mechanism of the bird respiratory system is incompletely understood but it is clear that the air sacs have a bellows-like action. Air enters when the thoracic cavity is increased in volume by the contraction of the external intercostal muscles causing the ribs to move outwards and the sternum to move downwards. The lowered pressure within the thoracico-abdominal cavity draws air into the system which traverses the lungs on its passage to the air sacs. During expiration air is forced back through the lungs on its way to the trachea. The lungs themselves are made up of a large number of bronchial tubes, the largest being the single mesobronchus which is a continuation of the primary bronchus. Arising from the mesobronchus of each lung are two sets of secondary bronchi some of which lead to air sacs. The anterior and posterior groups of secondary bronchi are joined together in the main body of the lung by a large number (1,000) of fine parabronchi. Leading off from the parabronchi are fine air capillaries which are profusely supplied with blood and are the site of the gaseous exchange. Air is circulated through the parabronchi in the same direction and flows from the posterior to the anterior secondary bronchi during both inspiration and expiration, so that there are scarcely any dead spaces within this system. Oxygen can diffuse rapidly from the parabronchi into the air capillaries. The distances

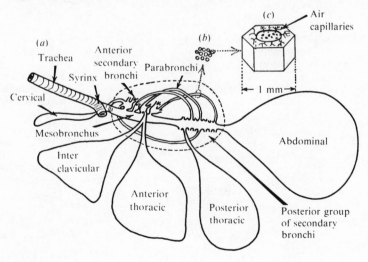

Fig. 4.17.
The main passages of a bird's lung (dashed area) and their connections with the air sacs.

89

are less than 0.5 millimetres and diffusion is sufficient to supply the oxygen required by the bird.

The air capillaries themselves across which the oxygenation of the blood occurs range from 3 to 10 micrometres in cross-section; in mammals the alveoli are from 300 to 1,000 micrometres in diameter.

This continuous unidirectional circulation of air through the parabronchi in birds is reminiscent of the fine adaptations for ventilation found in the gills of fishes.

4.5 Tissue respiration

4.5.1 *The role of energy transfer substances*

One of the fundamental properties of living matter is its constant demand for and utilisation of energy, which it requires to sustain its organisation and perform its various metabolic activities.

Living systems use energy in small quantities that are continuously made available as required in their cells. If all the potential energy of a fuel substrate were to be liberated at once most would be dissipated as heat and the delicate organic molecules of the cytoplasm irreversibly oxidised or coagulated. Thus we find that living systems release energy from their fuel substances by a series of down-grading reactions whereby the energy is handled in small packets. This idea is expressed diagrammatically in Fig. 4.18.

The most common of the energy carrier or transfer molecules in the cell is adenosine triphosphate (ATP), and even in slowly metabolising cells the total ATP pool may be turned over every second or two. This fact at

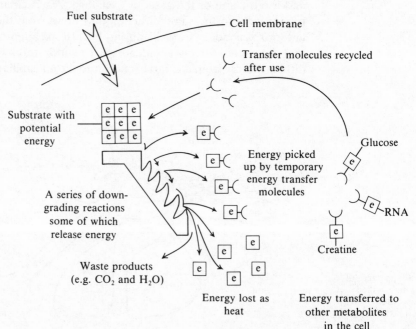

Fig. 4.18.
Diagram to show the method of energy transfer in cells.

90

once shows how transitory is the life of such energy transfer substances and it is not therefore surprising that muscles contain such very small amounts of energy stored in ATP at any one time. Muscles store their energy in glycogen and in the more stable creatine phosphate rather than in the much less stable ATP.

Despite the short life of ATP it still plays a vital role in most cell functions and so we will consider its properties in more detail.

4.5.2 ATP

The structure of adenosine and its various condensation products with phosphate is shown in Fig. 4.19. In fact at physiological pH the hydroxyl

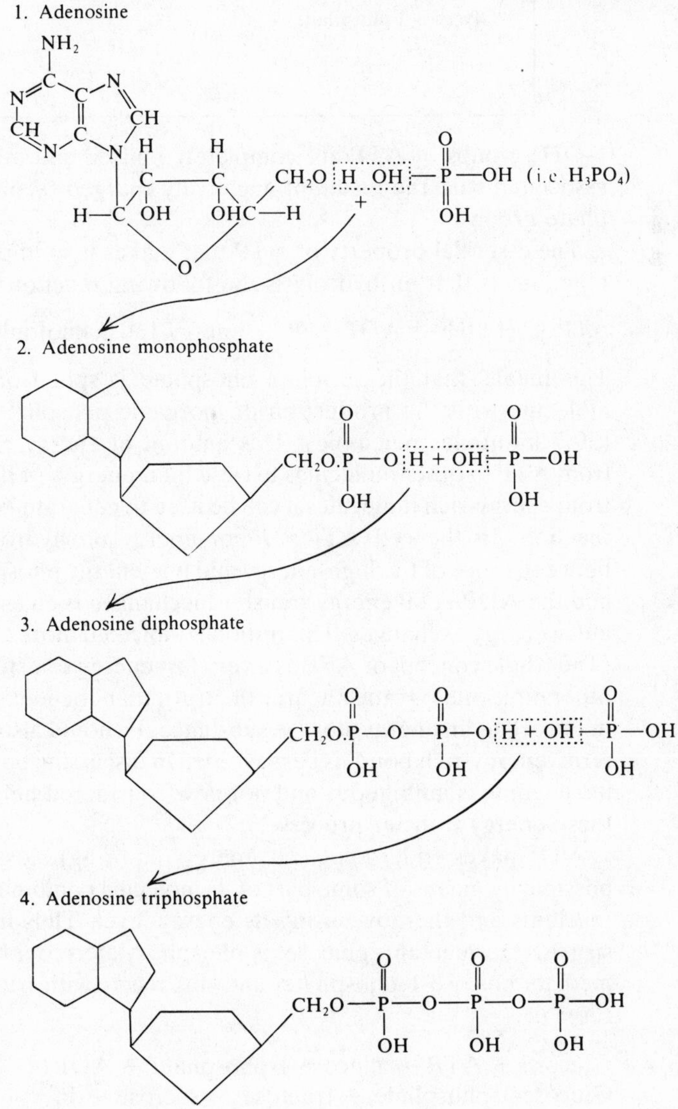

1. Adenosine

$CH_2O\ H\ OH\ \text{—}P\text{—}OH$ (i.e. H_3PO_4)

2. Adenosine monophosphate

$CH_2O.P\text{—}O\ H + OH\text{—}P\text{—}OH$

3. Adenosine diphosphate

$CH_2O.P\text{—}O\text{—}P\text{—}O\ H + OH\text{—}P\text{—}OH$

4. Adenosine triphosphate

$CH_2O\text{—}P\text{—}O\text{—}P\text{—}O\text{—}P\text{—}OH$

Fig. 4.19.
The structure of adenosine and its condensation products.

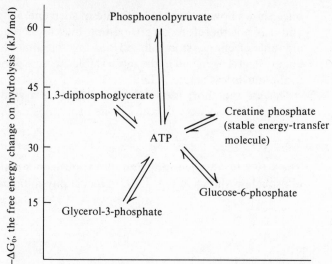

Fig. 4.20.
The energy content (measured as the free energy change on hydrolysis) of various phosphate compounds. ATP lies conveniently between that of the high-energy substances in the cell (e.g. phosphoenolpyruvate) and that of the low-energy substances (e.g. glycerol-3-phosphate) and is thus an ideal substance for energy transfer between molecules.

(–OH) groups of ATP are completely ionised and magnesium ions are associated with the resultant negatively charged second and third phosphate groups.

The essential property of ATP that makes it so important in so many reactions is that on hydrolysis the following reaction occurs:

$$ATP + H.OH \rightarrow ADP + P_i + energy \ (30.7 \ kilojoules \ per \ mole)$$

This means that the terminal phosphate is split from the rest of the molecule with the production of inorganic phosphate (P_i) and energy (30.7 kilojoules per mole). This amount of energy can be transferred from ATP to other molecules to raise their energy; or if received by ADP from energy-rich molecules it can be used to generate ATP by the reverse reaction. In the cell ATP's donor energy on hydrolysis lies midway between those of the high-energy and low-energy phosphate compounds, and the ADP/ATP energy transfer mechanism is an essential intermediate in energy exchanges. This notion is indicated more clearly in Fig. 4.20. (The whole concept of ATP as a very fast energy transfer system between other molecules is much nearer the truth than the idea that it lies around in the cell as an energy storage substance. It should also be noted that the term 'energy-rich bond' is best avoided in discussing energy exchanges as its meaning is ambiguous and not now considered helpful in visualising these energy transfer processes.)

ATP makes endergonic (i.e. energy-requiring) biosynthetic reactions possible by means of some part of its molecule combining with one of the reactants and thereby raising its energy level. Thus in the synthesis of sucrose the reactant, glucose, is phosphorylated to a high-energy intermediate glucose-1-phosphate, and this reacts with fructose to give sucrose: i.e.

Glucose + ATP → glucose-1-phosphate + ADP
Glucose-1-phosphate + fructose → sucrose + P_i

92

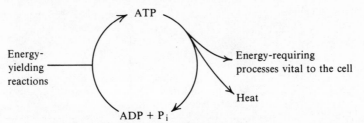

4.21.
role of ATP in
rgy transfer in cells.

This use of an intermediate high-energy substance produced by phosphorylation from ATP is found in many biosynthetic reactions. Sometimes it is other parts of the ATP molecule that are used, as for example in
the linking of transfer RNA (tRNA) to amino acids:

Amino acid + ATP → amino acid–AMP + $2P_i$
Amino acid–AMP + tRNA → tRNA–amino acid + AMP

(At the end of chapter 11 (p. 296) we shall consider the way in which
AMP plays a different role in the coordination of the cell's systems.)

Besides the part it plays in synthesis, ATP has many other functions in
the cell: the mechanical energy needed for muscle contraction, the
movement of cilia and flagella, and protoplasmic streaming all involve
energy transfer by ATP. This process is also essential to the membrane
pumps that shift ions or water against diffusion gradients and in the
generation of electrical impulses. In all sorts of reactions, therefore, ATP
plays a vital role, which can be summarised as in Fig. 4.21.

4.6 Carbohydrate respiration

4.6.1 *Glycolysis: the first stage of energy exchange*

This first part of the breakdown of carbohydrates and the transfer of their
energy to other molecules in the cell is termed anaerobic glycolysis as it
takes place without oxygen utilisation. Essentially the process is summarised as

Glucose + 2NAD + 2ADP + $2P_i$ → 2 pyruvate + $2NADH_2$ + 2ATP

but it takes place in a number of separate stages and involves a number of
enzymes all of which are found in the soluble part of the cytoplasm.

The equation shown above represents an energy exchange of only
some 60 kilojoules per mole whereas the glucose contains more than 30
times this amount of energy. Clearly if respiration ended here it would be
very inefficient! In some organisms the pyruvate passes through further
reduction reactions, passing hydrogen to organic hydrogen acceptors
such as NAD (nicotinamide adenine dinucleotide). This occurs, for
example, in yeast and tapeworms under conditions of low oxygen concentration, and it is termed fermentation. It may also occur in higher
plants, particularly during germination as the available oxygen is consumed. In most organisms, however, the pyruvate is oxidised to carbon
dioxide and water via pathways involving decarboxylation and phos-

93

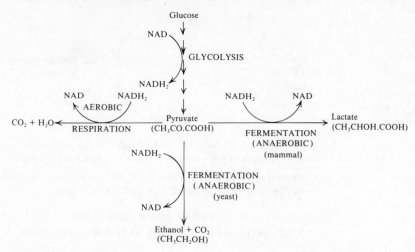

Fig. 4.22.
The relationship of
glycolysis to other
respiratory pathways.

phorylation, and it is in these terminal stages that the really significant exchange of energy occurs.

The relationship of glycolysis to other types of respiratory path is indicated in Fig. 4.22.

Substrate supply for glycolysis

The major energy store in most animal cells is glycogen (see p. 6), and it will be recalled that its molecules are especially suited to this purpose. The breakdown of glycogen takes place by the action of the enzyme glycogen phosphorylase which catalyses the cleavage of C-1 to C-4 glycosidic bonds:

Glycogen with n glucose units $+ P_i \rightarrow$ Glycogen with $n-1$ glucose units
$+$ glucose-1-phosphate

The glucose-1-phosphate is then converted to glucose-6-phosphate by the enzyme phosphoglucomutase, and it is this latter molecule that is the starting substrate for the process of glycolysis.

Where glucose itself is available in a cell it is directly phosphorylated by ATP and thus made into glucose-6-phosphate in a single step. It should be realised that the availability of this substrate depends on the rate of utilisation of ATP at any given time. Where this is high, because the cell is working, then there will be an increase in ADP and P_i, the degradation products of ATP. This in turn leads to phosphorylase activity so that more of the glucose-6-phosphate is made available for further ATP generation.

For glycolysis to occur, therefore, we need not only a supply of glucose but also ADP and P_i, and besides these NAD to accept hydrogen. The glucose is used up and the ADP and P_i are recycled from the breakdown of ATP. NAD is regenerated from $NADH_2$ by the transfer of the hydrogen to pyruvate to make ethanol or lactic acid in fermentation reactions, or to a phosphorylation sequence in aerobic respiration.

94

4.6 Carbohydrate respiration

The chemical stages of glycolysis

In mammalian muscle that is under work-load the end product of glycolysis is lactic acid, and between the glucose-6-phosphate stage and the lactic acid are 10 separate reactions. Essentially, however, the process of glycolysis is as follows. The first process is phosphorylation, which is followed by cleavage of one 6-carbon molecule to two 3-carbon ones. The next stage consists of oxidation (by the removal of hydrogen not by the addition of oxygen) and dephosphorylation. It is at this stage that energy is transferred to ADP, so this is a critical phase. The final stage in muscle is the reduction of pyruvic acid to lactic acid. The sequence is shown in Fig. 4.23. In this sequence there are two high-energy triose phosphate

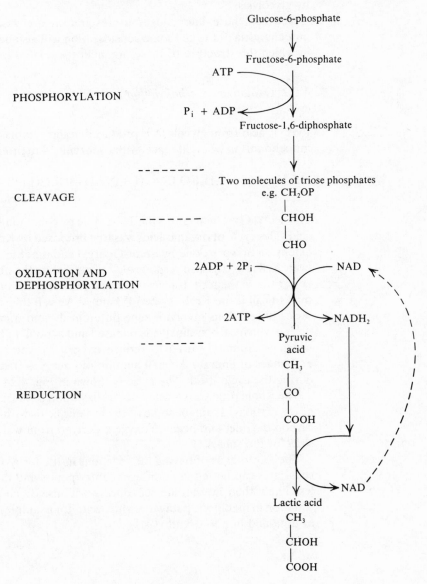

Fig. 4.23.
The stages of glycolysis in mammalian muscle.

molecules, namely 1,3-diphosphoglycerate and phosphoenolpyruvic acid, that have very high free energy released on hydrolysis, the former being some 50 kilojoules per mole and the latter above 60 kilojoules per mole. From the section on ATP (p. 91) it will be recalled that this has a free energy change on hydrolysis of 30.7 kilojoules per mole, so that it readily accepts energy from these triose phosphates.

Some small part of the original energy of the carbohydrate substrate has now been transferred to ATP, and thence to other molecules, as the result of glycolysis. The major part still resides in the pyruvic and lactic acid and the final phases of respiration consist of the means of the exchange of this energy via generation of much more ATP than was made by glycolysis.

Because these later stages of respiration are associated with the mitochondria of the cell some consideration will also be given to the way in which the structure of these organelles is related to their function.

4.6.2 *Oxidative decarboxylation and the tricarboxylic acid (TCA) cycle*

Pyruvic acid from glycolysis is produced in the cytoplasm and enters the mitochondrion where it reacts with coenzyme A to produce acetyl CoA:

$$R.CH_2-SH + CH_3CO.COOH + NAD \rightarrow R.CH_2-S-CO.CH_3 + CO_2 + NADH_2$$

The acetyl CoA then enters the TCA cycle by combining with oxaloacetic acid. This cycle of organic acids was first proposed by Krebs in 1937 as an extension of work done by Szent-Györgyi and other biochemists. (However, as Krebs was also responsible for suggesting the ornithine cycle it is better to use the term tricarboxylic acid (TCA) cycle, or citric acid cycle, than to call it the Krebs cycle.) Whatever we call this cycle it involves a series of reactions involving nine different di- and tricarboxycylic acids during which carbon dioxide is released and a 'pool' of hydrogen formed which is ultimately used to reduce oxygen to water with substantial exchanges of energy. A small amount of energy is transferred via ATP during the cycle itself. The cycle is shown in Fig. 4.24.

The whole process is termed decarboxylation because there is removal of the carboxyl group of some of the organic acids with the formation of carbon dioxide, and because oxygen, derived from water, is involved in some of the stages.

The advantages of having the reactions in the form of a cycle are that substrates can be fed in at different entry points and that blockages are less likely than in one-line sequence reactions. Biochemical cycles are common in metabolic pathways, and occur, for example, in the formation of urea and in photosynthesis.

4.6 Carbohydrate respiration

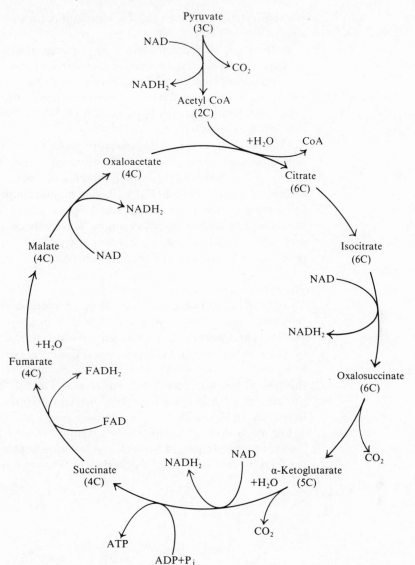

ig. 4.24.
he tricarboxylic acid
ycle. (Note that the
ydrogen acceptor for
our out of five of the
ehydrogenations is
AD, but that for the
age between succinic
cid and fumaric acid it
FAD. One molecule
f ATP is formed
uring the cycle,
etween α-ketoglutaric
cid and succinic acid.)

4.6.3 The respiratory chain: the process of greatest energy exchange

At the end of oxidative decarboxylation as described above there will be a pool of the hydrogen carriers $NADH_2$ and FADH derived from glycolysis and the TCA cycle. It is the transfer of the energy contained in the hydrogen carriers with the ultimate reduction of oxygen to yield water that yields by far the greatest amount of energy in the whole process of respiration.

This terminal pathway is called the respiratory chain or oxidative phosphorylation, and like the TCA cycle it occurs within the mitochondria of the cell (see p. 101).

In the chain of reactions that occurs electrons move along from the reduced carrier to oxygen because each carrier in the chain is a stronger oxidising agent than the previous one. Parts of this electron flow involve exergonic reactions and here the energy released can be transferred to other metabolites via the ubiquituous ATP, whose formation as we have already seen is highly endergonic.

Thus $NADH_2$ enters the respiratory chain and transfers electrons to dehydrogenase enzymes called flavoproteins, which are proteins using flavin as the prosthetic group. Examples of these are FAD (flavine adenine dinucleotide) and FMN (flavine mononucleotide). There seems to be good evidence for a further hydrogen transfer substance between the flavoproteins and the cytochromes. This has been termed ubiquinone or coenzyme Q(CoQ), and is a stronger oxidising agent than the flavoproteins and a weaker one than cytochrome *b*.

The cytochromes

Cytochromes are iron-containing chromoproteins which have the haem prosthetic group linked to proteins with molecular weights of about 16,000. There are several components in this system, the most important of which are cytochromes *b*, *c* and *a–a₃*. The latter enzymes are inhibited by cyanide and this is why 80–90 % of tissue respiration is abolished by this poison and with such disastrous results. The role of these carriers in the receipt of hydrogen from the breakdown products of glucose is illustrated in Fig. 4.25.

The electrons from the hydrogen are transferred first to cytochrome *b*. As a result this is changed from its oxidised form, that is where its iron is in the ferric (Fe^{3+}) form, to its reduced form, where its iron is in the ferrous

2H from hydrogen carrier (i.e. CoQ)

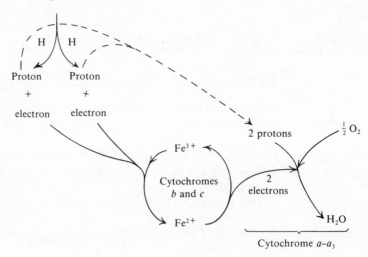

Fig. 4.25.
The role of
cytochromes in
oxidative
phosphorylation.

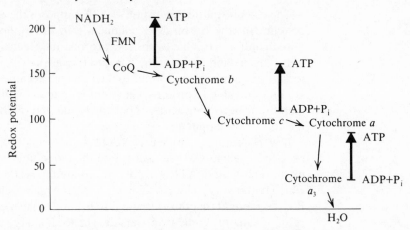

Fig. 4.26.
Energy changes in the
respiratory chain.

(Fe^{2+}) form. Thence the electrons continue via cytochrome c and $a-a_3$ in each case changing the oxidised (Fe^{3+}) form of the carrier to its reduced (Fe^{2+}) form. The final stage sees the formation of water by combination between two complete hydrogen atoms and a single oxygen atom. This reaction is catalysed by cytochrome $a-a_3$.

If we examine the energy changes in the various steps of the respiration chain then the picture that emerges is as shown in Fig. 4.26. There is sufficient energy generated between NAD and CoQ, between cytochrome b and c, and finally between cytochrome a and a_3 to allow formation of an ATP molecule from ADP and inorganic phosphorus (P_i). In fact the situation is somewhat more complex as some cytochromes, such as a_3, have sliding redox potentials. (In chemical reactions involving oxidations or reductions the energy change that takes place is related to the ability of the reactants to donate or accept electrons. This tendency is expressed quantitatively as the redox potential and for a given reactant it is calibrated against the potential of the hydrogen electrode, which at pH 7 is -0.42 volts. Thus cytochrome c has a redox potential of $+0.22$ volts while NAD has one of -0.42 volts, that is cytochrome c can donate electrons and NAD can accept electrons.)

The whole respiratory chain may thus be depicted

$$NADH_2 \longrightarrow FMN \longrightarrow CoQ \longrightarrow cyt.\ b \longrightarrow cyt.\ c \longrightarrow cyt.\ a \longrightarrow cyt.\ a_3 \longrightarrow H_2O$$

ADP   ATP   ADP   ATP   ADP   ATP
  i   +P_i +P_i

Mitchell's explanation of ATP formation by a chemiosmotic means

An alternative explanation as to how ATP is generated has been put forward by Mitchell. In 1979 Mitchell received a Nobel prize for his work in biochemistry, and his postulated mechanism of ATP formation in mitochondrial (and chloroplast) inner membranes seems to have found wide acceptance.

99

Respiration, gas exchange and transport

The chemiosmotic generation of ATP supposes the enzyme and carrier systems of the respiratory chain within the inner membranes of mitochondria to be able to abstract protons and electrons from substrates in the matrix and to direct the latter to the inside, where they are used to reduce oxygen, and the former to the outside. Because these membranes are impermeable to protons a gradient is created, with the outside of the membrane becoming acid and positively charged relative to the inside as the protons accumulate.

It is this gradient which drives ATP formation (and incidentally other processes such as Ca^{2+} transport into the mitochondrion) by causing the membrane-bound ATP synthase to condense ADP and P_i, a process which requires H^+ (since ATP bears less net negative charge than ADP + P_i). It is thought that the synthase is so located in the membrane that these protons can only come from outside – hence if a large enough gradient is formed it can drive ATP synthesis.

There is a good deal of evidence to support this view of ATP formation, including the properties of mitochondrial inner membranes with regard to protons and their behaviour on reversal of mitochondria, the effect on ATP generation of uncoupling agents such as dinitrophenol, and the ability of these membranes to generate ATP if an external gradient of charge is applied to them.

4.6.4 *Efficiency of energy exchange in respiration*

In the biochemical route that we have followed, the generation of ATP has come about in the various processes described.

Glycolysis
> Glucose + 2NAD + 2ADP + 2P$_i$ → 2 pyruvate + 2NADH$_2$ + 2ATP
> Subsequent phosphorylation of this 2NADH$_2$ gives 6ATP

The tricarboxylic acid cycle (for 2 pyruvate molecules)
> Between ketoglutaric and succinic, ADP + P$_i$ twice gives 2ATP
> Between succinic and fumaric, FAD → FADH$_2$ twice gives 2FADH$_2$
> Generation of 4NADH$_2$ twice gives 8NADH$_2$

Oxidative phosphorylation
2FADH$_2$ → 2FAD + 2H$_2$O, which yields 2 × 2 = 4 ATP
8NADH$_2$ → 8NAD + 8H$_2$O, which yields 8 × 3 = 24 ATP
Total ATP produced is therefore 38 molecules and we can compare the energy exchange here – which is some 1,170 (i.e. 30.7 × 39) kilojoules per mole of glucose – with the yield if the molecule had been combusted directly with oxygen. In this case the expected yield is some 2,856 kilojoules per mole. The efficiency of aerobic cellular respiration is therefore of the order of 1,170/2,856 × 100 = 41 %.* At first sight this

---

* Recent workers in cell thermo-dynamics have found an ATP/ADP ratio higher than previously estimated. This means the energy stored by ATP production is greater than 30.7 kilojoules per mole and that the efficiency of the respiration exchange could be much higher than 41 % – even as much as 70 % of the theoretical yield.

100

may seem rather a poor performance but it is worth remembering that steam engines are normally less than 5 % efficient. Besides this, in warm-blooded animals some of the heat losses in energy transfer are retained by the body of the animal (see Chapter 5).

4.6.5 *Mitochondria: their structure related to their function*

Mitochondria are associated with the oxidative decarboxylation reactions of the TCA cycle and with the final oxidative phosphorylation pathways of respiration. They tend to be most numerous near the source of an ATP-yielding substrate or a place of rapid ATP utilisation. A typical mitochondrion is some 0.5 to 1 micrometre across and 2 to 3 micrometres in length, but exceptionally long ones exceeding 100 micrometres are found. Externally the sausage-shaped mitochondrion is covered with a smooth membrane which is separated from an inner membrane system by a space. The inner membrane is thrown into numerous folds called cristae and surrounds a fluid central matrix. This arrangement gives each mitochondrion a very large internal surface area relative to its size, which has been calculated as being of the order of 40 square metres per gram of mitochondrial matter (Fig. 4.27).

Although mitochondria are visible in a light microscope, the detail of their structure is much more clearly revealed with the electron microscope, and by a variety of techniques it has been possible to relate the detailed structure to the functions of this organelle. Essentially the inner membrane holds many sets of enzymes bound to its surface, which are involved in the generation of ATP during oxidative phosphorylation, and as many as 5,000 individual sets of these enzymes may be present in each complete mitochondrion. As an active cell may contain up to 1,000 mitochondria, some 10_6 ATP-generating enzymes sets may be present altogether.

These organelles contain the enzyme systems for the final stages of energy exchange in the cell but they also contain their own nucleic acids and when the cell divides they divide also. This has caused some biologists to suggest that the mitochondria were at one time free-living systems

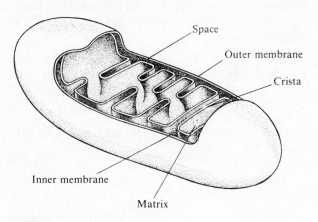

4.27.
nitochondrion with
ction cut away to
w its internal
cture.

Space

Outer membrane

Crista

Inner membrane

Matrix

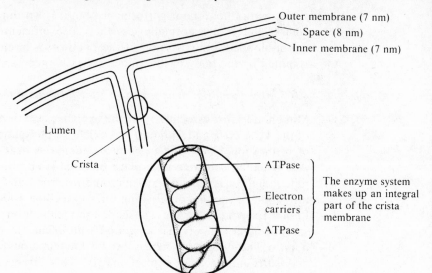

Fig. 4.28.
Section through the
membranes of a
mitochondrion,
showing the enzyme
systems bound to the
inner membrane.

which became associated in a symbiotic relationship with eukaryotic cells.

A more magnified view of a section of a mitochondrion is given in Fig. 4.28. The outer membrane, the intramitochrondrial space and the inner membrane are all around 7 nanometres in width. The outer membrane is permeable to most small molecules, but the inner membrane is impermeable to molecules of most sizes except for the very smallest, although it contains numerous carrier systems by which larger molecules may be transported across it. On the inner surface of this inner membrane have been discerned stalked bodies, sometimes called 'fundamental particles', but these are now seen to be artefacts due to a particular type of preparation of the section. In the living state it seems that the NAD, flavoproteins and cytochromes lie in the correct sequence on the surface of the inner membranes and cristae and that on the matrix side of these are the essential ATPase enzymes which are necessary for junction of the third phosphate to ADP. It is estimated that these enzymes take up 25 % of the protein of the membrane, the other 75 % being structural. The proportions of space occupied by the parts of the mitochondrion vary according to the physiological state. Thus in section an active mitochondrion might appear as

while in a non-active state it would be seen as

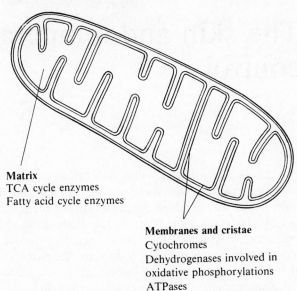

Matrix
TCA cycle enzymes
Fatty acid cycle enzymes

Membranes and cristae
Cytochromes
Dehydrogenases involved in
oxidative phosphorylations
ATPases

ₐ. 4.29.
ₑe locations of the
ₐin enzymes in a
tochondrion.

The orthodox presentation of the active mitochondrion and the enzyme location on the cristae is shown in Fig. 4.29.

In the matrix itself are the enzymes of the TCA cycle, such as the dehydrogenases of glutamate and malate, as well as acetyl CoA.

The intact mitochondrion takes in one ADP for each ATP it generates, neither more nor less, and the activity of the whole organelle depends on the ATP level in the cell. Where this is low the whole process of ATP synthesis in the mitochondrion is very high and conversely where ATP concentration in the cell is high then mitochondrial activity is very low.

Respiration of fats in mitochondria
Fatty acids combine with CoA at the surface of the mitochondrion and are taken into the organelle by transport systems in the inner membranes. All the steps of exchange of energy from the fatty acids to ATP are carried out by enzyme systems within the mitochondrial matrix. Small acetyl units are hydrolysed from the fatty acids one at a time and these combine with CoA. This naturally goes through the stages of oxidative decarboxylation of the TCA cycle yielding reduced NAD which in turn leads to ATP generation on the surface of the cristae.

5 The skin and temperature control

The skin of a mammal has a variety of functions to perform; marking as it does the boundary line between the animal and its environment, it controls the way in which the conditions of the latter affect the deeper tissues. Thus the skin protects the body from the entry of micro-organisms; it contains the sensory receptors which detect change of temperature, touch and pain; it prevents the passage of water into or out of the tissues; and it may be important in temperature control. Besides these functions the skin plays a vital role as a skeletal structure supporting the softer tissues of the body and, in the form of hair, nails and claws, preventing injury or aiding in the capture of food. Finally the skin may be coloured to protect the animal by a cryptic camouflage pattern.

5.1 The structure of mammalian skin

5.1.1 *The keratins and related molecules*

Chief among the chemical constituents of skin and its derivatives are the proteins related to keratin. They are of particular interest to the biochemist as being the proteins in which the typical α helix configuration of secondary protein structure was first recognised (by Linus Pauling in the 1940s).

α-Keratins contain many molecules of the sulphur-containing amino acids cysteine and cystine and have abundant sulphur cross-linkages between adjacent polypeptide chains. They are the major constituent of the skin, horn and nails, wool and hair of mammals. The α helices may themselves become coiled in groups about each other and this gives very considerable tensile strength to the molecules. β-Keratins by contrast have little or no sulphur or cross-bridging and are 'soft' proteins such as those found in the skins and scales of birds and reptiles.

Besides the keratins, vertebrate skin also contains another protein called collagen. This is a very important structural protein that has both flexibility and strength. Long-chain mucopolysaccharides related to the hyaluronic acid of connective tissue are also important cohesive elements in the outer covering of vertebrates. All the molecules described are characterised by their polymer-like construction, which gives them the tensile strength necessary for epidermal structures.

104

5.1 The structure of mammalian skin

5.1.2 The epidermis

The epidermis is the outer layer of skin and this layer can itself be subdivided into different functional and structural units. Directly above the dermis, which is mesodermal tissue below the epidermis, is the living Malpighian layer, composed of cells of cubical shape in which a large number of mitotic (normal cell division) figures can be found. These proliferating cells are well supplied with capillary networks from the dermis itself and may also carry the pigment granules responsible for the colour of the skin. Above the Malpighian layer, and derived from it, are the cells of the stratum granulosum, which are characterised by the presence of granules and contain a nucleus and well-marked boundary. The next layer is the stratum lucidum. Here the presence of the protein keratohyalin gives the layer its clear appearance, while the cell shapes have become irregular and no nuclei are visible. The upper layer is the stratum corneum, which contains the important protein keratin. Cells of this layer are squamous, that is, flattened, and with their individual structure completely lost.

In mammals the development of a lightweight waterproof layer in the form of the outer epidermis has been of great importance in their successful colonisation of a variety of inhospitable terrestrial environments. Keratin is quite impermeable to water and gases and its presence insulates the animal from water exchanges with its environment via the skin. While it is true that reptilian scales have the same effect, they are very much heavier than the mammalian epidermis. The actual depth of the epidermis varies over different parts of the body, and there is an adaptive increase in thickness over the soles of the feet and joints which protects against the wearing effects of the environment.

Variations in the epidermis from the type described above are found in two major forms: first the hairs, diagnostic of mammals, and secondly the thickening of the keratinised layer as claws, nails and hooves.

Hairs

These are epidermal in origin, originating from invaginations of the Malpighian layer into deep pits called follicles (Fig. 5.1). At the base of these follicles a mass of Malpighian cells proliferates, being well supplied with a plexus of blood capillaries, and elongated keratinised cells are budded off. The central cells of the hair contain air and the outer ones are usually pigmented, so that by inherited arrangements of appropriately coloured areas camouflage patterns are formed.

In general the shaft of the hair points backwards, minimising friction with the environment. Different species have their own particular hair patterns and it is interesting that the hair pattern of man is very similar to that of other anthropoids. Man has the same density of hair follicles as the great apes and his apparent hairlessness is due to the less-prominent hair shafts.

105

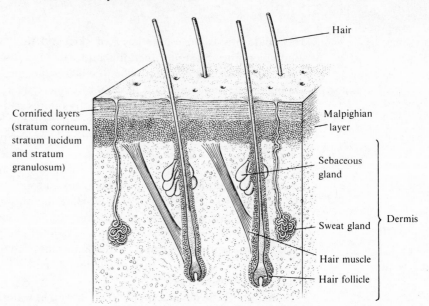

Hair

Cornified layers
(stratum corneum,
stratum lucidum
and stratum
granulosum)

Malpighian
layer

Sebaceous
gland

Sweat gland

Dermis

Hair muscle

Hair follicle

Fig. 5.1.
Section through human
skin.

By means of sensory cells at the base of the follicle the hair is able to act as an extended touch-receptor, and this capacity is particularly well developed in the vibrissae, found, for example, on a rat's snout. Threshold contact stimuli are magnified by the lever action of the hair on the sense organ when the former is touched. The main function of hairs, however, is to provide a thick insulating layer over the body which cuts down the loss of heat to the environment. Attached to the follicle is a small muscle, the erector pili, by contraction of which the individual hair can be erected. These hair muscles are under sympathetic control and not only operate to make the hairs retain a thicker layer of warm air next to the skin, but also cause them to be raised as an indication of aggression. 'Gooseflesh' seen in man is due to the contraction of hair muscles when cold and is a vestigial physiological response no longer serving in insulation. (A full treatment of the control of body temperature is found on pp. 111–15.)

Associated with the hairs are sebaceous glands, which lie some way up the shaft of the follicles. The sebum they secrete is an oily substance and makes the hairs more flexible and waterproof. The mammary glands are thought to be modified from sebaceous glands.

Claws

The claw, also found in birds and reptiles, is the basic epidermal modification from which the nails and hooves of mammals have been derived. Unlike nails the claw covers the whole end of the digit and consists of a germinal layer which proliferates a keratinised layer that continuously moves outwards towards the tip. On the underside of the claw a soft tissue represents the transition from claw to skin. Claws are important for defence of the body or for catching prey and are also used in climbing and grooming.

Nails

These are formed from a very thick stratum corneum, especially its deeper regions, and consequently have something of the characteristics of the stratum lucidum below (for example, the remains of nuclei are found). No stratum granulosum is present so the nail lies directly on the Malpighian layer. Nails are found at the ends of digits but only on the upper surface, assisting the animal in gripping objects and in grooming. The nail is a special type of claw.

Hooves

Like nails, hooves are adaptations peculiar to mammals, especially the ungulates, which walk on the ends of one or two lengthened digits. Hooves are much broader and less pointed than claws and are hemispherical with a hard keratinised layer surrounding a soft transition tissue in the middle. The latter may act as a shock absorber.

Horns

The horn is a thickening of the epidermis and the outer keratinised layer surrounds a bony outgrowth of the head. Horns are very strong though much lighter than bone and are used for protection and for fighting. (The horn of the rhinoceros is made of tightly packed hair.)

Other functions of the epidermis

The epidermis, in addition to the roles outlined above, also plays a part in the defence of the body against micro-organisms. Keratin is not easily digested by these organisms and they do not find it possible to establish themselves under normal conditions. The secretions of sebaceous and sweat glands are also inimical to their growth.

5.1.3 The dermis

This thick region below the epidermis is made largely of connective tissue, that is, it has the skeletal properties associated with the proteins collagen and elastin. The process of tanning that converts skin into leather involves making cross-linkages between these proteins by means of chemicals. Leather is an exceedingly tough biological substance, but even without tanning the dermis is very strong and supports the underlying tissues so as to maintain the characteristic shape of the animal.

Rich networks of capillaries permeate the dermis from vessels in the deeper tissues and by the contraction or dilation of these through the action of sympathetic nerve fibres the heat losses that take place directly from blood to environment can be controlled. The lower parts of the dermis form subcutaneous stores of fat and help to insulate the body.

Such insulation is found in man but, on the whole, is more typical of marine mammals.

Important structures of the dermis are the sweat glands, although much variation exists in the number and position of these among the mammals. The sweat gland consists of a secretory coiled tube fed by a capillary network and supplied with individual nerve fibres. The tube leads to the surface by a long sweat duct. Despite the fact that these glands are embedded in the dermis they actually derive from the Malpighian layer. The scents they emit may be important in the reproductive behaviour of the animal.

Various sorts of sensory cells are found in the dermis and the skin is an obvious site for receptors of touch, temperature and pain. Tactile corpuscles may be oval (the Pacinian bodies), while other receptors with different shapes respond to cold (Krause's organs) or heat (organs of Ruffini). There is a dendritic plexus associated with pain reception but the sensation of pain also results from the overstimulation of any type of receptor.

5.2 **Modifications of the skin in non-mammalian vertebrates**

5.2.1 *Elasmobranchs*

Elasmobranchs have characteristic placoid scales (Fig. 5.2) which are constructed on the same basis as a tooth and afford good protection to the animal. The scales are produced by certain cells from the dermis called odontoblasts, which form a pulp cavity and secrete a thick layer of dentine permeated by protoplasmic canals and surrounded by enamel. The scales originate at the junction of the dermis and epidermis and the latter is itself a thick keratinised layer which gives added protection. In the region of the jaws the dermal region produces along its edge types of structure transitional between teeth and scales, while further into the jaw long teeth are formed. The tooth is thus clearly seen to be homologous with the placoid scale.

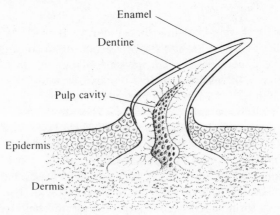

Fig. 5.2.
Section through a placoid scale of an elasmobranch.

108

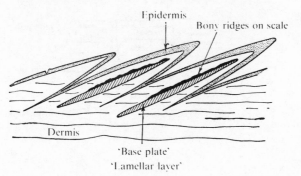

Epidermis

Bony ridges on scale

Dermis

'Base plate'
'Lamellar layer'

g. 5.3.
·ction through a
·cloid scale of a
·leost.

5.2.2 *Teleosts*

Teleosts have cycloid scales (Fig. 5.3) and these consist of flat 'bony' plates in the dermis which are covered by a keratinised epidermis. As the scales are only rooted at one end they allow flexibility of movement but, unlike the placoid scale, they do not penetrate the epidermis. In this cycloid scale, which is found in most present-day bony fishes, there is an outer calcified layer over a disc of fibrous connective tissue. Seasonal changes in growth rate are shown on the scale by a series of rings, so that the individual scale can be used to determine the age of the fish.

In both elasmobranchs and teleosts the skin is impermeable to the entry of water. All fish have mucus secreted over their scales which increases their resistance to the entry of water. This mucus is particularly important in eels, which live part of their life in freshwater and part in seawater and thus change their osmotic environment. The mucus film also reduces the drag on the fish as it swims.

5.2.3 *Amphibians*

Amphibians have a skin which is permeable to water and gases and in these vertebrates the skin is an important respiratory surface. One of the characteristic features of amphibian skin is the presence of mucous glands (Fig. 5.4), derived from the epidermis and opening by ducts to the exterior. The cells of the epidermis are partly keratinised though in modern amphibians there are no scales and periodically the skin must be shed to allow for growth. (It should be noted that the keratin layer is very thin and does not prevent the skin from being permeable.) Mineral salts are selectively assimilated through the skin from the environment.

The other type of gland found in the skin of amphibians (in some frogs and many species of toads and salamanders) is poison-secreting; these organs may be concentrated in definite regions, for example along the back in many species of frogs. In some species the mucus itself is poisonous. Such poisons are of benefit to their owners in preventing attack by predators and their presence is associated with bright warning colorations.

109

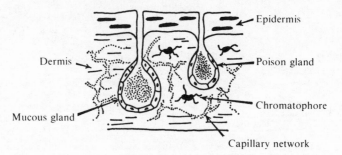

Fig. 5.4.
Section through
amphibian skin.

Amphibian skin may also be modified into horny tubercles, or thickened to provide nuptial pads with which the male holds the female during mating. It may contain scent glands. Finally, it may contain in its dermis various chromatophores which enable its owner to produce camouflage or other colorations.

5.2.4 *Reptiles*

Reptiles have scales which, unlike those of fishes, are formed mainly from epidermal layers. The scales have an outer dead horny part made largely of keratin and are periodically shed to allow for growth. In some reptiles, such as the snakes, the whole epidermis is sloughed off at one time, paralleling the moulting of arthropods. Reptile skins are almost impermeable to water, which has allowed their efficient colonisation of the land. Although their skins do contain glands of various types no sweat glands are present and the skin does not provide a means of temperature control. In some cases the scales are enlarged to form protective spines and horns while in others addition of dermal elements gives greater reinforcement (for example in crocodiles and turtles). As in the amphibians, chromatophores may be present which enable colour changes to be made (for example in the chameleon).

5.2.5 *Birds*

Birds have feathers, which are a diagnostic feature of the class. The feather develops from an outpocket of the skin called a papilla, which contains an inner Malpighian layer and outer periderm and stratum corneum. Within the Malpighian layer is a projection of the dermis with a good supply of blood capillaries and it is this region that pushes out the projections of the surrounding epidermis that form the feather (Fig. 5.5). The stratum corneum thus becomes arranged into a long rachis with extending barbs. From these barbs further projections called barbules arise which themselves carry small hooks or barbicels. The latter allow attachment between members giving a light airfoil, the vane, whose efficiency is maintained by the preening activity of the bird. An oily substance from the uropygial gland near the anus is used in preening and helps to keep the vane flexible.

110

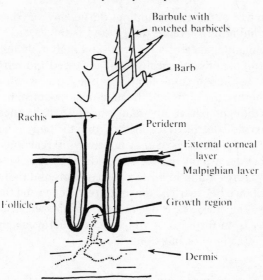

Barbule with
notched barbicels

Barb

Rachis

Periderm

External corneal
layer

Malpighian layer

Follicle

Growth region

Dermis

Fig. 5.5.
Section through the
skin and plume feather
of a bird.

After the feather structure has been laid down the internal dermal region of the Malpighian layer withdraws and forms a sunken follicle below the surface of the skin. As the blood vessels and nerves are associated with this layer, the distal part of the feather is no longer living after this withdrawal.

Variations on the flight feather described are found in the down feathers, which are mainly for insulation.

5.3 The control of body temperature

5.3.1 Terminology

Previously those animals that could maintain a constant internal temperature, such as the birds and the mammals, were called homoiotherms, while those whose temperature varied with their surroundings, such as the reptiles, were called poikilotherms. While these terms still find some use it is more correct (currently) to describe those animals that have a fast metabolism, generate much heat internally and have good thermal insulation as endotherms. Animals which do not produce much internal heat and whose outer layers provide little thermal insulation are now called ectotherms.

As will be seen the distinction is not absolute and among the endotherms there is considerable variation in the efficiency of temperature control, while many supposed ectotherms show some ability to remain at a constant temperature.

The term heterotherm is used to describe animals such as hedgehogs which maintain a high and constant temperature during summer but allow their temperature to fall to that of their surroundings when they hibernate in the winter. This term could also be applied to those animals such as bats whose temperature is held at a high level during night and

111

falls to that of their surroundings during day, or conversely to those such as humming birds where the reverse is the case.

The maintenance of a constant and high body temperature is characteristic of the physiologically most advanced and active of the vertebrates and is a fundamental factor in the success of the two classes of endotherms, the birds and the mammals. A constant high body temperature is thought to have advantages because of the effect of temperature on the enzyme-controlled reactions within the body. The optimum temperature at which the rates of enzyme-catalysed reactions within the body are most rapid is usually between 30 and 40 °C. This is also the range of body temperatures found in birds and mammals and there is evidently a delicate balance between the maximum efficiency and the rate of breakdown of these enzymes. The range of temperatures which animals can survive is roughly from 0 to 50 °C, which in ectotherms again coincides with the range of activity of most enzymes.

5.3.2 *Temperature control in ectothermic vertebrates*

Fishes have no means of maintaining a constant temperature. Their gills represent a large surface across which the temperature of the blood can come into equilibrium with that of the water. In very active and large fish

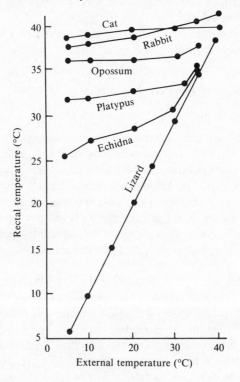

Fig. 5.6.
The relationship of external temperature to internal temperature in various vertebrates. Note that the eutherian mammals have more efficient temperature control than the marsupials or the monotremes.

such as tuna, however, there is evidence that a core temperature several degrees above that of the sea may be maintained during periods of activity. In all animals a large body size means a relatively small surface area and thus low heat loss, and conversely a small body size means very rapid heat loss through a relatively large surface area. Some fish, for example those living in hot lakes, show behavioural adaptations for temperature regulation.

Amphibians have no control over the passage of water through their permeable skins, and although on land this water loss will have some cooling effect it is not likely to be advantageous. In the normal aquatic environment of amphibians the water temperature will be the determining factor for their body temperature at any given time.

Discounting any behavioural adaptations we would expect the relationship of external to internal temperature for a range of higher vertebrates to be as shown in Fig. 5.6. While this relationship is broadly true a closer look at the behaviour of some of the ectothermic reptiles shows that their temperature is not always the same as that of their surroundings. Thus some reptiles show some degree of temperature selection by their behaviour and many lizards will burrow away from very high or very low ground temperatures. The lizards also exhibit postural behaviour to different ambient sources of heat, either turning broadside to the source when cold or end on when too hot. The large monitor lizards can control their temperature to some extent by increasing their basal metabolic rate at low temperatures and panting when it is hot. Marine iguanas slowly lose the body core heat that they acquire sitting in the sun on the land when they feed in the cold water of the sea. They therefore have to come out of the sea at intervals to warm up enough to be able to generate sufficient energy for swimming.

There is increasing evidence that many of the so-called cold-blooded reptiles of the Mesozoic era were in fact evolving towards or actually achieving a considerable degree of temperature control, although it is doubtful whether they ever achieved the efficiency found in their descendants the birds. The endothermy of many dinosaurs is indicated by the discovery that some of the plates on their backs could have acted as cooling devices; by skeletal proportions indicative of faster movement than they would have been capable of were they ectotherms; and by analysis of the ratio of carnivores to herbivores, which for some species is closer to that found nowadays in the mammals than in the ectothermic reptiles such as crocodiles.

5.3.3 *Temperature regulation in birds and mammals*

An endothermic animal will produce some 20 times the metabolic heat generated by an ectotherm, and this generation of heat correlates with the much higher food intake required by the endotherms. The following table shows the differences that exist between the two types, the heat generated being taken at the basal rates of metabolism.

Temperature regulation	Animal	Heat generated*
Ectotherm	Sturgeon	130
Ectotherm	Lizard	121
Endotherm	Canary	3,184
Endotherm	Rat	3,477

* In kilojoules per square metre per 24 hours.

In terms of their body weights the actual heat generated by all endotherms is very similar, but very large and very small endothermic animals have special difficulties in temperature regulation. Thus the small mammals and birds have very large surface areas relative to their volumes and as heat is lost via the surface they have the problem of retaining heat. Conversely the large endotherms such as elephants have relatively small surfaces compared with their volumes and thus have difficulty in getting rid of excess heat.

5.3.4 *Special problems of endotherms living in climatic extremes*

The most demanding environments in which endothermic animals have to regulate their body temperatures are clearly those of extreme heat and extreme cold, and the ability of certain species to survive in such hostile situations is remarkable.

The camel survives in the hottest deserts by having very considerable tolerance to the raising of the temperature of its body core, which in the day may be 6–7 deg C above that at night. Besides this the camel is unusual in being able to continue to live with as much as 30 % dehydration, whereas for most mammals the lethal figure is nearer 12 %. Other desert mammals may survive by heat evasion, burrowing down into the cooler layers of the sand during the day and only emerging during the night. All mammals living in very hot dry places have also to cope with the problem of water conservation (see p. 185). Another device found in desert endotherms is an extension of the surface area of the extremities such as the ears, as may be seen for example in the desert fox. While such extensions of the extremities may be efficient for cooling it could obviously involve the animal in great water loss by evaporation if the extremities were also supplied with sweat glands.

The other extreme is that of cold and here endotherms show adaptations such as reduction of extremities, fur with very high levels of insulation (up to four times those of temperate-dwelling species), and adjustments in the basal metabolic production of heat at very low environmental temperatures. An increase in metabolism in response to cold, which in a naked human starts below 28 °C, in animals such as arctic foxes may not commence until the surrounding temperature is as low as −40 °C. A further common adaptation in both birds and mammals living in very cold environments is to allow the temperatures of their extremities (feet, noses, etc.) to drop to just above freezing point and to provide these extremities with a counter-flow blood supply whereby most of the heat going out in arterial blood is transferred to a returning vein before reaching the exposed region of the body. Tissue damage to these extremities is prevented by periodic and transient admission of blood at the core temperature to those peripheral regions.

5.3.5 *Mechanisms of temperature regulation*

Heat may be generated for the body by a variety of mechanisms. Thus it is produced by muscular activity and also by shivering, which is an asynchronous contraction of portions of the skeletal muscles. It is also generated by all sorts of exothermic metabolic reactions, such as those which occur in the liver. Heat is produced by the calorigenic action of eating food (as much as 30 % of the energy content of protein is released to the body in this way 1 to 5 hours after digestion) and in young and hibernating mammals it may be released by oxidation of the free fatty acids mobilised from the brown fat store of the body. Man himself, as an adult, stores some 420,000 kilijoules as fat and at an absolute maximum output may produce up to 67,000 kilojoules per square metre of body surface, although such a heat production could not be long maintained.

Conversely heat may be lost from the body through evaporation of sweat and by conduction, convection and radiation from the body surface. In birds, and mammals which do not possess sweat glands, heat is lost mainly by shallow panting that causes a stream of cool air to pass over the moist surface of the tongue and buccal cavity. In dogs the panting rate may reach some 300 strokes per minute in extremes of heat and in some small birds it may go as high as 600 movements a minute.

There is conclusive evidence that the centre of temperature regulation in mammals is, like other regulatory centres, in the hypothalamus of the brain, and further that heat loss is controlled by the anterior region of this organ while its generation and retention are under control of the posterior part.

5.3.6 *Temperature control in man as a representative mammal*

Investigations of the effect of cooling the blood going to the brain, as compared with cooling the skin surface, clearly indicate that it is the temperature of the former that regulates heat production or loss. Changes in the temperature of the blood reaching the hypothalamus are continually monitored and this part of the brain acts as the thermostat of the body.

115

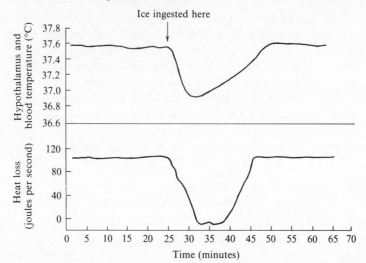

Fig. 5.7.
An experiment to demonstrate regulation of heat loss in man. For explanation see text.

Fig. 5.7 shows the results of an experiment in which a subject ingests ice at the point indicated. His blood temperature quickly drops and very shortly afterwards adjustment systems directed by the hypothalamus cut the loss of heat from the surface of the body. Within 20 minutes complete homeostatic adjustment has been achieved. Temperatures at the skin surface do not correlate with hypothalamic temperatures.

Should the temperature of the blood fall below the norm for the body then information goes out from the hypothalamus in several forms, all of which are designed to increase heat production and slow down heat loss. Thyrotropin releasing factor passes from the hypothalamus to the anterior pituitary where it causes the release of thyroid stimulating hormone (TSH). The subsequent release of thyroxine from the thyroid gland causes a general increase in the metabolic rate of many body tissues lasting from one hour to several days after the original stimulus. A short-term hormonal control route goes from the hypothalamus via the pituitary to the adrenal glands, which are stimulated to release adrenaline into the circulation. This causes immediate increases in metabolism though only for short periods (see section on hormones, p. 281).

Nervous connections from the hypothalamus to the higher brain centres initiate behavioural activity such as seeking warmth or putting on more clothes etc. At the same time connections with the autonomic nervous system cause changes in the periphery of the body. The circulation of the blood to the surface of the skin may be reduced by as much as 100 times via contraction of the precapillary sphincters in the capillaries feeding the superficial vessels.

The changes in heat distribution and blood supply at the surface of the body in man and other mammals can be well illustrated by Fig. 5.8.

In addition to these changes shivering may also be initiated from the hypothalamus via somatic nerve paths.

By these various physiological changes the body rapidly gains heat until the core temperature is restored to its normal level of around 37.6 °C.

5.3 The control of body temperature

Atmosphere

Surface capillaries

Subcutaneous
shunt vessels

Core of body

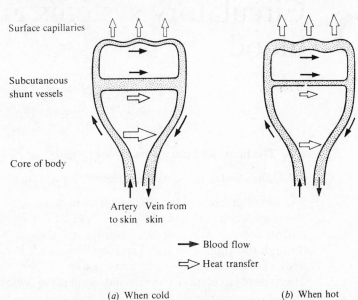

Artery Vein from
to skin skin

➡ Blood flow

⇨ Heat transfer

g. 5.8.
echanisms for
ntrolling heat loss
om the body surface
man.

(*a*) When cold   (*b*) When hot

The whole process is a good example of homeostasis in action.

Where the opposite situation arises and the body temperature rises above normal, events will be set in motion, again by the hypothalamus, which lead to its prompt reduction. Thus heat-generating metabolic hormones are not secreted and autonomically controlled sphincters allow increases in the flow of blood to the skin. This in turn leads to more heat loss, essentially, in man, through the evaporation of sweat. Man has some $2.5 \times 10_6$ sweat glands in his skin and although he will always lose some 40 grams of water per hour through insensible perspiration, this can be increased to 4 litres per hour during maximum sweating. Each gram of sweat requires some 2.45 kilojoules for its evaporation and much of this is taken from the surface of the body, leading to rapid cooling of the blood that circulates through the skin. From the higher centres behaviour is initiated which leads to seeking a cool environment or removing a layer of clothing. In animals extreme heat will also cause panting to occur, with consequent increases of heat loss from the upper airways.

117

6 Circulatory systems and the blood

6.1. **The heart and circulation of vertebrates**

6.1.1 *The double circulation of mammals and birds*

The speed of circulation is highest in mammals; in man the blood takes about 18 seconds to complete its circulation during maximum exercise, but 86 seconds when at rest. During this time it passes successively through the right ventricle, lung capillaries, left auricle, left ventricle, body capillaries, right auricle and back to the right ventricle (Fig. 6.1). This type of circulation, in which the blood passes twice through the heart during one complete cycle, is called a double circulation and was discovered by William Harvey in 1628. Rapid circulation is ensured because the pressure and the velocity of flow are raised by the heart after the blood has passed through the fine capillaries in the lungs. Circulation may therefore be divided into two parts – the systemic circuit to the body and the pulmonary circuit to the lungs. One essential condition for such a circulation is that equal volumes can pass round each circuit in any given time.

6.1.2 *Chambers of the heart*

The hearts of mammals and birds may be considered as two pumps (the right heart and the left heart) which act in series with one another although they are bound together as a single organ (Fig. 6.2). Each pump is made up of a relatively thin-walled auricle to which blood is returned from the capillary systems, and a thick-walled muscular ventricle which is responsible for raising the pressure in each circuit. The ventricles contract almost simultaneously and blood is forced into the aorta and pulmonary artery from the left and right sides respectively. During ventricular systole (i.e. contraction) blood is prevented from flowing back into the auricles by the auriculo-ventricular valves. These are one-way valves formed of flaps of tissue between auricle and ventricle which allow the blood to pass from the auricle to the ventricle but close when the pressure in the ventricle is greater than that in the auricle. Eversion of the valves into the auricle is prevented by fibrous strands (chordae tendinae) attached between the valves and the papillary muscles. The valve on the right side is called the tricuspid valve in man and that on the left side the bicuspid or mitral valve. The bicuspid valve is held in position by even

118

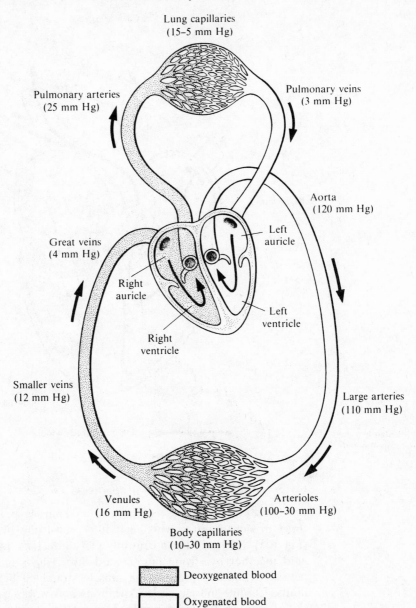

Lung capillaries
(15–5 mm Hg)

Pulmonary arteries
(25 mm Hg)

Pulmonary veins
(3 mm Hg)

Aorta
(120 mm Hg)

Great veins
(4 mm Hg)

Left
auricle

Right
auricle

Left
ventricle

Right
ventricle

Smaller veins
(12 mm Hg)

Large arteries
(110 mm Hg)

Fig. 6.1.
A simplified scheme of
double circulation. The
numbers indicate the
pressure (in mm Hg) of
the blood in different
parts of the human
circulation. (The
diagram is a ventral
view and hence the left
side of the body is on
the right-hand side of
the diagram.)

Venules
(16 mm Hg)

Arterioles
(100–30 mm Hg)

Body capillaries
(10–30 mm Hg)

Deoxygenated blood

Oxygenated blood

larger papillary muscles because the left ventricle is not only the largest
chamber of the mammalian heart but is also the most muscular. The left
side of the bird heart is like that of the mammal, but the right auriculo-
ventricular valve is a muscular ridge which almost entirely circles the
orifice and functions like a sphincter. Consequently there are no chordae
tendinae on the right side. The exit from the ventricles is protected by a
series of pocket, or semi-lunar, valves which prevent the return of blood
to the ventricles on the completion of systole.

119

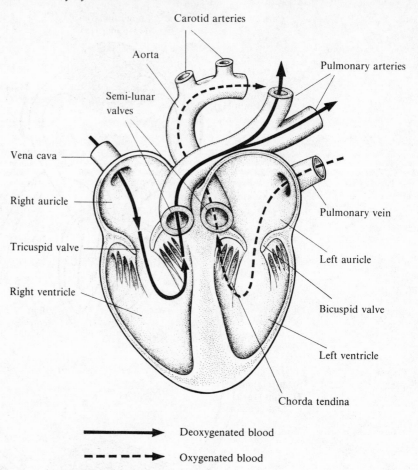

Carotid arteries

Aorta

Semi-lunar
valves

Pulmonary arteries

Vena cava

Right auricle

Tricuspid valve

Right ventricle

Pulmonary vein

Left auricle

Bicuspid valve

Left ventricle

Chorda tendina

Fig. 6.2.
Section through a
generalised mammalian
heart.

→ Deoxygenated blood

--→ Oxygenated blood

6.1.3 *Cardiac muscle and its properties*

Cardiac, or heart, muscle (compare striated muscle, p. 199) is made up of
short branched fibres with each fibre consisting of a chain of cells
(Fig. 6.3). Each cell has a centrally placed nucleus. Individual cells are
held together by strong intercalated discs which give extra adhesion
between cells and prevent the muscle stretching beyond its working
limits. Cardiac muscle shows striations and works essentially as does
skeletal muscle, by the making and subsequent breaking of cross-bridges
between molecules of the proteins actin and myosin. The muscle is very
rich in mitochondria and has a better capillary supply than skeletal
muscle. During exercise as much as a litre of blood may pass through the
human heart in a minute, and up to 100 millilitres of this through the
coronary blood vessels themselves. This is necessary because biochemi-
cally heart muscle is unable to carry out full anaerobic metabolism and
cannot run up oxygen debts. The copious arterial supply thus ensures that
adequate amounts of oxygen are supplied to the muscle, while the high

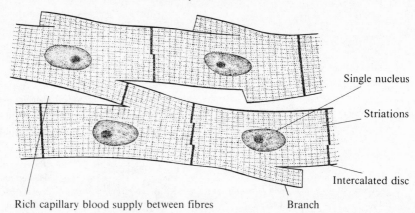

Single nucleus

Striations

Intercalated disc

Fig. 6.3.
Two adjacent fibres of
heart muscle.

Rich capillary blood supply between fibres   Branch

density of mitochondria is associated with aerobic oxidation. Heart muscle can bring about oxidation of lactic acid originating in other tissues.

Electron microscope studies indicate that the sarcomeres (contractile units) of cardiac muscle function over the range of their maximum efficiency of contraction, i.e. where most cross-bridges are made between the actin and myosin. This works out at a relaxed length of around 2.2 micrometres, and the properties of the surrounding cardia (see below) help to keep the muscle at this effective length. In this way heart muscle can always function at peak efficiency, and is able to increase its work according to the demands made upon it while remaining at optimum performance. This is unlike the situation in skeletal muscle, where performance declines at high work loads. On the other hand because of its restricted working range cardiac muscle has a smaller contraction percentage (percentage shortening from its relaxed length) than is found in skeletal muscle. The values are 32 % and 58 % respectively.

Lying outside the heart muscle is the pericardium, which is itself divided into two layers with a gap between. The outer layer is tough and fibrous and is attached to the sternum while the inner layer, which is connected directly to the surface of the muscle of the heart, has serous-producing cells which minimise friction. The movement of the heart takes place between these two layers of the pericardium. Lining the inside of the heart muscle and therefore in contact with the blood is the endocardium and this again has flattened cells with a surface lubricant to minimise friction. As already stated these cardia help to maintain the heart muscle fibres within their effective working lengths.

6.1.4 *The origin and conduction of the heart beat*

Rhythmicity is a characteristic feature of all hearts and in vertebrates it is myogenic in origin, that is, initiated from muscle tissue. In mammals the rhythm originates in a special region of the right auricle, known as the sinu-auricular node because it corresponds in position to the junction of the sinus venosus and auricle in the fish and ancestral heart. This

121

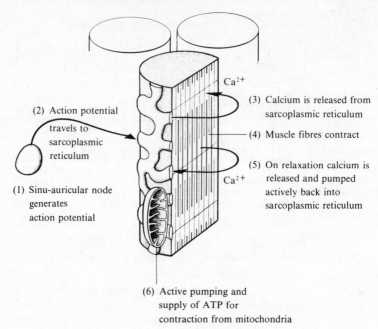

(2) Action potential
travels to
sarcoplasmic
reticulum

Ca^{2+}

(3) Calcium is released from
sarcoplasmic reticulum

(4) Muscle fibres contract

(5) On relaxation calcium is
released and pumped
actively back into
sarcoplasmic reticulum

Ca^{2+}

(1) Sinu-auricular node
generates
action potential

Fig. 6.4.
The mechanism for the
initiation of the heart
beat.

(6) Active pumping and
supply of ATP for
contraction from mitochondria

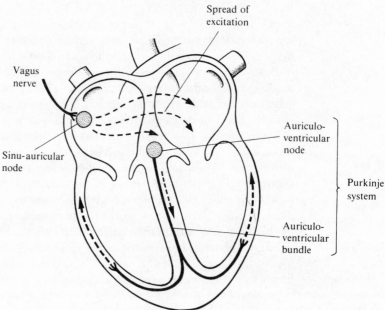

Spread of
excitation

Vagus
nerve

Sinu-auricular
node

Auriculo-
ventricular
node

Purkinje
system

Auriculo-
ventricular
bundle

Fig. 6.5.
Control of the heart
beat.

pacemaker region is made up of special muscle cells and is also the region
where the nerves controlling the heart have their endings.

The cells of the sinu-auricular node (Fig. 6.4) start their discharge cycle
with an internal resting potential of 90 millivolts. This potential gradually
leaks away owing to the slow diffusion of ions across the cell membranes.
When a certain level is reached an action potential is generated and this
spreads rapidly from the fastest leaking cells in the sinu-auricular node to

122

the others, and thence from the whole region to contractile cells in the auricles.

As with skeletal muscle, the actual contraction mechanism is linked to the release of calcium from the sarcoplasmic reticulum which adheres closely to the individual fibrils. The arrival of the action potential causes the membrane of the sarcoplasmic reticulum to become permeable to calcium ions, and these promote the formation of cross-bridges between actin and myosin molecules. When the muscle relaxes the calcium is released and pumped back across the sarcoplasmic membranes. The beat originates spontaneously and spreads relatively slowly (1 metre per second) through the whole of the auricular muscles. It is especially important that the two ventricles should contract simultaneously and this is ensured by a special conducting system formed by modified cardiac muscle fibres (Purkinje tissue). This takes its origin at the base of the inter-auricular septum at the so-called auriculo-ventricular node and is vital for transmitting the heart beat through the non-conductile connective tissue which separates the auricles from the ventricles. The wave of excitation which spreads out from the sinu-auricular node (Fig. 6.5) reaches the auriculo-ventricular node which becomes excited and conducts relatively slowly (0.2 metres per second), but after this delay the excitation is transmitted by the auriculo-ventricular bundle which spreads out posteriorly along the inner wall of both sides of the inter-ventricular septum. The wave of excitation is conducted at 5 metres per second along these fibres and in the absence of any sinu-auricular pacemaker they may develop their own beat spontaneously. The passage of the wave of contraction follows the electrical events after a shorter delay than in skeletal muscle. The electrical phenomena can be recorded from electrodes placed at different points on the body surface. This electrocardiogram (e.c.g.) gives very valuable information concerning the state of the heart (Fig. 6.6). It appears that there are three phases in the spread of ventricular activity – first, from right to left in the inter-ventricular septum, then from the inside to the outside of the ventricular walls, and finally from the tip (i.e. posterior) of

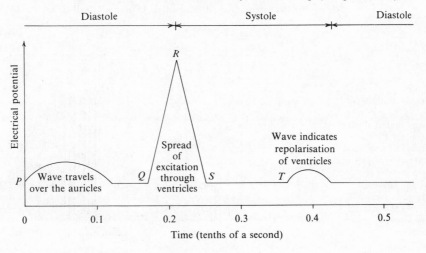

Fig. 6.6.
The electrocardiogram (or ECG) which shows the changes in electrical activity of the heart during a heart beat.

123

the ventricle towards its base, as this region is not supplied by conductile tissue. As a result of this activity both ventricles contract almost simultaneously, beginning first of all with the inner layers of the more posterior regions of the ventricles. During ventricular systole the papillary muscles also contract and maintain the position of the auriculo-ventricular valve. All of these adaptations ensure a more powerful beat of the heart.

6.1.5 *Pressures within the circulatory system*

The aortic pressure produced during systole is about 120 mm Hg, falling to 80 mm Hg during diastole (relaxation of the heart). The pressure difference between these two during each contraction is the pulse pressure. Pressures in the pulmonary circuit are much lower (systolic 27 mm Hg, diastolic 10 mm Hg), because the pulmonary circulation is much more distensible than the systemic (see Fig. 6.7). During its passage through the systemic circulation the pulse, as well as the systolic and diastolic pressures, decreases progressively. The smoothing out of the pulse is due to the elasticity of the arterial walls. The general fall in pressure is not great in the arteries because of their relatively large diameter, but in the small arterioles and capillaries the greater resistance of the fine tubes reduces the pressure because of the increased friction at their walls (Fig. 6.7). By the time blood reaches the veins the pressure is extremely low (less than 10 mm Hg) and its return to the heart is aided by other features of the circulation. These are absolutely vital because within the large veins the pressure may be as low as 5 mm Hg. The tone of the body muscles and especially the increase in muscular contraction within the extremities during exercise play a very important role. Within the veins there are valves preventing the backflow of blood and hence any compression causes flow towards the heart. Another feature aiding the venous return is the lowered pressure within the thoracic cavity which results from the lung elasticity as described previously. The negative pressure in the thoracic cavity of −6 mm Hg falls to −2.5 mm Hg during expiration, but during both phases it will draw blood towards the heart.

Fig. 6.7.
Diagram to illustrate pressures in different parts of mammalian circulation. Pressure fluctuations produced by three heartbeats are shown in each section. Note the fall in pressure as blood traverses the arterioles of the systemic circulation.

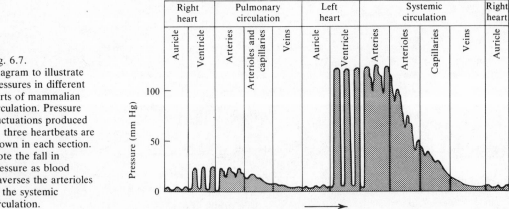

124

6.1.6 *Heart output and its regulation*

Essentially, then, a circulatory system consists of a pump and a series of tubes of varying diameter which convey the blood to the tissue capillaries and then back to the pump again. The passage of blood from the pump is intermittent but its movement through the capillaries is continuous. The elasticity of the thick-walled arteries plays a vital part in the maintenance of this continuous supply of blood to the tissues. During systole energy is stored in the artery walls and can be released during diastole to help squeeze the blood through the arterioles. Clearly the volume of blood passing to the tissues in a given time depends on the amount pumped by the heart and the resistance to its flow through the capillary system. The cardiac output may be varied either by changes in frequency or output per beat. Frequency varies a great deal among different mammals; in general the larger the animal the slower its heart rate. In small shrews and in birds the frequency may be between 800 and 1,000 beats per minute. For a given animal there appears to be a limit to which the frequency can be raised and in man it is about 190–200 per minute. At rest, the output of a man's heart is about 5 litres per minute and it may be increased by as much as 7 or 8 times during severe exercise. In trained athletes this maximum output can be maintained at lower frequencies of heart beat. The heart functions more efficiently at these frequencies, partly because there is greater time for recovery between each beat. Many athletes at rest have pulses less than 50 per minute, the lowest recorded being about 39, as compared with the usual frequency of around 70 per minute. The output per stroke is greater in athletes and those accustomed to strenuous work and in some cases is associated with hypertrophy (enlargement) of the heart.

The arterial blood pressure is monitored by receptors in the carotid sinus and aorta which send sensory fibres to the medulla in, respectively, the glossopharyngeal nerve (IX) and depressor branch of the vagus (X). Excitation of this pathway leads to a reduced cardiac output and consequent fall in the arterial pressure. Other important reflex mechanisms include those which ensure that the heart becomes accelerated when the large veins and auricles are distended. An intrinsic regulation is also produced by the so-called law of the heart whereby the tension exerted by the individual muscle fibres increases with their length. Changes in heart frequency arise mainly by variations in the so-called vagal tone. The endings of the vagus nerve near the sinu-auricular node secrete acetylcholine (originally called 'Vagusstoff') which has an inhibitory effect upon the pacemaker region. Most of the effects of afferent stimulation result from an increase or decrease in this tone. These vagal fibres form part of the parasympathetic system and the reflexes in which they are involved are fairly specific. The sympathetic innervation has an accelerating effect due to the local liberation of adrenaline but it is often associated with liberation of adrenaline from the adrenal medulla as part of a general response changing the condition of the body in an emergency. Not only does it produce an increase in heart rate but it also speeds up the blood flow to the muscles by causing the smooth muscles of the arteriole walls to relax (vasodilation). Altering this peripheral resistance of the vascular system is an important way in which the blood flow is regulated. Not only the total output of the heart, but also its precise distribution can be regulated by varying the diameter of the different capillary systems.

The extrinsic factors regulating heart output are summarised in Table 6.1.

TABLE 6.1. *Summary of extrinsic regulation mechanisms for maintaining constant heart output*

Stimulus	Receptor	Input to medulla	Change in output	Effect
Increase in return of venous blood to the right auricle	Stretch receptors in great veins	Vagus X	Decrease in parasympathetic and increase in sympathetic supply to heart	Increased heart rate
Increased arterial blood pressure	Stretch receptors in aorta and carotid sinus	Vagus X and glossopharyngeal IX	Stimulation of parasympathetic and inhibition of sympathetic supply to heart	Decreased heart rate

6.1.7 *Blood flow in the capillary beds*

The total length of the capillaries in a human is several thousand kilometres and altogether they have a vast surface area of some 700 square metres (this is approximately 100 times greater than the surface area of the body itself). At any one time only about a quarter of this network is in operation.

The resistance of organs like the brain and the liver to blood flow is constant, and thus a fairly constant volume of blood passes through them at all times. The skin and the muscles, however, have capillary networks whose resistance can decrease by some 75 % as they become active, and in such organs blood flow will be dependent on metabolic activity.

The changes in the resistance of an organ to blood flow are largely due to the opening or closing or precapillary sphincters (see Fig. 6.8), which control the bore of the terminal arterioles entering the capillary beds. The sphincters are innervated by vasoconstrictor nerves from the sympathetic system which secrete noradrenaline. In any one capillary system, however, it should be noted that an accumulation of toxic wastes leads to relaxation of the controlling sphincter and thus a washing away of such metabolites by the blood. The system is thus self-regulating.

In shock the body may suffer a general relaxation of the precapillary sphincters and in this case the blood pressure is likely to fall drastically.

6.1 The heart and circulation of vertebrates

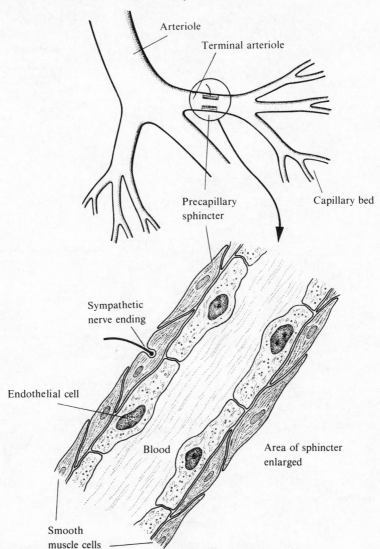

Fig. 6.8.
Diagrams to show the position of precapillary sphincters.

6.1.8 Development of the aortic arches

Lower vertebrates

The double circulation of mammals and birds is very distinct from the single circulation of fishes and in the absence of other evidence it would be difficult to envisage how one could evolve from the other. Fortunately, however, many intermediate stages can be seen among living vertebrates, both in their adult and developmental stages. A study of the latter shows that in all vertebrates six aortic arches join the ventral aorta with the lateral dorsal aorta on each side (Figs. 6.9 and 6.12). In fishes these embryonic arches become broken up into the capillary networks of the

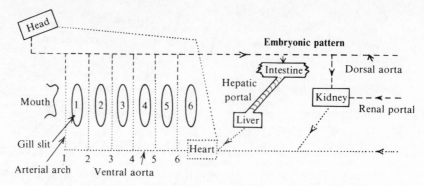

Fig. 6.9.
The basic type of
vertebrate circulation.

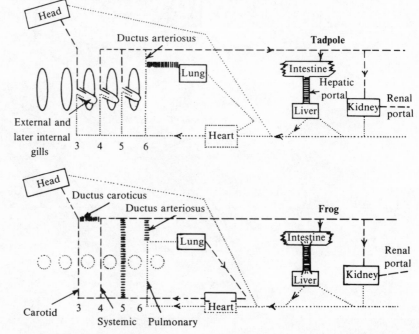

Fig. 6.10.
The aortic arches of
the tadpole and frog,
showing how they can
be derived from the
basic 'fish condition'.
Dotted lines represent
deoxygenated blood,
dashed lines
oxygenated blood, and
hatched lines vestigial
tissue.

gills. In adult tetrapods those which persist are continuous between the
ventral and lateral aortae. Usually these represent the third, fourth and
sixth embryonic arches and are known respectively as the carotid, systemic and pulmonary arches in adult forms. The carotid supplies the
head, the systemics take blood to the rest of the body, and the pulmonary
arteries go to the lungs. The pulmonary artery may retain its connection
with the lateral dorsal aorta. This connection, found during development
and also in some adult amphibians, is called the ductus arteriosus or
ductus Botalli (Fig. 6.10). Blood returning from the lungs enters the left
auricle in all lung-breathing forms. In some amphibians (e.g. salamanders
and newts) the fifth aortic arch persists and in some reptiles (e.g. lizards)
the connection between the systemic and carotid arches remains as the
ductus caroticus.

During development, tadpoles pass through a stage in which the arches

128

form gill capillaries and may also supply the external gills. Later the gill circulations are lost and the adult pattern develops. In mammals the six homologous arches are represented but are never all present at the same time.

The foetal circulation

There are two remarkable features of the circulation of the mammalian foetus associated with the way it receives oxygen. The first is that the lung is non-functional and therefore little blood can pass round the pulmonary circuit. Secondly, blood oxygenated at the placenta passes directly into the inferior vena cava. The way in which these features are combined in the foetal circulation and the changes which occur at birth provide excellent examples of the adaptation of the circulatory system to the means of gaseous exchange. The following account is based on the foetal lamb, but a similar change occurs in man. Of especial importance is the ductus arteriosus, which functions as a shunt allowing blood to pass from the pulmonary artery into the systemic circulation. In this way a greater volume of blood is passed to the placenta for oxygenation before its return to the right auricle. As so little blood circulates through the poorly

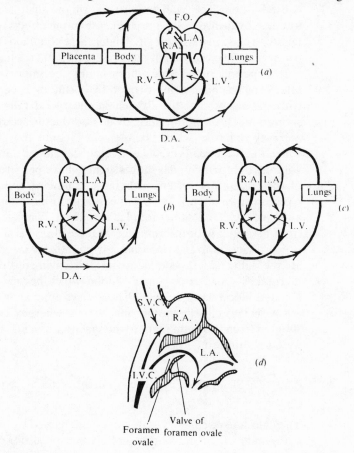

Fig. 6.11.
Diagram indicating changes in the circulation during the development of the mammal. (a) Foetal, (b) neonatal, (c) adult, (d) foetal circulation showing entry of vena cava to the right and left auricles.

developed lung capillaries, the left auricle receives only a small quantity of blood. This is supplemented, however, by about 40 % of the blood returning to the right auricle, which passes through the foramen ovale into the left auricle (Fig. 6.11).

At birth the umbilical cord is severed and the volume of blood in the systemic circuit is reduced; its oxygen content falls and carbon dioxide tension rises. This stimulates respiratory neurones in the medulla and leads to the expansion of the lungs. Consequently, the resistance to flow in the pulmonary circuit is reduced, and more blood passes by this route. The increased flow of blood into the left auricle from the lungs and the decreased flow into the right auricle from the systemic circulation have the important effect of shutting the valve mechanism which covers the foramen ovale. This closes up subsequently by tissue growth. At this stage, very shortly after birth, the ductus arteriosus remains open and in fact the direction of blood flow through it reverses (Fig. 6.11*b*). In this way a larger volume of the cardiac output passes to the pulmonary circuit than to the systemic circuit and the relative inefficiency of the gaseous exchange in the lung is compensated for. Within the next 24 hours, the ductus arteriosus becomes occluded and the complete double circulation is established. The vital importance of the ductus arteriosus and the foramen ovale are well known because of the effects resulting from their failure to close in development. Under these conditions the oxygenated and deoxygenated blood do not remain separate and a 'blue baby' may result.

In experiments with new-born lambs the importance of the ductus arteriosus has been demonstrated by occluding it very soon after birth, with the result that the oxygen in the blood falls from 20 to 10 % saturation. Both the foramen ovale and ductus arteriosus act as shunts and play vital roles which tide the young mammal over the period during which the two circuits are incompletely developed and cannot take equal volumes of blood. Similar shunts seem to have persisted in different parts of the circulatory system of living amphibians and reptiles. We have already drawn attention to the ductus arteriosus in some amphibians, and a diagnostic character of reptiles is that the conus is divided into three from its origin with the ventricles. One of these is the pulmonary arch, which divides into the right and left pulmonary arteries, and the others are the right and left systemic arches. In all living reptiles the left systemic arch takes some deoxygenated blood which becomes mixed with oxygenated blood when it joins the right systemic arch to form the dorsal aorta. This is clearly another mechanism whereby excess blood can be shunted from the pulmonary to the systemic circuit, and other shunts are found within reptile hearts.

6.1.9 *The heart and circulatory systems of fishes, with particular reference to the dogfish*

The chordate circulatory system is characterised by the direction of flow being anterior in the ventral vessels and posterior in the main dorsal vessels. Blood is pumped forward from the ventral heart behind the gills into the ventral aorta, from which afferent branchial vessels supply the gills. Oxygenated blood collects in the efferent vessels, which communicate with the paired lateral dorsal aortae that convey blood backwards to the single dorsal aorta from whence it is distributed to the rest of the body (Fig. 6.12). After its passage through the tissue capillary system the blood returns to the heart via the small veins which lead into the main veins entering the auricle. The single circulation contrasts with that of mammals and birds in that all the blood must traverse at least two capillary systems before it returns to the heart. Consequently the pressure of blood supplying the tissues is lower than in the systemic circulation of birds and mammals. The circulation time has not been measured for many fishes; in

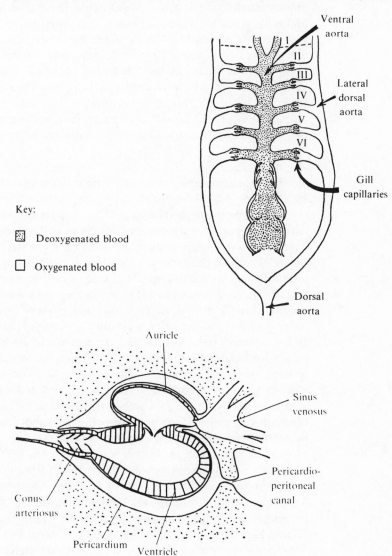

Key:

▨ Deoxygenated blood

▢ Oxygenated blood

Fig. 6.12.
The heart and arterial arches of a fish.

Fig. 6.13.
Medial section through a dogfish heart.

131

an eel it may take 2 minutes and in a dogfish probably longer. Fish hearts, except those of lungfishes, only contain deoxygenated blood and receive their oxygen from a separate coronary circulation which takes its origin from the efferent side of the branchial circulation. The heart of a fish (e.g. dogfish, Fig. 6.13) functions as a single pump and is made up of four distinct chambers in series, each separated from the next by valves. The original ventral tube has become S-shaped as seen in side view. Blood from the venous sinuses collects into a sinus venosus from which it enters the auricle after passing the paired sinu-auricular valves. The auricle is dorsal and thin-walled and when it contracts the blood passes into the thick-walled ventricle. From the ventricle, blood is forced into the ventral aorta after having passed through a conus arteriosus or, in teleosts, a bulbus arteriosus. The conus is contractile, being made of cardiac muscle, and it has three longitudinal rows of semi-lunar valves, up to six in each row. In more advanced fishes the conus becomes reduced, but in all cases at least one pair of valves persists. The swelling in front of the ventricle in bony fishes is a fibrous non-contractile expansion at the base of the ventral aorta which lies outside the pericardium. The semi-lunar valves of the conus prevent reflux of blood into the ventricle during diastole. The bulbus, being elastic, evens out the pressure wave so that blood flows through the gill capillaries at a more uniform velocity.

The dogfish circulation

The heart is contained within the pericardial cavity, coelomic in origin, which, in the dogfish and other cartilaginous fishes, is contained within a box formed by the pectoral girdle. This is important functionally because, with the exception of a small duct (the pericardio-peritoneal duct) which contains a valve, it is completely isolated from the main perivisceral (literally 'around the guts') coelom. Because of the single circulation and especially the large cross-sectional area of the venous sinuses, the problem of returning venous blood to the heart is very much more acute in a dogfish than in mammals. Fishes do not possess valves in their veins although muscular activities must aid the venous return. But the major factor assisting the return is the suction produced by the heart. This results from the incompressibility of the fluid contents of the pericardium, which must maintain constant volume. When the ventricle contracts there will be a tendency for the volume to be reduced which can only be compensated by the flow of an equal volume of blood into the pericardium through the sinus venosus. Fluid cannot enter through the pericardio-peritoneal canal because its valve does not allow the passage of fluid in this direction. In this way the dogfish heart functions as an aspiratory pump which draws blood into itself from the venous sinuses.

Pressures within different parts of the circulatory system of the dogfish have been measured and shown to be much lower than those of mammals and other tetrapods. The systolic pressure within the ventral aorta is 23 mm Hg and falls to 15 mm Hg during diastole. This pulse pressure of 8 mm Hg is reduced after the blood's passage through the branchial capil-

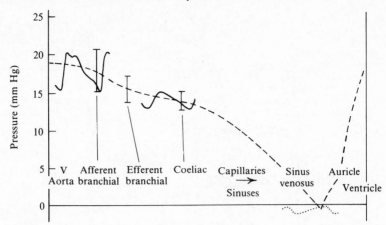

Fig. 6.14.
Vascular pressures at
various points in the
dogfish circulation.

laries (Fig. 6.14) and even further reduced as it passes into the finer arteries. The pressure within the venous system is extremely low and in the large sinuses may even be negative. This is especially true in the region of the heart, where blood in the sinuses will be subjected to the reduced pressure resulting from the aspiratory action of the heart. Circulation in bony fishes appears to be much more rapid and the pressures are higher at all parts of the system. The volume of blood in most bony fishes is distinctly less than that of the dogfish, which has approximately the same relative volume as a mammal (70–80 millilitres per kilogram body weight).

Regulation of the frequency and output of the heart of fishes is controlled by the sympathetic and parasympathetic systems as in mammals. Changes in blood pressure are detected by receptors in the branchial arches in regions homologous with the carotid sinus of mammals.

6.1.10 *The heart of frogs*

The frog heart seems an ideal intermediate between the fish heart and that of mammal and bird, for it contains a sinus venosus and two auricles but only a single undivided ventricle. Aortic arches 3, 4 and 6 persist as the carotid, systemic and pulmonary arteries (Fig. 6.15). The evolution of the lung as an air-breathing organ was an important step in the conquest of the land. But it was equally important that mechanisms should evolve which enable the best use to be made of the oxygenated blood returning from the lung in the pulmonary veins. An essential stage which occurred during the evolution of all lung-breathing forms is that the auricle became divided, and the pulmonary veins returned blood to the left side. Blood from the rest of the body collects, as it does in fishes, in the sinus venosus, which opens into the right auricle. The subsequent fate of blood on the left and right sides varies in different tetrapods but some degree of separation is maintained in most of them. (It should be noted that the lungfishes, which might be thought to have a circulation intermediate between fishes and amphibians, have undergone a separate evolution

133

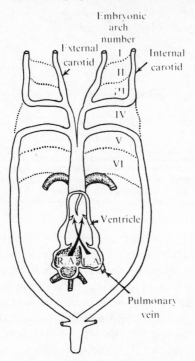

Fig. 6.15.
The heart and arterial
arches of a frog.

from modern amphibians for many millions of years. These fishes have special structural adaptations for the separation of venous and arterial blood which are not homologous with the circulatory changes seen in other vertebrates.)

The main features of the classical (Brücke and Sabatier) description of separation of systemic and pulmonary circulations in the frog heart are as follows:

(*a*) Separation is present in the ventricle because of its spongy nature and the high viscosity of the blood. Consequently at the completion of auricular systole, blood from the right auricle lies on the right side of the ventricle and that from the left auricle on the left side.

(*b*) When the ventricle contracts, the blood from the right side is the first to leave because it is from this region that the conus takes its origin. Deoxygenated blood, mixed blood and, lastly, oxygenated blood from the left auricle would therefore pass successively up the conus arteriosus.

(*c*) Because the resistance to flow in the pulmonary circuit is least the first lot of blood would flow into this channel. As the pressure rises in the pulmonary circulation, the second portion of blood (i.e. mixed blood) would pass into the systemic arch and only the last lot of blood (oxygenated) would be forced into the high-pressure system of the carotid arch. The high resistance of the carotid arches was thought to be due to the presence of the carotid bodies.

Some parts of this description have been established experimentally but others are now known to be incorrect, at least for the common frog (*Rana*). (*a*) One feature which appears to be true is that there is little

134

mixing in the ventricle. This can be demonstrated by injection experiments, so long as only small quantities of dye (0.001 millilitres Evans Blue) are injected into the blood returning to the right or left auricle. (*b*) During ventricular systole there is no successive movement of the blood into the conus arteriosus as blood from the right and left sides pass up it simultaneously. Nevertheless, blood from different parts of the ventricle maintain their separateness when passing up the conus, partly because of the action of the spiral valve. (*c*) Further evidence in support of the simultaneous passage of the different streams of blood up the conus comes from measurements of the pulse pressures, which are simultaneous within the three arches. There appears to be little difference in the resistance to flow in the three arches. (*d*) Distribution of the blood into the aortic arches depends upon the morphological relationships of their exits at the anterior end of the conus. In *Rana* and *Bufo*, blood from the right auricle mainly enters the pulmo-cutaneous arch but some also enters the left systemic. Blood from the left auricle is distributed to the carotids and mainly to the right systemic, and never goes to the pulmo-cutaneous arch (Fig. 6.16).

As the oxygen content of the blood in all the arches has not been measured, it is not possible to say whether there are significant differences between the carotid and systemic arches. It must be remembered that blood oxygenated at the skin forms an important proportion of the blood returning to the right auricle. Consequently, the difference in

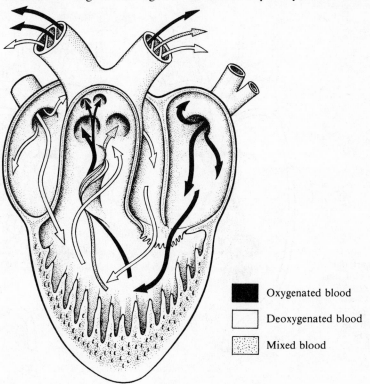

Fig. 6.16.
The frog heart. The path of the blood is indicated by injection experiments. Blood from the pulmonary vein (black arrow) passes to the carotid arches and the right systemic, blood from the vena cava (white arrow) enters the pulmo-cutaneous arch. Mixed blood (stippled arrow) passes to the left systemic.

■ Oxygenated blood

□ Deoxygenated blood

▨ Mixed blood

135

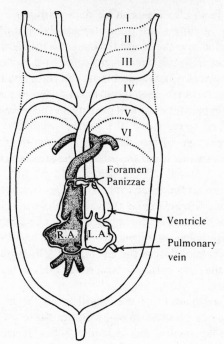

Fig. 6.17.
The heart and arterial
arches of a crocodile.

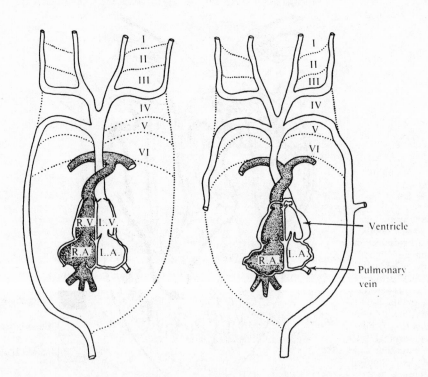

Fig. 6.18.
The heart and arterial
arches of (a) a bird and
(b) a mammal. (Note
that these are views
with the heart pulled
forwards, so left and
right appear to be
transposed.)

oxygen content of the right and left auricular blood may not be as great as might be imagined. In forms in which cutaneous respiration plays an even more important part than in *Rana* and *Bufo*, a greater degree of mixing of the bloods seems to occur; for example, in *Xenopus* the left auricular blood enters all three aortic arches. In urodeles a spiral valve is absent and complete mixing of the blood takes place in the conus. In some urodeles mixing even occurs in the auricle, which is undivided.

There are, however, important changes in the patterns of distribution of blood when an amphibian is using lung ventilation, because of a resistance to blood flow in the pulmonary circuit.

6.1.11 *The heart of modern reptiles*

In reptiles there are two auricles, but division of the ventricle is incomplete in all groups except the crocodiles (the crocodiles are close descendants of the stock from which birds arose). Another interesting feature of the crocodile heart is that shortly after the left and right systemic arches leave the ventricles they communicate with one another through a small foramen Panizzae (Fig. 6.17). During diving crocodiles are able to shunt blood past the lungs, re-routing blood from the right ventricle to the left aortic arch, and hence allowing a portion of the systemic blood to be recirculated.

6.1.12 *The heart of birds and mammals compared*

The circulation of a bird (Fig. 6.18*a*) can be derived from that of a crocodile by the loss of the left systemic arch. The mammalian type of circulation (Fig. 6.18*b*) cannot be derived from that of any living reptile. This is because the carotids of all modern reptiles arise from the right systemic arch, whereas in mammals they come from the persistent left systemic arch. Secondly, completion of the septum in any modern reptile except crocodiles would not result in a double circulation because both auricles open into the same part of the incompletely divided ventricle. Despite the possibilities for mixing apparent in these systems, injection experiments have shown that a precise distribution of the blood occurs, for example, in the lizard heart. Blood passing into the ventricle from the right auricle is distributed to the pulmonary arteries, whereas that which enters from the left auricle is pumped to the carotid and systemic arches. Mixed blood passes into the left systemic arch and becomes mixed with the right systemic blood when they both join to form the dorsal aorta.

6.1.13 *Phylogenetic considerations*

The persistence of so many apparently inefficient types of circulation in which oxygenated and deoxygenated blood are not kept completely separate suggests that they are related to some aspect of the functioning of the cardiovascular and respiratory systems. One suggestion is that the

137

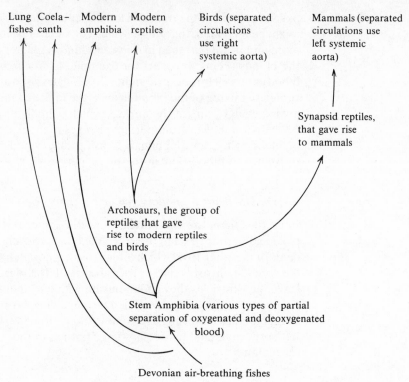

Fig. 6.19.
Phylogenetic
interpretation of
vertebrate hearts and
circulatory systems.

lungs have not evolved sufficiently to take a volume of blood equal to that passing through the systemic circulation. It is equally possible, however, that a greater volume of blood is passed through the pulmonary circulation than the systemic in order to compensate for the relative inefficiency of the oxygen exchange mechanism.

The relationships between the different classes of vertebrates that can be deduced from studies of the heart and arterial arches are amply confirmed by reference to palaeontological data (Fig. 6.19). The birds and crocodiles both evolved from the same group of reptiles – the Archosaurs. Other modern reptiles (the diapsids) are on a different line of evolution but share with the crocodiles and bird ancestors the possession of two temporal vacuities in the skull (hence their name). The reptiles which gave rise to the mammals (the synapsids) were on an entirely separate line of evolution for they possessed only a single vacuity in the temporal region of their skulls. There are no living representatives of this group of reptiles and it is not surprising therefore that the mammalian type of circulation cannot be derived from that of any modern reptile. The circulation of the modern amphibians is doubtless specialised relative to that of the ancestral forms which first came on to land, but nevertheless it is from a type similar to this that the mammal circulation is most easily derived. The first stage in this evolution would probably have been the separation of the pulmonary arch from the systemic and carotid arches in its origin from the ventricle. At a later stage the right systemic arch would disappear and this would only have been possible had both carotid arches taken their origin from the left systemic arch. By whatever means it was achieved, however, the double circulation of both birds and mammals is a very efficient mechanism and ensures that blood from the left side of the heart which has been oxygenated at the lung is pumped to the body. Whether it goes to the left or right in the systemic arch is relatively immaterial.

6.2 **The blood**

The blood, lymph and coelomic fluids form the internal medium of the body, which is regulated in a precise way with respect to its ionic content, osmotic pressure, gas content and temperature. This relative fixity of the internal medium is one of the most striking features of vertebrates, especially mammals and birds, but it is found, to a lesser extent, in all animals. This tendency to maintain a constant internal environment is called homeostasis. The efficiency of the controlling mechanisms is greatly aided by the rapid communication between different parts of the organism which the circulatory system provides.

6.2.1 *The structure of blood*

Blood is a tissue, in the same way that nerve and muscle are tissues; that is to say it is a collection of similar cells specialised to perform a given function in the body. Unlike most tissues, however, blood is a liquid, consisting of 'solid' corpuscles suspended in a fluid plasma.

The red blood corpuscles (erythrocytes)
There are between 5 and 6 million erythrocytes in each cubic millilitre of human blood and they are the most numerous type of corpuscle found in blood. The numbers fluctuate, but within small limits; certain organs, such as the spleen, act as stores and liberate more red corpuscles when they are required. In structure the red corpuscles are biconcave discs which are 7.2 micrometres in diameter with a thickness of 2.2 micrometres. Mammalian red corpuscles have no nucleus (this is a diagnostic character of the class).

In the foetus red corpuscles are formed in the liver but in the adult they form in the marrow at the ends of bones, especially the ribs and vertebrae. It has been demonstrated that the hormone erythropoietin stimulates their formation in all vertebrates. At high altitudes the production of red corpuscles is increased in response to a rise in the level of this hormone which is secreted by endocrine cells in the kidney cortex. It appears that it is the kidney that senses sustained low oxygen tensions in the blood. (When applied to a culture of liver cells from a foetal mouse, erythropoietin was found to cause an increase in DNA in 20 minutes and in RNA and haemoglobin synthesis within 2 hours. Here then is an example of a hormone 'switching on' a specific gene; see also p. 293.)

139

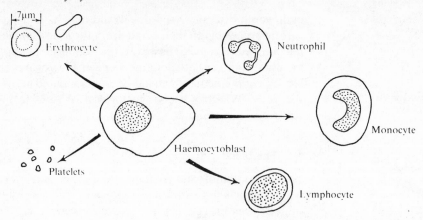

Fig. 6.20.
The main types of
blood corpuscle.

In man red corpuscles proliferate at a high and continuous rate of the order of 10^6 per second, and are, of course, destroyed at the same rate after living from 100 to 120 days in the circulation. (Other vertebrates, such as birds and reptiles, have nucleated red blood cells, which while they live much longer than mammalian red cells have an appreciable consumption of oxygen themselves.) As with other types of blood corpuscle erythrocytes appear to develop from a particular type of mother cell, called a haemocytoblast (Fig. 6.20). This goes through a number of stages, including the complete degeneration of the nucleus, until the mature red cell is virtually only a membrane system filled with haemoglobin. For the formation of haemoglobin, vitamin B_{12} and folic acid are required. These are two ingredients of diet which are assimilated in the stomach if the intrinsic factor of Castle is present in the mucosa (see p. 19). All the necessary constituents of haemoglobin must also be present, plus copper as an intermediate.

After their time in the circulation is completed the red cells are phagocytosed (engulfed) into the lining cells of spleen, liver, connective tissues and bones. The products of their breakdown are returned to the liver and some parts discharged into the bile canaliculi from the sinuses of the latter. The breakdown product of haemoglobin is bilirubin ($C_{33}H_{36}N_4O_6$), some of which is oxidised to biliverdin; both are excreted. Meanwhile most of the iron is released and stored in the liver for future use.

The functions of the red corpuscles are to transport oxygen, in the form of oxyhaemoglobin, and carbon dioxide as a carbamino compound, with haemoglobin. The oxygenation of haemoglobin also plays an important part in the release of carbon dioxide from the bicarbonate of the plasma (see p. 00). Finally it should be noted that, despite their lack of a nucleus, red corpuscles are able to pump out any excess sodium ions and to concentrate potassium ions across their membranes.

The white corpuscles (leucocytes)
There are only $5-6 \times 10_3$ of these per millilitre but like the red corpuscles

140

their numbers, and in this case their types, vary considerably under different conditions. Leucocytes are produced variously by reticulo-endothelial cells in the liver, spleen and lymph ducts, or by the bone marrow, and live for some 2 to 3 weeks in the circulation (or in some cases a good deal longer). Other than by their white colour they can be distinguished from red corpuscles by the presence of a nucleus. As at least five types of white corpuscle, the appearance and properties of each type will be dealt with separately. In general they all play a part in the defence of the body against infection.

Neutrophils (or polymorphs) are distinguished by their lobed nuclei; they make up 70 % of all the white corpuscles. Neutrophils are attracted out of the general circulation at any site of infection, and despite the fact that they are larger than red corpuscles they are able to pass through the walls of the capillaries by a process called diapedesis. This involves the white corpuscle changing shape into a thin pseudopodium which can pass through a pore in the capillary wall; the polymorph resumes its normal shape on the further side of the wall. They secrete certain enzymes but there is no doubt that their effectiveness in resisting infection is helped by secretions of antibodies from other types of white cells (the lymphocytes).

Once the neutrophil has found bacteria or other foreign bodies it engulfs them in the same way that amoeba takes in its food and, once inside, the invading organism is digested. The process is called phagocytosis. Many neutrophils, however, succumb to the poisons (toxins) of the germs and are killed, and these dead corpuscles form the bulk of the pus that collects about an infection.

Monocytes are larger than red cells, being 10–15 micrometres in diameter, and have a bean-shaped nucleus. They act in conjunction with neutrophils, described above, and are strongly phagocytic.

Lymphocytes are white cells that have entered the blood from the lymphatic system in which they originate. Their function seems to be to provide globulin proteins, the class of protein that includes antibodies, and like the first two types they collect at the site of bacterial invasions. The nucleus is spherical and they are not phagocytic (more will be said about these under lymph: see p. 145).

The remaining types of white corpuscle are *basophils* and *eosinophils* and these are normally few and far between in the blood. Not much is known about them except that the latter increase in certain types of infection and allergic conditions. They both have lobed nuclei.

The integration of the activity of these types of corpuscle and their place in the defence of the body will best be considered when the reticulo-endothelial system and the lymph have been described.

Platelets and the prevention of bleeding

Platelets are the other solid constituents of the blood and there are some 250,000 of them per millilitre. They are smaller than the red corpuscles and are formed from the disintegration of very large megakaryocytes of the red bone marrow. Platelets have no nucleus and consist of lumps of granular cytoplasm. They function in the prevention of bleeding, both by releasing vasoconstrictor hormones that assist in the automatic sealing of a cut vessel and by their involvement in the complex clotting mechanism of the blood.

The process of blood clotting is dependent on the change of the soluble plasma protein fibrinogen to insoluble fibrin, which provides the mesh on which other solid particles, including the platelets, build a clot. This change is due to the hydrolysis of peptides in the fibrinogen by the enzyme thrombin.

The detailed path of thrombin formation is still controversial, but it seems likely that contact of an inactive factor with the wound surface activates it and so sets off a 'cascade' of reactions involving at least six different factors. (The significance of such a cascade system is that it acts as an amplifier, with a small initial change at each step triggering off progressively greater ones.) Activation of each of these factors from its predecessor in turn causes the newly activated factor to act enzymically on the next one and so on until the last factor catalyses the conversion of inactive prothrombin to thrombin. Some of these factors need calcium ions to operate as enzymes, the availability of calcium appears never to be limiting in a normal clotting situation as the amount required is so small. Most of the factors have names, the best known being No. VIII, the anti-haemophilic factor that is absent from people suffering from haemophilia.

In addition to all these events that lead to thrombin (and hence fibrin)

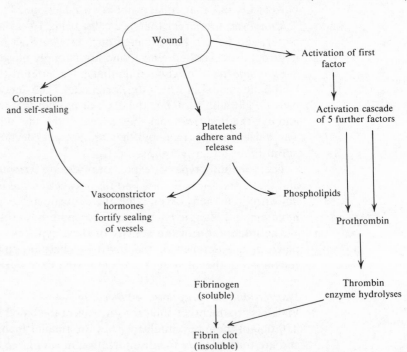

Fig. 6.21.
Summary of the
mechanism of
prevention of blood
loss from wounds.

formation the release of phospholipids is required from the platelets. If this does not happen no clotting results, so that it can be seen that while the role attributed to platelets is different from that in older schemes of how blood clots, they are still thought to play an essential role in clotting. It is not easy to see why such an elaborate system is necessary; it may be to ensure that clotting does not occur too readily within the circulation, but on the other hand these cascade systems are more subject to genetic mutation and failure than are single-factor systems. Another possible advantage is the amplification effect described above, which could not be obtained by a single factor system.

The plasma

This is a straw-coloured fluid making up about 50 % of the blood volume. It contains 90 % water and 10 % dissolved substances, including proteins, food substances, salts and excretory products.

Proteins in the plasma make up most of the solid material and have a number of specific functions. The main protein is albumen, while others include globulin and fibrinogen; together these produce a colloid osmotic pressure (about 30–40 mm Hg) which in turn helps to control blood volume and water balance in the body. The globulin is concerned with the synthesis of antibodies in the blood and is probably released by lymphocytes and reticulo-endothelial cells, while the function of the fibrinogen has been described above (i.e. formation of the fibrin clot). These proteins are all manufactured in the body and may be differentiated from the small percentage of amino acids that are transported by the blood – for example, from the ileum to the liver and from the latter to other tissues.

Besides amino acids, fatty acids and sugar (in the form of monosaccharides) are also transported in the plasma from the gut to the liver and thence to all parts of the body. There are some 0.6 grams of fat and 0.1 grams of sugars in 100 millilitres of blood. The quantities are strictly regulated because outside a certain range of concentrations these substances have deleterious effects on the body, the former on pH and the latter osmotic. The main mechanism controlling the blood sugar depends on insulin secreted by the pancreas (p. 286).

Other organic substances in the plasma are excretory urea, creatinine and uric acid, which are found collectively at concentrations of 0.3 grams per 100 millilitres of blood, together with traces of lactate and hormones.

The inorganic ions include K^+ (5 millimoles per kilogram) Na^+ (143), Cl^- (103), Mg^{2+} (2.2), and Ca^{2+} (5), together with traces of sulphate, phosphate and carbonate.

The functions of these ions have been described elsewhere (p. 165), but the alkali present has an important buffering function and together with the haemoglobin and plasma proteins forms the alkali reserve of the body. From the physiological point of view the most important of these buffers is bicarbonate, because the nervous mechanisms regulating respiration are tuned to maintain a constant blood pH or carbon dioxide content. As a chemical buffer bicarbonate mops up hydrogen ions from organic acids as follows:

$$CH_3CHOH.COOH \rightarrow CH_3CHOH.COO^- + H^+$$
(lactic acid)

(Sodium and bicarbonate ions in blood) $\rightarrow Na^+ \quad HCO_3^-$

$$CH_3CHOH.COONa \quad H_2CO_3 \rightarrow H_2O + CO_2$$

Phosphates and haemoglobin in the red corpuscles and plasma proteins are also important chemical buffering systems.

Small amounts of oxygen (0.25 %) and carbon dioxide (3 %) are carried in solution in the plasma, but the oxygen would not be adequate for the needs of the mammal, and most is carried in association with haemoglobin.

6.2.2 *Exchange between the capillaries and the tissues*

The blood, with the composition described above, passes from the small arterioles of the arterial system into the capillaries. These tiny vessels (some 8–10 micrometres in diameter) permeate all the tissues of the body. It is across the capillary walls that the cells of the organism must take up the substances they require (Table 6.2).

Oxygen and carbon dioxide are readily exchanged by diffusion across the walls of the capillaries. Lipids also pass through the membranes and are taken up by the numerous pinocytotic vesicles characteristic of the capillary surfaces (Fig. 6.22). Large proteins probably also pass across capillaries by pinocytotic action. Water and water-soluble molecules, however, pass in and out of the capillaries via the pores that lie between adjacent endothelial cells, being effectively squeezed out at the arterial end and sucked back at the venous end.

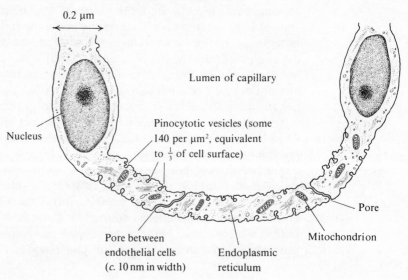

Fig. 6.22.
A drawing based on an electron micrograph of a capillary (with one particular type of endothelial lining).

0.2 μm

Lumen of capillary

Nucleus

Pinocytotic vesicles (some 140 per μm², equivalent to $\frac{1}{3}$ of cell surface)

Pore

Pore between endothelial cells (*c.* 10 nm in width)

Endoplasmic reticulum

Mitochondrion

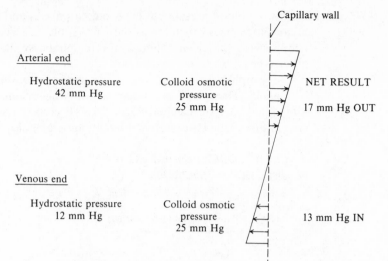

ig. 6.23.
he forces causing
ovement of water
nd water-soluble
olecules into and out
f capillaries.

TABLE 6.2. *Permeability and molecular sizes of some blood constituents that pass across capillary walls*

Substance	Radius of molecule (nm)	Permeability (cm per second $\times 10^{-6}$)
Water	0.46	0.54
Sodium chloride	0.24	0.33
Urea	0.27	0.26
Glucose	0.36	0.09
(cf. Haemoglobin	3.25	0.00)

The force that causes the movement of water and water-soluble molecules into or out of the capillaries is the net result of the hydrostatic pressure of the blood, which tends to force water out, and the colloid osmotic pressure of the plasma proteins (especially albumen), which tends to suck water in. The hydrostatic pressure decreases along the capillary from the arterial to the venous end with increasing distance from the heart; the colloid osmotic pressure, being proportional to the plasma protein concentration, remains more or less constant. The net result is that at the arterial end the hydrostatic pressure is greater than the colloid osmotic pressure and so water is forced out of the capillary, while at the venous end the situation is reversed and water is sucked in. The situation is shown diagrammatically in Fig. 6.23. It can be seen that on balance there is more fluid leaving the capillary than is drawn back in, and this excess forms the basis of the lymph.

6.2.3 *The lymph*

As described above, many substances leave the capillaries and enter the tissue spaces. The fluid thus formed is called lymph and is in immediate contact with the cells of the tissues. Lymph has much the same composition as the blood plasma but lacks red corpuscles and has various other minor differences which are as shown in Table 6.3. The lymph has more lymphocytes, which originate in the lymphatic system, but less of the other type of white corpuscles.

The lymph passes from the tissue spaces into small vessels called lymphatics. These lead into larger vessels which eventually drain into a tube called the cisterna chyli (Fig. 6.24). This tube passes into the venous system just as the latter enters the right auricle, that is, at the base of the

TABLE 6.3. *Differences in composition between blood plasma and lymph*

Substance	Lymph (as compared with plasma)
Ca^{2+}	Less
PO_4^{3-}	Less
K^+	Less (by 3 %)
Protein	Less (by 38 %)
Fat	Less
Glucose	More (about 20 %)

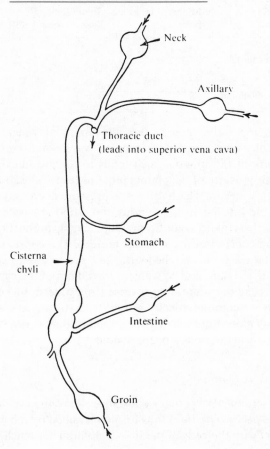

Fig. 6.24.
Layout of the lymphatic system. (The round areas represent regions where many nodes are found.)

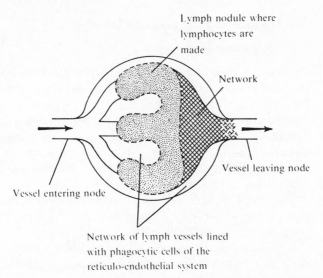

Lymph nodule where lymphocytes are made

Network

Vessel leaving node

Vessel entering node

Fig. 6.25.
Schematic diagram of a lymph node.

Network of lymph vessels lined with phagocytic cells of the reticulo-endothelial system

jugular vein, by the thoracic duct. The blood pressure here is only 5 mm Hg, which is less than the lymph pressure, and thus the lymph fluid returns to the general circulation. Lymph vessels are similar to the veins and depend for their flow on muscular activity and valves which prevent backward movement of the fluid. At various points along these vessels are lymph nodes, essentially filters, concerned with the defence mechanisms of the body.

These nodes (Fig. 6.25) are prominent in the intestines and in certain regions such as the back, groin, and under the arm. The lymph returning to the heart enters the node in a sort of pool; it then passes into the body of the node where great masses of lymphocytes are proliferated from the reticulo-endothelial system (see below), and thus into the efferent vessel. It is known from experiments on dogs that a lymph node is able to remove great numbers of foreign bodies such as germs from the circulating fluid and this is their main function. In cases of infection the lymph nodes (or glands) become swollen. These nodes are only found in warm-blooded vertebrates.

6.2.4 *The reticulo-endothelial system*

The cells of this system line the sinuses of the liver, spleen, lymph and connective tissues of the body. They are partly sessile (fixed) and partly mobile and are concerned in the defences of the body. Sessile members are able to ingest foreign matter as it passes and to destroy it, while the large mobile macrophages engulf germs and ingest them. The cells of the reticulo-endothelial system also produce antibodies and antitoxins in response to specific stimuli and these circulate in the body fluids acting against foreign bodies. Monocytes (p. 141) are budded off from the reticulo-endothelial system and one of the normal functions of this system is the destruction of old red blood cells.

6.3 **The body's defences against infection**

6.3.1 *The causative agents*

Disease is a state of deleterious departure from the normal condition of the body. This is a wide definition and covers genetical abnormalities of form and function, sometimes called 'inborn errors of metabolism', functional diseases, and the presence of parasitic organisms. It also covers the deficiency diseases which result from the body being deprived of some essential constituent of diet. It is only the defences of the body against infectious agents that concern us here, and the role of the blood and reticulo-endothelial system.

The infectious agents of disease comprise viruses, bacteria, protozoans and the macro-parasites. Examples of these as well as of other diseased states are shown in Table 6.4.

Viruses are entirely parasitic and can only reproduce within the body of their host, but as far as most micro-organisms are concerned the mammalian body is by no means an ideal environment and it is not surprising that so few of the many thousands of species of bacteria and protozoans have been able to live within the tissues. On the whole, the mammalian body is too hot for the most efficient functioning of micro-organisms (hence the old-fashioned method of curing syphilis by giving the patient malaria, which raises the body temperature and kills the syphilis bacteria) and the chemical environment of the body is inimical to their growth. It is

TABLE 6.4. *Human diseases due to micro-organisms, larger parasites and other causes*

Bacteria

Pneumonia (some types)	Dysentery (one type)
Tetanus	Boils
Typhoid	Tuberculosis
Diphtheria	Scarlet fever
Cholera	

Virus

Mumps	Rabies
Influenza	Yellow fever
Smallpox	Poliomyelitis
Measles	

Protozoa

Dysentery (amoebic)	Sleeping sickness
Malaria	

Other parasites

Elephantiasis (nematode)	Liver fluke (platyhelminth)
Bilharzia (platyhelminth)	Tinea (fungus)

Diseases due to deficiency of necessary food substances

Simple goitre	Beriberi
Scurvy	Pellagra
Rickets	

Functional diseases due to failure of organs

Heart conditions: angina, etc.

not practical for a parasite to destroy its host, and the usual state that exists between parasite and host is one of adaptation rather than violent reaction.

We all carry great numbers of bacteria on our bodies and on the linings of the respiratory tracts, as well as in the gut. Some of these are potential pathogens but the establishment of our normal surface species of bacteria does seem to prevent invasion by other harmful species. But if our defence mechanisms fail these organisms multiply and bring about disease, as happened with the enteric infections at Hiroshima when radiation temporarily suspended the natural relining of the gut by cell divisions.

Infectious micro-organisms must first enter the tissues of the body, and this is not easy as the skin limits such entry by the continuous sloughing off of dead keratinised epidermis. Inside the body the various tracts are protected by wandering white cells and by secretions, both of which destroy infectious agents, as well as by having a commensal flora which resists the settling of new species. The nose filters out foreign matter from the air taken into the body and the acidity of the stomach does much to limit bacteria entering the small intestine. The reproductive and respiratory tracts are lined with cilia beating towards the openings of the body which convey foreign particles to be expelled. If the body is wounded a blood clot seals off the tissues from micro-organisms.

In other words, it is neither easy for a pathogen to enter nor to survive within the body and this is why we remain relatively free of disease most of the time.

Viruses

These are between 12 and 400 nanometres in size and are composed of nucleoprotein, usually surrounded by a protein wall. They have no internal organelles and although they possess a genetic mechanism they require the protein replication system of the host to express this. Viruses lack an energy-exchange mechanism. For this reason they must be parasitic, using the energy-exchange units of their host cell to satisfy their metabolic requirements.

Viruses enter the cells of their host by injecting their nucleoprotein core into the cytoplasm, the protein case remaining outside the cell membrane. The core of the virus reproduces rapidly, taking over the protein replication mechanism of the host and using it to synthesise its own nucleoprotein. In a matter of minutes the host cell may be completely converted to virus material, giving an increase in the latter of some 200 times, and this new generation of viruses is released to attack further cells. Sometimes viruses remain latent in a cell, only reproducing when subjected to abnormal conditions such as heat or radiation. An example in man is the skin infection herpes simplex, which is activated by the ultraviolet radiation in sunlight and produces cold sores. Because viruses are contained within the cells of the host, antibiotics (which circulate in the blood) are not effective against them and the body has to rely on its

TABLE 6.5. *Pathogenic micro-organisms and man*

Tissue affected	Disease	Transmission
(a) Viruses		
Skin and mucous membranes	Smallpox	Contact, sputum, etc.
	Measles	Contact
Nervous tissue	Poliomyelitis	Faeces, sputum
	Rabies	Animal bites
Respiratory tract	Influenza	Nasal and oral discharges
	Viral pneumonia	Sputum
Intestinal	Epidemic diarrhoea	Faeces
General infections	Yellow fever	Mosquitoes
	Dengue	Mosquitoes
(b) Bacteria		
Intestinal	Salmonella	Contact, faeces
	Typhoid	Faeces
	Cholera	Infected food and water
Lung	TB	Droplet infection
	Whooping cough	Droplet infection
Genitalia	Syphilis	Contact
	Gonorrhoea	Contact
Skin	Leprosy	Prolonged contact

own defences. Recently artificial anti-viral chemicals have been synthesised and are likely to find increasing use in the treatment of disease. In the mammal, viruses cause infections of the skin, nervous tissue, respiratory tract and alimentary canal, as well as general disorders (see Table 6.5).

Bacteria

These are unicellular organisms ranging from 1 to 8 micrometres in size (although exceptional forms may be much larger than this), which have a single ring chromosome not enclosed in a nuclear membrane. They have a form of sexual reproduction and a cell wall of polysaccharide somewhat resembling that found in the fungi. In adverse conditions some bacteria form resistant spores, which remain viable and potentially infective for long periods.

Bacteria fall into three classes. First there are cocci, which may be subdivided into double cell types (such as the pneumonia species), chain forms (streptococci) and cluster-forming species (staphylococci). The second type of bacteria is the rod-shaped bacilli (bacillus means rod-shaped), and the remaining type the spiral-shaped spirochaetes.

When the pathogenic forms invade the body they may cause damage by the destruction of cells or by their accumulation, as in diphtheria. It is, however, far more usual for the harmful effects to be due to the metabolic wastes or toxins that they produce. Some of these toxins are among the most potent poisons known; as little as 0.0001 millilitres of botulin toxin can cause the death of a guinea-pig. In some cases, such as tetanus and gangrene, the bacteria feed on living cells and dead wound tissues but the

150

powerful toxins they produce cause death by diffusing into uninfected tissues.

In nature, bacteria compete with many other micro-organisms and some of these produce substances which destroy the bacteria. Well-known examples are the fungal extracts penicillin and streptomycin, which are used as antibiotics to combat infectious bacteria in the body.

6.3.2 Antigens and antibodies

Antigens are substances that promote antibody formation by the body, the antibody being specific to the antigen with which it reacts. Pathogenic micro-organisms and their products are the most common antigens but any foreign protein or tissue, for example the peptide-carbohydrate surface component of grass pollen, may act as an antigen.

In recent years there has been a marked increase in our knowledge of the detailed structure of antibodies and the way in which they are produced, as well as the way in which they combine with antigens.

All antibodies belong to the class of proteins called immunoglobulins (Ig). While the structure of specific antibodies has much in common, differences in chain length etc. exist and a number of types are recognised (e.g. IgA, IgD, IgE, IgG, IgM). The immunoglobulin whose structure was first elucidated is gamma-globulin or IgG and this is the one about which most is known, owing to the researches of Porter and Edelman in the early 1960s.

Gamma-globulin is made up of four polypeptide chains; two of these are long and are called heavy chains and two are shorter and termed light chains. The four chains are held together by disulphide links. As the

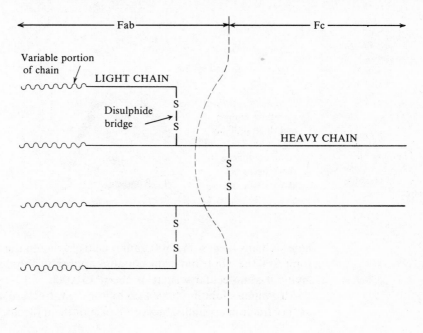

Fig. 6.26. Diagrammatic representation of a gamma-globulin molecule.

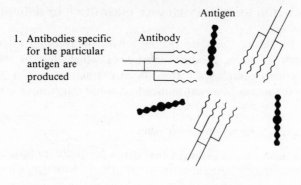

1. Antibodies specific for the particular antigen are produced

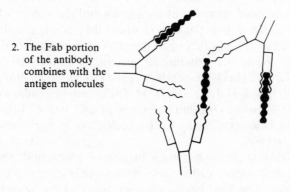

2. The Fab portion of the antibody combines with the antigen molecules

Fig. 6.27.
One of a number of possible ways in which antibodies can combine with antigens.

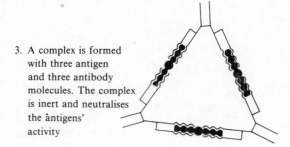

3. A complex is formed with three antigen and three antibody molecules. The complex is inert and neutralises the àntigens' activity

heavy chains have some 500 amino acids each and the light chains have some 200 the whole molecule contains approximately 1,400 amino acids, giving it a molecular weight of about 140,000.

All gamma-globulins have a common stem structure, and this is known as the 'fraction crystallisable' or Fc part of the molecule. The ends of the

152

four chains, however, vary in structure and form the site of binding with the antigen. For this reason this part of the molecule is called 'fraction antigen binding' or Fab. The variability in the amino acid sequences in the Fab part of the antibody molecule allow it to combine successfully with a great many different antigens and it has been suggested that the human body can produce at least 100,000 different antibodies. Fig. 6.26 is a diagrammatic representation of a gamma-globulin molecule.

When antibody and antigen combine the Fab portions of the antibody bind onto the active sites of the antigen to form an inactive complex, thus effectively neutralising the antigen (Fig. 6.27).

Evolution of vertebrate immunoglobulins

Immune responses essentially similar to those seen in mammals and involving proteins of the immunoglobulin type have been detected in all vertebrates down to the level of the agnathous lamprey.

Chemical analysis shows the protein of the fish type (for example that of sharks) is a single form, seemingly a precursor of all the forms of immunoglobulin seen in higher vertebrates. It may be that the colonisation of the land required the development of more complex antibodies; modern amphibians have two distinct types of immunoglobulins. Thereafter it appears that parallel changes in chain complexity took place separately in the reptile–bird line and in the mammal line.

It is interesting to compare the rates of evolution of the vertebrate immunoglobulins with other proteins such as their cytochromes and haemoglobins. Such data as exist suggest that the non-variable parts of the chains have evolved faster at a rate of 0.2 mutations per 100 amino acids each million years, which is faster than other proteins studied. It has been estimated from differences in immunoglobulin structure between men and monkeys that the time of divergence of the hominoid and monkey lines was 20 to 30 million years ago, which is consistent with the palaeontological evidence.

6.3.3 *The immune response*

As the mammalian embryo grows in the uterus of its mother its immunity system becomes competent to respond to antigens, although, in its sheltered environment no antigenic material will be present. Immediately after birth however the young mammal becomes exposed to numbers of antigens from the air, its food and from other sources.

This post-natal period is a time of great activity for the B cells, which are derived from the bone marrow. An individual is programmed genetically to produce a large number of different types of B cell but whether or not a particular type of B cell will proliferate depends on its exposure to antigen. Antigens with which the very young animal comes in contact result in triggering off the multiplication of the specific B cells tailored to meet the antigen with an appropriate antibody.

During this time the T cells from the thymus are also active and assist

153

the process described above. It seems that the T cell is a coarse adjustment system to the antigen, which becomes attached to it at one point making it easier for the appropriate B cell to respond.

This can be shown as follows:

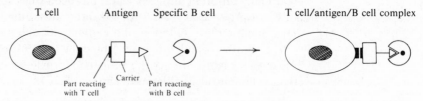

At an early stage of its life, therefore, the mammal becomes armed with a large number of antibody-producing cells which will confer immunity to a wide range of antigens with which it is likely to come in contact. Of course a very large number of genetically programmed B cells may never be needed if the individual does not ever come in contact with the antigen to which they are set to respond.

After this early phase in the setting up of immunity the activity of the thymus declines and continues to do so. By the age of 20 in man it has virtually ceased to function.

When antigens enter the body it is thought that the immune system responds in the following way (Fig. 6.28):

(1) Antigen molecules are picked up from the blood, lymph or tissues by macrophage cells. These are large phagocytic cells with numerous cytoplasmic granules that are found scattered throughout the body but particularly in the lymph nodes, the liver, the lining of the blood vessels, the spleen and the bone marrow. In the bloodstream monocytes and lymphocytes may also be transformed into macrophages.

(2) Once attached to the surface of a macrophage the antigen comes into contact with the T cells and B cells.

(3) At this stage the critical interaction of the three cell types – T cell, B cell and macrophage – takes place with the antigenic molecule, as described above. The T cells are then able to switch on genes in the nuclei of the appropriate B cells which causes these to proliferate rapidly into plasma cells. It is the plasma cells which actually secrete the antibody molecules.

(4) Antigen and antibody then combine chemically in the manner already described and the resulting neutral complex can be phagocytosed from the bloodstream or the tissues by appropriate white corpuscles. The activity of these latter is stimulated by the T cells directly.

It will be noted that if this method of antibody formation is true it requires vast numbers of different types of B cells, each making a different type of antibody, to be present in the first place. Only the B cell suited to the antigen will be triggered to respond to the instructions of the T cells recognising this antigen. Improbable as this may appear it had wide acceptance as the most likely mechanism, and much experimental work is in progress in this area.

154

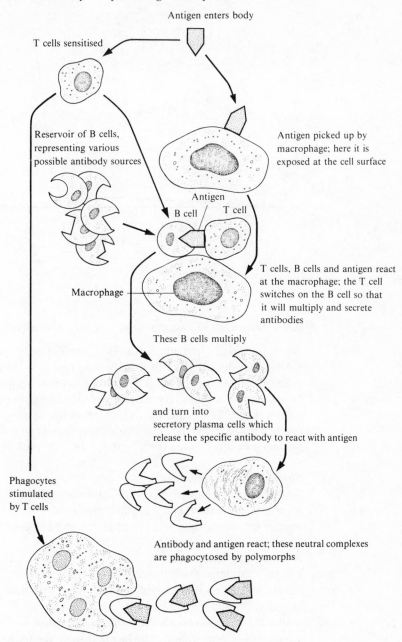

Antigen enters body

T cells sensitised

Reservoir of B cells, representing various possible antibody sources

Antigen picked up by macrophage; here it is exposed at the cell surface

Antigen

B cell   T cell

T cells, B cells and antigen react at the macrophage; the T cell switches on the B cell so that it will multiply and secrete antibodies

Macrophage

These B cells multiply

and turn into secretory plasma cells which release the specific antibody to react with antigen

Phagocytes stimulated by T cells

Antibody and antigen react; these neutral complexes are phagocytosed by polymorphs

Fig. 6.28.
A proposed scheme for the immune response.

The role of complement

Antibodies are not the only proteins involved in combating antigens and other enzyme-like blood proteins called complement (so-called because they *complement* the action of the antibody) may act with the antibody. Complement is particularly important in the lysis of bacterial cells and viruses and is also involved in tissue reactions to some forms of tumours.

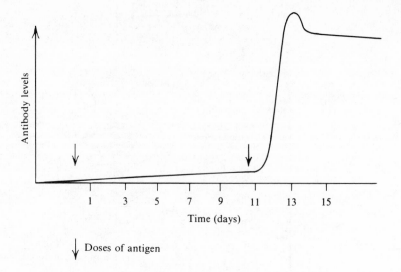

Fig. 6.29.
The antibody response
to two successive doses
of antigen.

6.3.4 *Sensitisation to antigens and types of immunity*

When an antigen first enters the body it gives rise to specific antibodies in
the manner described. This process may take several days and to a small
dose of antigen only a small amount of antibody is produced. If another
dose of antigen follows the first, however, a massive secretion of antibody
occurs (Fig. 6.29). While the exact basis of this response is not clear, it can
be seen that it gives the body protection against any subsequent invasion
by antigens after it has been sensitised by an initial contact.

The role of antitoxins
Antitoxins are special types of antibodies that act against the toxins
formed by infectious agents, which they neutralise. They act in the same
way as the antibodies and are made by the same sort of cells.

Protection against virus infections
The virus attacks inside the cell and therefore is not readily accessible to
the circulating antibodies or the phagocytes, or indeed to chemicals, such
as the antibiotics, introduced into the body.
   In 1957 it was shown that a cell produces against the virus that attacks it
a substance, now called interferon, which seems to prevent the virus
making use of the energy mechanism of the cell. (It will be remembered
that the virus needs to use the cell's energy supply system in order to
reproduce.) This interferon has a very similar composition and molecular
weight wherever it is found and there is some hope it may be possible to
produce it for clinical use. Of course the mammal also makes antibodies
against viruses and these appear before interferon and are an essential
part of the body's defence. Cells infected with viruses undergo changes to

156

their surface chemistry and are recognised as foreign by other body cells.

Natural immunity and active and passive immunisation
The defence system of the body is stimulated to produce antibodies not only when a disease is present but under a number of other conditions. Such a production of antibodies may give rise to natural immunity, that is, an immunity to a disease which has not been acquired by medical means. During life the body is exposed to many potentially harmful micro-organisms, such as the polio virus or tuberculosis bacteria, which may enter the body and stimulate its antibody system. The latter can some-times resist and destroy the infectious agent without any clinical symp-toms of the disease appearing, and the antibodies formed provide further immunity against subsequent infections. In other cases the disease will manifest itself (e.g. mumps, measles, whooping cough) but on recovery the body retains antibodies, as well as the 'memory' of their synthesis, which will provide a lifelong natural immunity. Viruses, however, are able to change their antigenic identity more readily than bacteria, and this is one of the reasons why prolonged immunity to such virus diseases as colds and influenza does not occur.

Artificial immunity by immunisation consists in the formation of anti-bodies to a disease when this process has been stimulated by the deliber-ate introduction of antigenic substances into the body. In some cases weakened or related micro-organisms are introduced (e.g. rabies, small-pox immunisations), in others toxoids (weakened toxins) are used (e.g. tetanus). Dead antigenic material also stimulates antibody production without fear of infection. Such a form of immunisation is called active because it relies on the response of the body's own defence mechanisms and in most cases leads to prolonged immunity. This whole system of artificial immunisation is usually attributed to Jenner, who 200 years ago showed that infection with the mild disease cowpox gave immunity to the much more serious disease of smallpox.

Another form of artificial immunity is called passive and this is used where an infection has already taken place. Well-known examples are the tetanus antitoxin given after a deep wound and the snake-bite antitoxin administered after a bite. These antitoxins are prepared by the inocula-tion of an animal, such as a horse, with increasing doses of antigenic material and the collection of the antitoxins it forms in its blood. The method is passive because, although it assists the body to neutralise the infection, it does not stimulate the cells to produce antibodies and thus does not confer more than a temporary resistance.

By the use of these techniques, together with public health measures and the use of antibiotics, the incidence of infectious diseases in many parts of the world has decreased. Thus smallpox has been virtually eliminated from Asia and hopefully is not far off total extinction. That many such diseases are only kept under control and not permanently destroyed is shown when our modern health measures have temporarily collapsed, as in the Congo in 1959, and in India in 1949 when, during the

157

pilgrimage to the Ganges, an epidemic of smallpox broke out owing to the failure to ensure a widespread programme of immunisation.

Antibody reactions not associated with micro-organisms

There are a number of ways in which foreign protein may be introduced into the body other than by micro-organisms. Such foreign proteins can evoke antibody responses by the body which are not only of importance to the physician but also illustrate the very wide scope of protective reactions.

Many individuals are sensitive to a variety of animal and plant proteins which produce allergic responses in the cells with which they come in contact. Grass and other plant pollens may stimulate antibodies and increase secretion of histamine (a hormone involved in defence mechanisms) in the cells of the respiratory tract, eye membranes, etc., a condition known as hay fever. Other types of protein may cause asthmatic attacks. A serious type of allergy is where the body has previously been made hypersensitive to a protein, for example, the venom of a bee, and a subsequent dose causes a general reaction in the tissues. This exaggerated response can lead to the death of the individual, but despite its disadvantageous nature it is only an extreme form of antibody action.

Another manifestation of the sensitivity of the body to foreign protein occurs in blood transfusions. Certain factors in the red blood corpuscles act as antigens when transfused into a person with a different genetic make-up. Antibodies to these antigens are naturally present and do not need to be induced, being a result of the genotype of the individual concerned. In man the antigen–antibody pairs most commonly found are those associated with blood groups A, B, AB, and O. The letters A and B represent antigens and are clotted by the opposite antibody, that is, α and β. Thus A blood is clotted by both B and O recipients, but not by A or AB groups, while O blood can be transfused safely into any blood group. The compatibility of the various blood groups is shown in Table 6.6. This transfer of protein from one individual to another does not happen in nature except in the case of the mammalian placenta where the foetal and maternal tissues are in contact. The evolution of this organ had to overcome rejection of foreign proteins built into the parental tissue through

TABLE 6.6. *The compatibility of blood groups on mixing*

Blood group (and antibodies) of recipient	Blood group (and antigens) of donor			
	A (A)	B (B)	AB (AB)	O (none)
A (α)	–	×	×	–
B (β)	×	–	×	–
AB (none)	–	–	–	–
O ($\alpha\beta$)	×	×	×	–

–, compatible; ×, clots.
From the table it can be seen why blood group O is termed the universal donor while AB is the universal recipient.

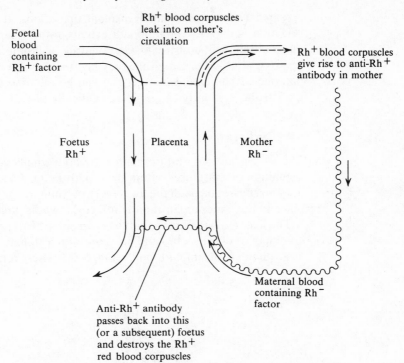

Foetal blood containing Rh⁺ factor

Rh⁺ blood corpuscles leak into mother's circulation

Rh⁺ blood corpuscles give rise to anti-Rh⁺ antibody in mother

Foetus Rh⁺

Placenta

Mother Rh⁻

Maternal blood containing Rh⁻ factor

Fig. 6.30.
The possible antigenic reaction between a Rh⁻ mother and an Rh⁺ foetus.

Anti-Rh⁺ antibody passes back into this (or a subsequent) foetus and destroys the Rh⁺ red blood corpuscles

the immune response. The sort of difficulties that can arise are seen in the Rhesus factor of man (Fig. 6.30).

The Rhesus factor can exist in two allelomorphic forms (i.e. genes situated at the same point on homologous chromosomes), Rh⁺ and Rh⁻, these two forms being antagonistic to each other. Let us suppose that the mother has Rh⁻ blood while the foetal blood is Rh⁺ by parental inheritance, Rh⁺ being the dominant gene. If a small leakage of foetal blood occurs across the placenta then antibodies will be generated in the maternal bloodstream against the Rh⁺ antigens in the foetal blood. When these pass across the placenta into the foetal circulation they will destroy the Rh⁺ red blood cells and then a complete blood transfusion is necessary immediately after birth if the child is to survive. In fact the most likely time for the leakage of foetal blood into the mother's circulation to occur is at birth. This means that while the first baby is not normally affected, because the anti-Rh⁺ antibodies remain in the mother's circulation then subsequent foetuses are progressively at risk.

The ABO and Rhesus systems may interact in certain situations and the whole problem of tissue matching (blood is a tissue) is very complex.

A final type of antibody reaction takes place when living cells of skin, or of some organ, are grafted from one individual to another. Unless the donor and patient are genetically identical (i.e. are identical twins) the foreign proteins in the donated tissue will act as antigens and cause an immune response in the patient that will tend to destroy the graft. The basis of tissue rejection has become an important field of modern

159

research because of organ transplant operations. It seems that tissue rejection is due to lymphoid cell activity rather than processes taking place within the bloodstream. Certain drugs will suppress these rejection responses but at the same time they lay the body open to infection from pathogens. The extent of rejection can be minimised by matching the graft tissue as closely as possible genetically to that of the patient.

6.4 **The liver**

The liver is an important organ of the body vitally concerned with the regulation of substances within the bloodstream. As all the blood of the mammal passes through the liver every two minutes or so, changes taking place in this large organ rapidly affect the whole circulation.

The liver is a large organ that takes up one-fifth of the whole viscera. It is composed of four lobes and has a uniform histology, being made up of large numbers of similar lobules (Fig. 6.31). These tend to be hexagonal

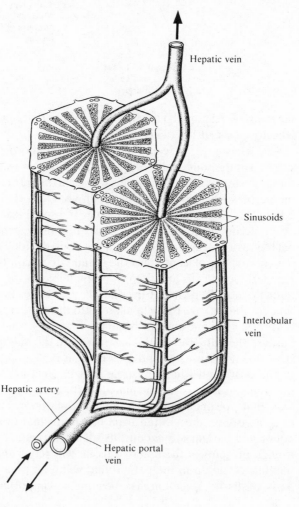

Fig. 6.31.
Two lobules of the
liver.

and from a central vessel radiate out 50 or so small sinusoids along which blood passes from the peripheral vessel. It is in its passage through these sinuses that exchanges take place.

As with other organs, the liver receives an arterial supply (the hepatic artery from the aorta) and drains into a vein, the hepatic vein, which passes into the posterior (or inferior) vena cava. In addition to this the liver also receives the terminal branches of the portal system from the ileum and other parts of the intestines. Detailed arrangement of these three systems in relation to the lobule is shown in Fig. 6.31. Branches of the hepatic portal system pass into the edges of the lobule while the hepatic vein originates from its centre. Hepatic arteries are also at the periphery.

Besides the vascular system the liver has a complex meshwork of bile canals leading from the intercellular canaliculi and entering the gall bladder.

The functions of the liver are as follows:

(*a*) the reception and storage of food substances assimilated in the intestines;

(*b*) the deamination of proteins;

(*c*) the production of bile;

(*d*) the removal of waste substances and foreign matter and general detoxication of the blood;

(*e*) the production of heat.

It therefore completes and integrates many of the metabolic processes, such as digestion, that have already been described.

The reception and storage of food is by uptake from the hepatic portal vein of substances absorbed from the ileum. Monosaccharides, usually in the form of glucose, are absorbed along cells of the liver sinusoid and transformed by a condensation reaction into the storage compound glycogen. This is a polysaccharide similar to starch and the cells of a normal liver are full of glycogen granules. The condensation involves a phosphorylation and requires energy. It is stimulated, both in the liver and the tissues, by the action of insulin, which seems to facilitate the passage of the sugar across the cell walls. When the blood glucose level falls, glycogen is hydrolysed back into glucose which is passed into the blood for transport to the body cells that require it. This process is controlled by hormones secreted by the anterior lobe of the pituitary, pancreatic islets and the adrenal cortex and medulla. These two mechanisms are finely balanced so that the blood glucose level remains within narrow limits at all times.

Fats and fatty acids entering the liver, either by the hepatic portal system or the general circulation, are metabolised in there. Much fat is stored in its cells and the breakdown and oxidation of fat (Fig. 6.32) also takes place in the liver. Amino acids are taken up and either stored or passed into the circulation, in the cases of essential ones, or deaminated, and respired or condensed. This deamination is an important process and consists of the removal of nitrogen from excess amino acids. The nitrogen

161

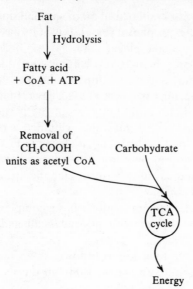

Fig. 6.32.
Fat oxidation.

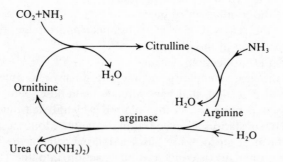

Fig. 6.33.
The ornithine (or urea)
cycle.

in the form of ammonia is converted to urea by reacting with carbon dioxide (Fig. 6.33). Other constituents of amino acids, including phosphorus and sulphur, are also removed in the liver and eliminated as waste materials, leaving carbon, hydrogen and oxygen compounds.

The mechanism of bile production has been described on p. 48. The bile canaliculi run parallel with the sinusoid of the liver and bile substances are continually passed into the canaliculi from the surrounding cells. At the edges of the lobule are found the longer bile ducts which eventually empty into the gall bladder.

Part of the reticulo-endothelial system (see above) lines the sinuses of the liver lobules and this gives the liver the ability to pick up and destroy foreign bodies. In addition to this, these cells in the liver seem to have a wide range of detoxicating effects and are able to down-grade and excrete a number of toxic substances, for example alcohol, that are taken into the bloodstream.

7 Excretion

The high degree of efficiency to which the physiological processes of mammals have evolved underlies the whole success and activity of the group. A part of this efficiency lies in the maintenance of a stable internal (hence cellular) environment in which metabolic reactions can proceed under optimum conditions.

Two of the ways in which stability is achieved have already been described – the regulation of temperature and of the pH of the blood. No less exact is the control of the concentration and ionic composition of the body fluids. Metabolism means, essentially, the reactions between protein and other systems which make up all the activities of protoplasm. Protein reactions and properties depend not only on the pH and temperature but on the degree of hydration and the ionic medium in which they operate. Such important processes as contraction, conduction and membrane permeability are particularly dependent on the ionic medium; also, ions act in conjunction with many enzymes. The end-products of metabolic activities are often toxic to the organism. This is particularly true of nitrogenous substances produced during protein deamination, and the removal of these from the body is one of the main functions of the excretory organs.

The regulation of all these factors makes up the physiological function of osmotic and ionic regulation and excretion. Its main agents are the kidney, lungs and skin, and the substances involved are salts, water, carbon dioxide and nitrogenous waste. Although these organ systems have evolved independently for the mammal, efficient regulation as well as temperature control depend on their interaction.

A number of terms are used to describe the concentration of the body fluids and urine. It is important to know these in order to understand the following account of regulation. First there is the important concept of osmosis. This is the tendency of water to pass across a semi-permeable membrane (i.e. one which water molecules alone can pass through) to a medium of higher concentration. All living membranes are not truly semi-permeable because they allow the passage of ions and molecules other than water. The osmotic concentration of the body fluids is largely the result of small molecules or ions; large protein molecules are less significant per unit length in their contribution to osmotic pressure. Two solutions are said to be isotonic if there is no tendency for water to pass from one to the other when they are separated by a semi-permeable

163

membrane. If there is passage of water across the membrane then the solution that gains the water is hypertonic relative to the other and the solution that loses the water is hypotonic. These terms do not necessarily indicate the relative osmotic concentrations of the two solutions; these are given by the terms iso-osmotic, hyperosmotic and hypo-osmotic respectively. It would be possible, because of the properties of the membrane, for water to move between two iso-osmotic solutions if they had different tonicities.

7.1 **Water balance**

For most mammals water is an essential constituent of their diet (though a small number of desert-dwellers can derive sufficient water from the metabolism of foodstuffs). Whatever the source, the body of a mammal is some two-thirds water (man 63–67 %), and active protoplasm is always hydrated to this sort of proportion. Protoplasm is a colloid whose continuous phase is water, and as the cell membrane is permeable to water and some solutes, the concentration of the body fluids surrounding the cell must be similar to that of the cell contents in order to avoid a net flow of water into or out of the cell. Water is used in the elimination of toxic nitrogenous waste, and although the urine of a mammal is usually hypertonic to the body fluids, salts cannot be concentrated beyond a certain point, so that salt excretion also needs water. A further use of water is in temperature control, where excess heat is lost by the evaporation of water from the skin or lung.

Over a period of 24 hours a man needs anything between 1 and 7 litres of water. If this need is not met the body becomes dehydrated; death will occur when some 20 % of the body's water content has been lost, owing to failure to transmit heat to the skin and to the strain on the cardiovascular system as a result of the increased viscosity of the blood.

The usual source of water is the diet, either directly as a liquid or as a constituent of solid foodstuffs. Most vegetables are 80–90 % water and even dry foods, such as bread, contain as much as 10 % water. The water is taken up along the whole length of the alimentary canal, probably by the osmotic pull of the lining epithelium (unless the osmotic pressure of the gut contents happens to be higher than that of the epithelia, as it may be under certain conditions). The skin of mammals is almost impermeable to water, unlike that of the frog, so this is the only means of uptake.

TABLE 7.1. *Relative water losses from different organs*

Organ	Water loss per 24 hours (ml)
Kidney	600–2,000
From gut in faeces	50–200
Skin	400–4,000
Lung	350–400

Excess water taken in is rapidly eliminated from the body by the kidney. The mechanism of this regulation is described later. Table 7.1 shows the different ways in which water loss from the body is accounted for.

7.2 **Inorganic ions**

Much of the effective osmotic pressure of the cell and body fluid is due to the presence of inorganic ions. Colloid osmotic pressure resulting from large organic molecules is very much less than the pressure exerted by free ions. From the time of its early evolution in the sea the cell has been surrounded by a chemical environment and it has needed to select certain ions, important in its functioning, and eliminate others. If the cell membrane is not to have a permanent charge the total sum of positively charged ions (cations) and negatively charged ions (anions) must be the same on either side of the membrane. The cell has to use a number of devices to achieve a cytoplasmic environment which is isotonic with its surroundings. This need not depend on having the same concentration of osmotically active substances inside and outside the cell. Some of the devices for producing isotonicity are as follows:

(*a*) Differential permeability to various ions.

(*b*) The joining of an inorganic ion (e.g. K^+) with an oppositely charged organic ion which cannot pass through the membrane (e.g. an amino acid, $NH_3CH_2COO^-$).

TABLE 7.2. *The ions of the body fluids*

Ion	Where found	Use
K^+	Main cation of cell	(*a*) Contributes to osmotic pressure of cell (*b*) Provides potential across nerve cells (*c*) Activates enzymes (including phosphorylases)
Na^+	Main cation of body fluid	(*a*) Works with K^+ in producing potential across cell membrane (*b*) Contributes major osmotic pressure of body fluid
Ca^{2+}	In body fluids mainly	(*a*) Affects permeability of cell, possibly involved in Na^+ pump action (it accumulates in over-stimulated nerve) (*b*) Involved in muscle contraction through activation of 'myosin' enzyme (*c*) Involved in the functioning of 'cement' holding cells together
Mg^{2+}	Both in fluid and cell	(*a*) Decreases permeability of membranes but otherwise acts contra-osmotically to Ca^{2+} (*b*) Decreases excitability of muscle (*c*) Has activating effect on some oxidising enzymes
PO_4^{3-} HCO_3^-	Both in body fluid and cell	(*a*) Use in buffering changes of pH
Cl^-	Main anion of body fluids	(*a*) Affects permeability (*b*) Balances cation concentration within cell

TABLE 7.3. *The ionic composition of the blood plasma and urine of man compared with that of seawater (millimoles per kilogram)*

	Na^+	K^+	Ca^{2+}	Mg^{2+}	Cl^-	SO_4^{2-}
Plasma	143	5	5	2.2	103	1
Urine	143	35	6	9	136	20–60
Seawater	465	9.9	10.2	53	542	35

(It should be noted that seawater varies slightly in composition and urine very widely.)

(*c*) The active transport and removal of one ion in exchange for another.

It is mechanism (*c*) which is used a great deal in animals, and the best-known example is the sodium pump. This is a mechanism that was demonstrated first in the giant nerve fibres of the squid, and is an active pumping out of Na^+ that requires metabolic energy. The removal of this cation from the cell allows other cations such as K^+ to enter – though ions may be concentrated within the cell by actively pumping them in. The mechanism is found in nearly all mammalian cells as well as in other groups. One can suppose that the early chemical environment in which living organisms arose was over-rich in Na^+ ions and that in order to acquire other essential ions inside the cell, the ever-present Na^+ had to be pumped out. On the other hand the sodium pump may have evolved in the cell as a means of reducing its osmotic problems, the removal of the common Na^+ ion lowering the osmotic pressure in the cell and reducing the tendency of water to enter the cell by osmosis.

Thus the cytoplasmic colloids became adapted to the other ions brought into the cell. Compared with sea-water the mammalian body and cell fluids are low in Na^+, Cl^- and Mg^{2+} (see Table 7.3).

The ions of the body fluids are taken in with the food and water of the diet. They are selectively absorbed by the cells of the alimentary canal but the main method of uptake is by diffusion across a gradient. Some have the specific uses outlined in the discussion on the diet (that is, as straight-forward building materials – for example, in bone structure), but others have more complex roles and these are briefly summarised in Table 7.2.

The composition of seawater and of the urine and blood plasma of man is shown in Table 7.3.

In a very general sense the urine eliminates from the body the excess of those ions in which the composition of the body fluids most departs from that of seawater. However, the concentrations of these ions are by no means constant in the body even if some major differences from seawater can be seen. A human red blood corpuscle, for example, contains relative amounts of ions (millimoles per kilogram) as follows: Na^+, 10; K^+, 105; Ca^{2+}, 0; Mg^{2+}, 5.5; Cl^-, 80; SO_4^{2-}, 0. In other words the Na^+, Ca^{2+} and Cl^- are actively kept out while K^+ and Mg^{2+} are concentrated. This is for specific reasons to do with the functioning of the corpuscle, but it shows what a large local variation in ionic composition the body can produce.

7.3 **The kidney**

The mammalian kidneys are made up of some 2,000,000 filtration units, called nephrons; the total length of these amounts to some 80 kilometres in an adult man. Although we are mainly concerned here with the mammals, some vestiges of early excretory systems are found in this group, so some account of the origin of the mammalian kidney should be given.

7.3.1 *Origin of the kidney*

The nephrons arise from the mesoderm surrounding the nephrocoel (or segmental excretory cavities of the embryo); there is one pair of tubules to each segment and they open into a duct which leads to the cloaca. Blood is supplied by the aorta to the nephrocoels and forms a capillary network – this is the glomerulus; the capillaries join up again and lead into the posterior cardinal veins. The tissue immediately surrounding the glomerulus is the Bowman's capsule and this, together with the glomerulus, makes up the Malpighian body. The whole excretory body plus its tubule forms the nephron.

This segmental plan is very primitive and does not exist in the adults of modern vertebrates. In mammals the excretory bodies and their tubules form in three areas: the pronephros is formed in the anterior area, the metanephros in the posterior area and the mesonephros is intermediate. They are not separate organs but all arise from homologous embryological tissues. The pronephros functions as the embryonic kidney but does not persist, though its ducts become incorporated into the collecting ducts of the mesonephric adult kidney. It is, however, found as the functional kidney of the tadpole. The mesonephros is the functional kidney of fishes and amphibians. It loses its segmental arrangement as more tubules and Malpighian bodies develop. A part of it is found in the amniotes in the collecting ducts (or vas efferentia) from the testes. The metanephros is only found in the reptiles, birds and mammals, where it is the functional kidney. It has its own collecting duct, the ureter, and a great many extra excretory units, the nephrons, whose tubules often become very complicated.

The ureter leads down to a sac or bladder originally derived from the walls of the cloaca. Here the urine passed down from the kidneys collects, and is passed out of the body from time to time via the urethra. Except in young mammals the sphincter muscle at the bladder is under conscious control, but emptying of the bladder is normally stimulated by the pressure of urine on proprioceptors within the bladder muscles.

7.3.2 *How the kidney operates*

In gross structure the mammalian kidney can be seen to consist of two main layers – an outer cortex and a central medulla. A variable portion of

167

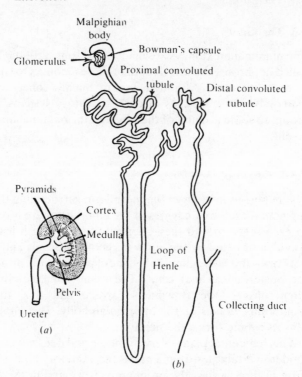

Fig. 7.1.
(*a*) Longitudinal
section of the kidney.
(*b*) A nephron.

the medulla is made up of the pyramid of large collecting ducts which open into the ureter (Fig. 7.1). The nephrons (i.e. Malpighian bodies and tubules) are arranged so that the filtration unit, i.e. the Malpighian body, lies in the cortex while most of the tubule is in the medulla. The tubule system is much modified from the primitive anamniote condition, and a loop of Henle is present between the proximal and distal convoluted tubules. This loop functions in the complex processes of salt concentration and reabsorption of useful substances filtered off from the blood (as does the whole tubule). The descending and ascending limbs of the loop differ in their histological appearance in transverse section.

The fact that the two limbs of the loop of Henle run parallel to one another and to the collecting duct has been shown to be of physiological significance. The length of the loop is correlated with the size of the pyramids, which are most distinct in desert-dwellers and other mammals that live in conditions of water shortage.

7.3.3 *The blood supply*

The blood supply of the tubules is very complicated, as can be seen from Fig. 7.2. In any one minute as much as a quarter of all the blood in the body is passed through the kidney, from which the glomeruli filter about 120 millilitres (one-tenth of the total volume). The capillaries of the glomeruli are particularly permeable (see below) and in all they provide

168

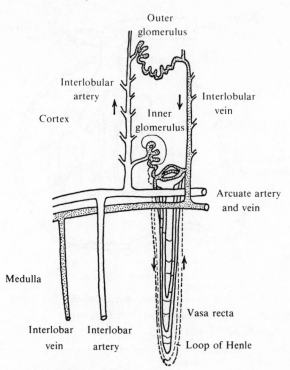

Fig. 7.2.
Blood supply to a
kidney tubule.

1–2 square metres of surface-area across which the smaller molecules and ions can be passed.

The main artery to the kidneys is the renal artery from the dorsal aorta. This branches into interlobar arteries and these in turn into circumferential arcuate arteries (Fig. 7.2). The latter give off small interlobular arteries and from these the glomeruli branch off. The glomeruli have an afferent vessel, taking blood to them from the interlobular artery, and an efferent vessel collecting up the blood and passing it to interlobular veins and hence, via interlobar and arcuate vins, to the main renal vein and thus back into the circulation.

Whereas the outlying glomeruli have the simple circulation described above, those lying near the medulla (the juxtamedullary glomeruli) have quite different connections. The efferent vessel on leaving the glomerulus loops down in a series of U-shaped bends, called the vasa recta, and these come in close communication with the convoluted tubules and loop of Henle (Fig. 7.2).

7.3.4 The glomerulus

The walls of the capillaries which make up the glomerulus are permeable to substances of low molecular weight; these include water, nitrogenous waste such as urea, inorganic ions, and some useful substances such as sugar and occasionally protein. The sort of molecule which can pass

169

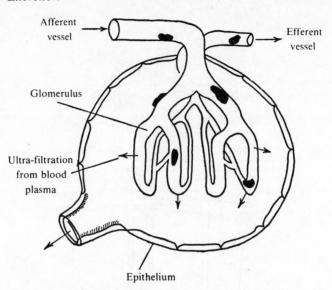

Fig. 7.3.
Malpighian body.

TABLE 7.4. *A comparison between the amounts of various substances in glomerular filtrate and urine*

Substance	Amount passed in filtrate of glomerulus/day	Amount passed in urine/day
Na^+	600 grams	6 grams
K^+	35 grams	2 grams
Ca^{2+}	5 grams	0.2 grams
Glucose	200 grams	Trace
Water	180 litres	1.5 litres
Urea	60 grams	35 grams

through the filter has a molecular weight which is below 67,000; albumen is a border-line size for filtration. The passage of each filterable substance across the membranes of the glomerulus will vary according to its concentration in the blood, but by and large the composition of the filtrate for these smaller molecules and ions is the same as that of the blood plasma. However, by the time the urine reaches the bladder the amount of these substances is quite different, as shown in Table 7.4, as useful substances are reabsorbed (p. 173).

The afferent vessel entering the glomerulus is wider than the efferent one that leaves it (Fig. 7.3). This increases the filtration pressure, just as constricting the nozzle of a garden hose increases the force of the jet it produces.

The hydrostatic pressure of the blood in the glomerulus is some 70 mm Hg, but the colloids it contains exert a back colloidal osmotic pressure of about 30 mm Hg. The resultant filtration pressure is thus $70 - 30 = 40$ mm Hg, although this may be reduced further by back pressure from the lumen of the glomerulus. In a normal adult some 120 millilitres of filtrate are produced from both kidneys per minute giving a total about 170 litres every 24 hours.

Fine structure of the glomerular capillaries

Examination of the glomerular capillaries with the electron microscope shows a specialised arrangement of the cells for the process of ultrafiltration (Fig. 7.4).

On the blood side of the capillary is a layer of endothelial cells stretched out in flat plates. The cells have transverse pores and the layer is sometimes called the lamina fenestrata. After fixation these pores are around 20 nanometres in diameter but there is good reason to believe that they are not more than 4 nanometres in the living state.

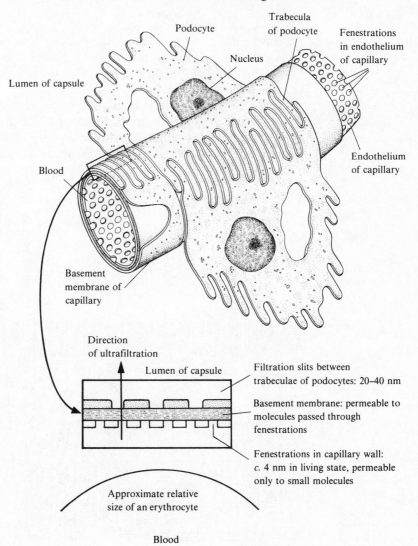

Fig. 7.4.
The ultrastructure of a glomerulus.

Excretion

Next to these endothelial cells is the non-cellular layer of the basement membrane to which the cells on both side are attached. The material of the basement membrane is made up of microfibrils and is presumably readily permeable to substances allowed through by adjacent cells.

The outer layer of cells, i.e. those cells in contact with the lumen of the capsule along their outer sides, is the fenestrated podocytes of the epithelium. Such cells are raised up above the basement membrane on numerous projections of their cytoplasm called trabeculae, which end in pedicels in contact with the membrane below. Between the pedicels are filtration slits some 20 to 40 nanometres across.

Here therefore we have an elaborate arrangement of cell layers producing a fine sieve across which small molecules and ions can be forced by hydrostatic pressure.

The proximal convoluted tubule

On leaving the Bowman's capsule the filtrate passes into the long, coiled proximal tubule of the nephron. Here the main exchanges take place that are to result in the formation of the urine and the reabsorption of useful substances.

The cells of the proximal tubule are reminiscent of those lining the ileum, having pronounced brush borders with microvilli and the large numbers of mitochondria that are associated with a high metabolic rate (Fig. 7.5). Their function is actively to secrete into the lumen of the tubule

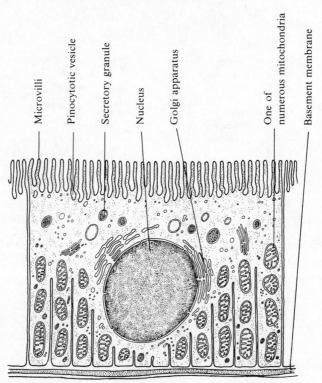

Fig. 7.5.
The ultrastructure of a cell from the proximal convoluted tubule.

waste nitrogenous products such as creatinine, choline and urates, and to reabsorb useful substances such as amino acids, glucose, bicarbonates, phosphates, potassium, protein, sodium, chloride, sulphate and water. Thus all of the 1.8 kilograms of glucose that passes through the glomeruli of a normal kidney every 24 hours is reabsorbed in the proximal tubules, and seven-eighths of the Na^+ and HCO_3^- ions are also returned to the blood.

The total surface area of the proximal tubules exceeds 5 million square millimetres, and they represent the main working area of the kidney, the other regions being the means, albeit very important, of fine adjustment of urine composition. It is in the proximal tubules that part of the essential acid–base regulation of the blood fluids occurs (see p. 175 for discussion).

The loop of Henle

The role of the loop of Henle in osmoregulation in the mammal has been firmly established since the researches of Wirz and others in the mid-1950s. Essentially regulation is achieved by building up high osmolarities of NaCl at the lower part of the loop which, according to the osmolarity of the urine, may or may not cause water to be reabsorbed from the final urine passing down the collecting duct. This initial concentration of salt is achieved by a system of counter-current multiplication.

The filtrate passes from the proximal tubule into the descending limb of the loop of Henle (Fig. 7.1). Here little active transport occurs, but as the filtrate enters the ascending limb sodium pumps in the cells lining this limb begin to extract Na^+ ions from the passing fluid. These are actively transported out into the extracellular fluid and then passively re-enter the descending limb as well as entering the surrounding capillaries of the vasa recta. Cl^- ions follow the Na^+ passively out of the ascending part of the loop owing to the charge left across its membranes by active movement of the Na^+.

After the pumping mechanism has been operating for a short period the osmolarities in a typical human loop of Henle might be as represented in Fig. 7.6.

The filtrate now leaves the ascending limb for the distal convoluted tubule and at this stage, owing to the removal of the NaCl, it will be hypotonic to the body fluids. But the high osmotic concentrations at the bottom of the loop of Henle are also represented in the vasa recta and external fluids in contact with the final collecting tubule as it emerges from the distal convoluted region (because the flow of blood through the vasa capillaries is sufficiently slow not to disperse the concentration gradients produced by the active transport mechanisms), so the urine will go through the final stage of its concentration in the collecting tubule. If the body needs to make a hypertonic urine, water is withdrawn across the walls of the collecting duct by osmosis. How much water is removed and thus the actual concentration of the urine is determined by hormones, as is the proportion of salts lost.

It can thus be understood why the region of Henle takes the form of a

173

Excretion

loop and why the vessels of the vasa recta take the form of loops. Increases in length of the limbs of these loops lead to higher concentrations in the glomerular filtrate, and in desert mammals the loop of Henle is very long. Such mammals can produce urine that is more than twice the concentration of seawater, a remarkable physiological feat, and most mammals are able to produce urine somewhat more concentrated than seawater. The human kidney has relatively short loops of Henle and is not able to make urine as strong as this, which is why man cannot use salt water as his only source of liquid.

Table 7.5 shows the relative thicknesses of the medulla of the kidney (the region where the loops of Henle are situated) relative to the maximum concentration of urine that can be produced. It can be seen how closely the loop length, as indicated by the thickness of the medulla, is related to the concentration of the urine, and this in turn to the water conservation problems of the environments in which each of these mammal species live.

The ability of the human kidney to concentrate excretory substances including salts, when this is necessary, is shown in Table 7.6, and it is obvious from this that the kidney is a highly effective excretory and

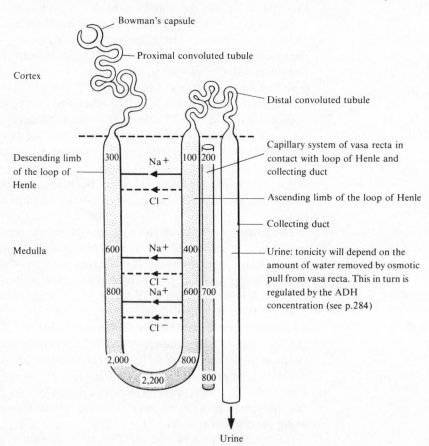

Fig. 7.6.
The counter-current multiplier system of the loop of Henle. Figures indicate the osmolarities of NaCl.

174

TABLE 7.5. *The relationship between thickness of medulla and concentration of urine*

Species	Environment	Thickness of medulla (relative to beaver)	Freezing point depression* of urine (Δ °C)
Beaver	Freshwater	1	0.96
Man	Land	2.6	2.6
Cat	Land	4.2	5.8
Kangaroo rat	Desert	7.8	10.4

* The more concentrated the urine the greater is the depression of its freezing point.

TABLE 7.6. *The maximum concentrations of various substances in human urine relative to their concentration in the body fluid*

Substance	Ratio urine: body fluid
Urea	60:1
Water	1:1
Creatinine	100:1
Na^+	1:1
K^+	7:1
SO_4^{2-}	60:1
Ammonia	400:1

osmoregulatory organ. The ability to excrete hypertonic urine is essential for successful terrestrial life and has also developed in reptiles, birds and insects.

The distal convoluted tubule

The distal tubule actively secretes ammonia, H^+ and K^+ may selectively reabsorb Ca^{2+}, Mg^{2+}, K^+, NaCl, and water. Its functioning in urine formation is not so significant as that of the proximal tubule although it has a special part to play in the maintenance of acid–base equilibria of the body fluids (see below).

7.3.5 Acid–base regulation and the part played by the kidney in maintenance of a constant pH

As a result of its metabolic activities the body produces a large amount of acid. This acid, by changing the pH of the body fluids, would affect the performance of protein molecules such as enzymes and lead to death if not rapidly removed or buffered.

One way in which acid is removed is via the lungs where HCO_3^- and H^+ ions recombine to give carbonic acid, which in turn dissociates into carbon dioxide and water which are exhaled. The other organ involved in removal of acid and conservation of the body's base reserves are the kidneys, and they may act in a number of complementary ways.

Essentially H$^+$ ions produced during respiration and other metabolic processes diffuse into the bloodstream where they are picked up and buffered by the body's bicarbonate reserve, i.e.

$$H^+ + HCO_3^- \rightarrow H_2CO_3$$

H$^+$ ions may also be picked up and transported by alkaline phosphates in the bloodstream

$$H^+ + Na_2HPO_4 \rightarrow NaH_2PO_4 + Na^+$$

The problem is to remove these H$^+$ ions in the urine while at the same time not depleting the body's reserves of alkaline buffer, namely bicarbonate. The main methods found in the kidneys are as follows:

(1) Should there be a loss of HCO$_3^-$ ions in the glomerular filtrate then base may be recovered, mainly by the cells of the proximal convoluted tubule, as follows

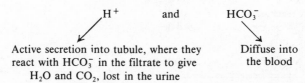

Tubule cells metabolise H$_2$CO$_3$
(in the presence of carbonic anhydrase)

This dissociates into

H$^+$ and   HCO$_3^-$

Active secretion into tubule, where they react with HCO$_3^-$ in the filtrate to give H$_2$O and CO$_2$, lost in the urine   Diffuse into the blood

Thus H$^+$ ions picked up from the metabolism of the body and combined to give bicarbonates are effectively lost from the kidney *while at the same time* for every H$^+$ lost in this way a basic bicarbonate ion is returned to the bloodstream.

(2) Alkaline phosphates passing down the tubule in the filtrate may pick up actively excreted H$^+$ from tubule cells, thus becoming acid phosphate which is lost in the urine. At the same time (as shown above) tubule cells regenerate HCO$_3^-$ ions which pass into the bloodstream. Once again acid H$^+$ ions are lost without loss of base from the bloodstream.

(3) A further process that removes acid while conserving basic ions is associated with metabolic processes taking place in distal convoluted tubule cells. These may catalyse the dissociation of glutamine (an amino acid) into glutamic acid and ammonia. The ammonia is actively passed out into the tubule and there combines with H$^+$ ions also secreted by tubule cells. The hydrogen passes out in the urine as NH$_4$Cl while HCO$_3^-$ ions are again able to diffuse from tubule cells to the blood.

These systems are flexible and if few H$^+$ ions are being produced then HCO$_3^-$ will not be recovered from the filtrate (this applies especially to system 1 above), and will be lost in the urine which may become alkaline. Generally, however, this state does not last for long and normal urine has a pH around 6.

7.3 The kidney

7.3.6 Control of kidney output

In general the output of the kidney will depend on such variables as the blood pressure, which directly influences the rate of filtration, on the state of hydration of the body, and on the load of surplus salts and nitrogenous wastes to be removed at any one time. In recent years some of the complex endocrine controls of kidney output have been established, and this is very much an active area of current investigation.

Antidiuretic hormone and water conservation

Antidiuretic (literally 'against passing of urine') hormone (ADH), also called vasopressin, is a small molecule consisting of only eight amino acids. It is thus more properly described as a peptide than a full-scale protein and in this respect is similar in structure (but not effects) to the hormone oxytocin (see p. 284).

ADH is actually synthesised in the neurosecretory cells of the hypothalamus, from where it passes down the hypophysial tract to the posterior pituitary where it is stored. When the viscosity of the blood (as indicated by its osmotic pressure) rises, sensitive osmoreceptors in the hypothalamus monitor this change and bring about release of the stored hormone from the nearby posterior pituitary.

The ADH travels in the blood to the region of the collecting ducts of the kidney, whose permeability it seems to increase (Fig. 7.7), possibly by the enlargement of cell pores. Because of this increase in permeability the water is drawn out of the tubules more rapidly by the high osmotic

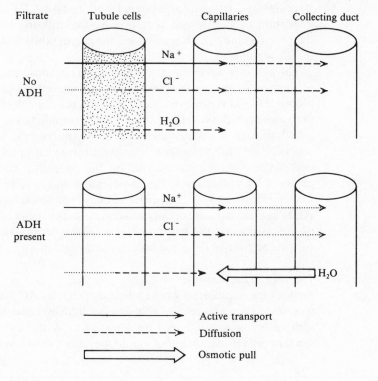

Fig. 7.7.
The effect of ADH on water retention by the kidney. ADH makes the collecting ducts more permeable to water so that water is drawn by osmosis from the urine and returned to the bloodstream.

177

concentrations of salt in the surrounding fluids and capillaries (see p. 163). Thus the effect of ADH on the urine passing out of the kidney is to render it much more hypertonic; in fact it may enter the collecting duct as hypotonic to the blood and leave it as hypertonic.

In its normal state of hydration the body recovers about 99 % of the water passing out of the glomeruli, mostly during its passage through the proximal tubules. But the vital fine control on the osmotic stability of the blood is provided by ADH and the response of the collecting ducts to this hormone.

The role of ADH in the retention of water is supported by the action of aldosterone in the way described below. Recently a new class of very rapidly metabolised hormones called prostaglandins have been described. It appears that one of these, produced in the kidney itself, may be antagonistic to ADH.

Aldosterone and salt regulation

It is convenient to divide the adrenal cortex hormones into those that are concerned with regulation of glucose (glucocorticoids) and those concerned with salt regulation (mineralocorticoids). The distinction is somewhat arbitrary as the two groups are complementary as regards osmotic and ionic control.

Aldosterone is produced from the adrenal cortex partly as a response to ACTH (adrenocorticotrophic hormone) from the anterior pituitary, the ACTH itself being released in response to a cortical releasing factor (CRF) from the hypothalamus. The levels of aldosterone are kept constant through a monitoring system involving both CRF and ACTH. More important in aldosterone secretion is glomerulotrophic hormone from the pineal gland, which is an endocrine organ now known to be involved in salt regulation.

The pineal is directly stimulated (via the nervous system) by the extent of dilation of the great veins, which in turn reflects the volume of the blood. If blood volume decreases then the resulting decreased dilation of the veins stimulates release of glomerulotrophic hormone. This in turn causes release of aldosterone from the adrenal cortex. Aldosterone itself acts on the kidney tubules to stimulate increased uptake of Na^+, which makes the osmotic pull exerted on the collecting duct higher. In the presence of aldosterone, therefore, more water is removed from the kidney filtrate and so the volume of blood restored. Suppression of both ADH and aldosterone secretion is controlled by the increasing expansion of the auricles of the heart as blood volume increases. Thus there is a reciprocal control ensuring homeostatic regulation.

The role of angiotensin

Besides the stimulation of the adrenal cortex by ACTH and glomerulotrophic hormone to release aldosterone, a further system for monitoring sodium levels exists, based on a response by the kidneys themselves. Certain cells in the region of the glomerulus will, when the sodium level

ig. 7.8.
ne hormonal
echanisms involved in
smoregulation.

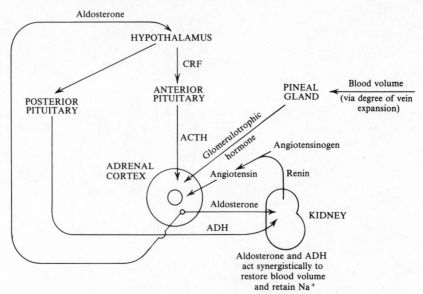

falls, release an enzyme renin into the blood. Here it acts on a precursor plasma protein angiotensinogen, hydrolysing it to the active hormone angiotensin. One target area of angiotensin is the adrenal cortex, where it promotes aldosterone secretion and therefore, via the tubule cells, Na^+ uptake. Angiotensin also has other target areas, for example the kidney arterioles, where it ensures a regular flow of blood through the whole organ.

The complex interaction of all these hormones in osmoregulation and salt retention is shown in Fig. 7.8.

Vitamin D and parathormone in calcium and phosphate regulation
Vitamin D, which, if made in the body, may be considered a hormone, causes the assimilation of Ca^{2+} and PO_4^{3-} from the gut and its release from bone. These ions are also regulated by parathormone from the parathyroids, small endocrine glands near the thyroid, which when stimulated by lack of Ca^{2+} in the blood cause an increase in its absorption from the gut and its retention in the kidneys. Parathormone also causes release of both Ca^{2+} and PO_4^{3-} from bony tissues, but as it encourages the elimination of the latter ion from the kidney tubules it actually produces a rise only in Ca^{2+} ions.

7.4 **The source of nitrogenous excretory substances in the urine of mammals**

7.4.1 *Urea*

Urea, $CO(NH_2)_2$, is formed from the deamination of amino acids taken in with the diet. It is thus derived from exogenous protein (i.e. protein originating outside the body). Urea is formed in a cyclic process called the ornithine or urea cycle. Carbon dioxide and ammonia combine with the

Does not drink sea water

Hypertonic NaCl from rectal gland

Isotonic urine

Drinks sea water

Secretes salt from gills

Isotonic urine

Water enters via gut and gills

Much hypotonic urine

Drinks sea water

Hypertonic tears

Isotonic urine

Does not drink sea water

Strongly hypertonic urine

Drinks sea water

Hypertonic nasal secretion

Weakly hypertonic urine

	Liability to osmotic water loss	Liability to evaporative water loss	Blood concentration relative to medium concentration	Urine concentration relative to blood concentration
Elasmobranchs	×	×	Isotonic	Isotonic
Seawater teleosts	+	×	Hypotonic	Isotonic
Freshwater teleosts	×	×	Hypertonic	Hypotonic
Reptiles	×	+	Hypotonic	Isotonic
Mammals	×	+	Hypotonic	Hypertonic
Birds	×	+	Hypotonic	Hypertonic

× = No liability
+ = Liability

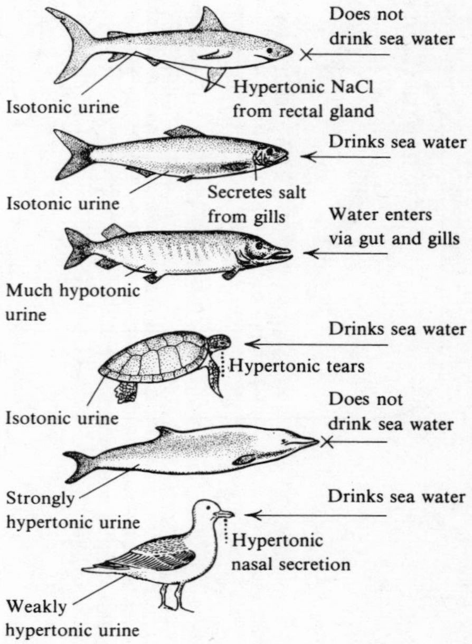

Fig. 7.9.
The problems of salt and water balance for different marine vertebrates and the ways in which they have solved them.

amino acid ornithine to form the amino acid citrulline which in turn takes up a further molecule of ammonia to give arginine. Hydrolysis of this by the enzyme arginase yields a molecule of urea and a new molecule of ornithine which recommences the cycle. Urea is found in mammalian blood at a concentration of some 30 milligrams per 100 millilitres and is the main nitrogenous excretory substance.

7.4.2 *Ammonia*

Ammonia, NH_3, is very toxic and is not released into the blood in any quantity in mammals (concentration approx. 0.02 milligrams per 100 millilitres). It is made by deamination of amino acids in the liver, where it is rapidly converted to urea by the route described above. It is also made in the kidney, whence it enters the urine in small quantities.

7.4.3 *Creatinine*

Creatinine is a complex nitrogen-containing substance derived from the breakdown of endogenous protein (i.e. protein derived from the body tissues).

7.5 **A comparative account of water and salt regulation in the vertebrates**

The osmotic pressures of the body fluids of animals are customarily measured by the depression of the freezing point they cause. This is expressed by the symbol $\triangle$, the internal medium being represented by $\triangle_i$ and the external medium as $\triangle_o$.

Fig. 7.9 shows the problems of salt and water balance for different vertebrates and the ways in which they have solved them. These are discussed in more detail below.

7.5.1 *Fishes*

Freshwater fishes
Fishes first evolved in freshwater and all present-day seawater species have a freshwater ancestry. For fish living in freshwater the external medium has a lower osmotic pressure than their internal fluids ($\triangle_i = 0.57$, $\triangle_o = 0.03$), so water will tend to diffuse in across permeable surfaces (mainly the gills and gut) and elimination of excess water will tend to carry out valuable salts in the dilute urine. The osmotic and ionic problems of freshwater fish are thus to eliminate excess water and to conserve salts. The excess water is filtered out by the glomeruli of the kidneys, and freshwater fish have large Malpighian bodies to deal with the passage of considerable quantities of water. Ionic balance is maintained by replenishing salts by way of the diet, reabsorbing salts in the kidney and absorbing salts through special cells on the gills (Fig. 7.10).

181

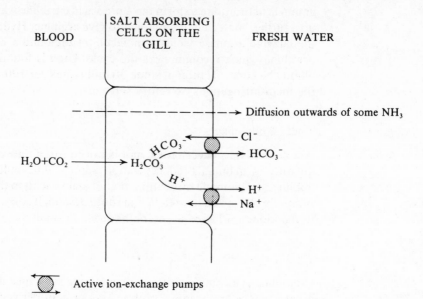

BLOOD

SALT ABSORBING
CELLS ON THE
GILL

FRESH WATER

- - - - - - - - → Diffusion outwards of some NH_3

H_2O+CO_2 ────→ H_2CO_3

HCO_3^-

H^+

Cl⁻

HCO_3^-

H^+

Na^+

Fig. 7.10.
Ion exchange in a
chloride cell of a
freshwater fish.

Active ion-exchange pumps

Marine fishes

(*a*) *Teleosts.* Seawater has an osmotic pressure greater than that of the blood of marine teleosts, whose body fluids ($\triangle_i = 0.6$) tend to have a similar composition to those of their freshwater relatives. The osmotic pressure of the sea corresponds to $\triangle_o = 1.7$. Marine teleosts thus tend to lose water through their gills, gut or any permeable surfaces, and their osmotic and ionic need is to dilute the internal medium. In order to do this the fish swallow large quantities of salt water, absorbing both the water and the salts from the gut. They conserve the water, producing only small quantities of urine, and eliminate excess salts via special secretory cells on the gills as well as in the urine. Some of the excretion of nitrogenous waste takes place through the gut lining. Because of the small amount of water eliminated the capsule and glomeruli of the kidneys are very small or even absent.

(*b*) *Elasmobranchs* that live in the sea are unusual in using the retention of urea in the body as a means of raising the internal osmotic pressure (a similar adaptation is found in some marine frogs). In these fishes the $\triangle_i \simeq$ 1.8 and that of seawater $\triangle_o \simeq 1.7$. For this reason the marine elasmobranch has no tendency to lose water to the medium and its kidney has a large capsule and well-developed tubule for selective reabsorption of salts (see also p. 175). The young elasmobranch is provided with a supply of urea from its mother. Freshwater elasmobranchs descended from marine ancestors retain this characteristic of urea in the bloodstream. And it is interesting that modern dipnoans and coelacanths, although more closely related to teleosts than elasmobranchs, also show this same osmoregulatory adaptation.

182

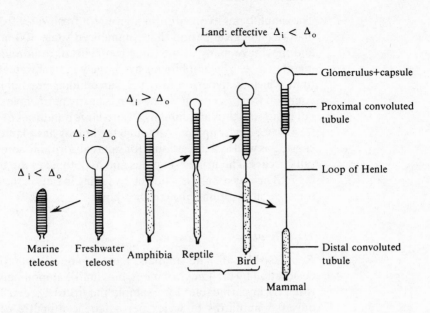

Land: effective $\Delta_i < \Delta_o$

$\Delta_i > \Delta_o$

$\Delta_i > \Delta_o$

$\Delta_i < \Delta_o$

Glomerulus+capsule

Proximal convoluted tubule

Loop of Henle

Distal convoluted tubule

Marine teleost

Freshwater teleost

Amphibia

Reptile

Bird

Mammal

Hypotonic urine in small quantities

Hypotonic urine in large quantities

Hypotonic urine in large quantities

Hypertonic urine in small quanties

Hypertonic urine in larger quantities than reptiles and birds

Fig. 7.11.
The relationships of vertebrate kidneys.

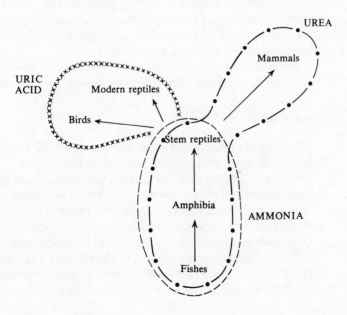

UREA

URIC ACID

Mammals

Modern reptiles

Birds

Stem reptiles

Amphibia

AMMONIA

Fishes

Fig. 7.12.
Relation of main nitrogeneous excretory product to vertebrate phylogeny.

183

7.5.2 *Amphibians*

The amphibians evolved from a group of freshwater fishes in the Devonian (a geological period that commenced some 300 million years ago) and they have many of the same problems of osmoregulation and ionic maintenance. The amphibians live largely in freshwater and have a body fluid which is hypertonic to the surrounding medium (for *Rana*, the common frog, $\triangle_i = 0.56$). The skin is partly permeable to water as is the gut lining and thus amphibians absorb large quantities of water from their freshwater environment. Although the kidney has a large glomerulus and capsule as well as a long tubule for salt reabsorption, some leaching out of salts occurs. The urine is copious and hypotonic to the body fluids ($\triangle_i = 0.17$). The loss of salts is made up by taking in salts in the diet as well as by selective absorption through the skin.

7.5.3 *Reptiles*

Reptiles, like all land animals, have a potentially serious problem of desiccation by their environment, but, unlike amphibians, their scaly skin is nearly impermeable: for example the lizard *Lacerta* loses on average only 0.5 millilitres of water per square centimetre of skin per hour, compared with the frog *Rana* that loses 8.3 millilitres.

The supply of water available is limited and some water must be lost in the elimination of toxic waste products. Water is taken in with the diet and there is only a small glomerulus and capsule so that there is a low filtration rate into the tubule. The urine produced is concentrated ($\triangle_o \simeq 0.6$) and may be semi-solid. The ability to produce such concentrated urine lies in the reptiles' use of the fairly insoluble and non-toxic uric acid as the main nitrogenous excretory product. This substance requires less water than ammonia or urea for its elimination. (Fig. 7.12 shows the relation of the main nitrogenous excretory product of different groups of vertebrates to vertebrate phylogeny.)

Some reptiles for example the turtle, have colonised the sea, and these have similar osmotic adaptations to the sea-birds described below.

7.5.4 *Birds*

The birds are closely related to reptiles and have the same problems of osmoregulation and excretion. Unlike the former they have a large glomerulus and capsule and rely on the uptake from the long tubule to return water and salts to the bloodstream. A new element is present in the tubule of birds and is also present in mammals; this is called the loop of Henle and is an integral part of salt and water uptake in these two classes. It is much less developed in the birds, however, than in the mammals.

Water is also reabsorbed from the urine as it passes through the cloaca, and like the reptiles birds use uric acid as their main nitrogenous excretory product, producing a semi-solid and hypertonic urine. In the eggs of

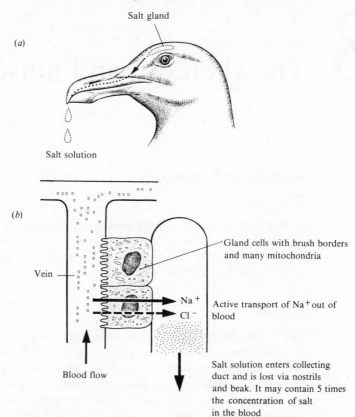

Salt gland

(*a*)

Salt solution

(*b*)

Gland cells with brush borders
and many mitochondria

Vein

Na$^+$
Cl$^-$

Active transport of Na$^+$ out of
blood

Blood flow

Salt solution enters collecting
duct and is lost via nostrils
and beak. It may contain 5 times
the concentration of salt
in the blood

Fig. 7.13.
(*a*) The position of the
salt gland in a herring
gull. (*b*) Diagram
showing a portion of a
salt gland. There is a
counter-current
multiplier system
analogous to that
found in the
mammalian kidney.

both reptiles and birds the nitrogenous waste products of the embryo are
stored as the insoluble, non-toxic allantoic acid. The evolution of such a
substance was essential for metabolism to be possible in eggs with shells.

Marine birds and marine reptiles have an internal osmotic pressure
which is less than that of their environment, so that their problems are
much the same as those of marine teleosts. When they either drink
seawater or feed on animals whose body fluids are isotonic with seawater,
marine reptiles and birds have an excessive intake of salts. The surplus
salt is eliminated from a secretory gland on the head called the salt gland
(Fig. 7.13). In marine birds this gland is some 10 to 100 times larger than
its homologue in terrestrial forms. The salt gland exudes a salt solution
more concentrated than seawater and thus eliminates any excess salts
taken into the body; the exudate from the gland gives marine birds and
turtles the appearance of shedding tears.

7.5.5 *Mammals*

The mammals have already been dealt with earlier in this chapter. They
can secrete a hypertonic urine and have a large glomerulus and capsule
with a long tubule and loop of Henle for reabsorption of water and salts in
the tubule.

185

8 The skeleton and muscles

8.1 **Bone and the skeleton**

8.1.1 *Function*

Bone forms the main skeletal element of the body. It provides not only support but also protection, and gives a jointed structure of struts and levers which is operated by the muscles. In the red marrow of the bone erythrocytes and certain of the white corpuscles are made.

8.1.2 *Structure*

In the adult mammal, bone consists of numerous trabeculae, or lines of bone-forming cells, and bone matrix oriented with respect to the forces acting on the bone. Both spongy and compact bone is full of the vessel-carrying Haversian canals from which the osteoblasts (bone-forming cells) derive food and oxygen.* These osteoblasts lie in lacunae or spaces, arranged concentrically round a central canal. Each osteoblast produces a fine canal system so that there is no part of the bone matrix far from a strand of living protoplasm (Fig. 8.1). Around the bone is the periosteum; here ligaments are attached and tendons from muscles 'integrate strongly' into periosteal cells. The bone-forming osteoblasts also originate from the periosteum.

The long bones of mammals ossify from three centres. The main shaft of the bone, or diaphysis, is initially separated from an epiphysis at each end, but during development the three parts grow towards each other until finally they fuse and growth stops.

During bone synthesis it is thought that mineralisation is a result of the calcium and phosphates that are present in high concentrations in the body fluids 'salting out' as crystals of hydroxyapatite ($Ca_{10}(PO_4)_6(OH)_2$). This is due to hydrolysis by the bone-forming cells of pyrophosphate, which is normally present in the body fluids and prevents such precipitation. There are certain key spots in the bone where this mineralisation takes place very rapidly during growth. In addition to the hydroxyapatite

---

* Spongy bone is derived from the first ossification process, and is fluid-filled and therefore less dense than compact bone. It is usually found in the centre of the ends of long bones. Compact bone is stronger and has dense Haversian systems. (Haversian systems are prominent in human bone and that of many other mammals but are rarer in small mammal species.)

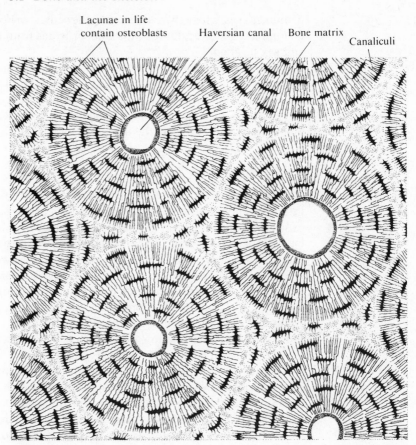

Lacunae in life
contain osteoblasts   Haversian canal   Bone matrix
                                                        Canaliculi

Fig. 8.1.
Structure of bone in
transverse section.

crystals, bone substance also consists of collagen fibres and a mucopolysaccharide cement substance. At the same time as the osteoblast causes the precipitation of the mineral it secretes these two other components. Mineral crystal and collagen fibres are linked together chemically, the former making up some 60 % of the total. The combination of the properties of the two substances gives bone some unique characteristics (see p. 188).

Bone substance is continuously turned over in the body; in the adult human some 0.05 % of the skeleton is replaced per day. The removal of bone is effected by certain osteoclast cells, which secrete organic acids to dissolve the apatite crystals and enzymes to digest the collagen and cement.

8.1.3 *Endocrine control of bone synthesis*

In the young mammal bone growth is dependent on the action of the growth hormone somatotrophin, from the anterior pituitary. This stimulates activity of the bone-forming cells of the periosteum, particularly in the main areas of ossification in the epiphyses. Somatotrophin works in

187

conjunction with high thyroxine levels; the activity of both these hormones is suppressed in the adult by mineralocorticoids from the adrenal cortex and sex hormones.

Homeostasis of calcium levels in the bone and the blood is brought about by the antagonistic action of two other hormones. If the blood calcium level falls then parathormone from the parathyroids brings about its release from bone substance. Conversely if the level is high in the blood then calcitonin from the thyroid stimulates further bone synthesis.

8.1.4 *Adaptations of bone*

Bone is a tissue designed to withstand tension and compression. It must be strong enough to take the standing weight of the body and the forces exerted by that weight in moving. These forces may be very great – for example, a man weighing 300 kilograms may, whilst running, exert on the shaft of the femur some 35×10^5 newtons per square metre.

The forces acting on bone are met in ways similar to those used by engineers in the construction of girders, cranes, bridges, etc. It can be seen (Fig. 8.2) that forces applied eccentrically to solid struts produce compression and tension in the outer rim of the strut. A solid strut can thus be replaced by a hollow one, which has the effect of lightening the skeletal material without reducing its strength.

Another device serving to reduce the forces acting on a bone is the use of braces of muscle or tendon, such as the biceps which run from shoulder to forearm, or the ilio-tibial tract along the outside of the thigh. These act as a bow-string and balance some of the forces of tension acting on the side of the strut opposite the weight, and also reduce to a lesser extent the compression forces on the strut below the weight (Fig. 8.3).

Because the brittle crystalline structure of bone is 'integrated' with a flexible fibrous one the tendency of bone to break under sudden stress is much reduced. A crack in the mineral will not spread easily beyond a collagen 'crack-stopping' region; stress is also taken up in all the fluid-filled spaces of the spongy bone and it may also be absorbed at a molecular level by movement of the mineral and collagen relative to each other.

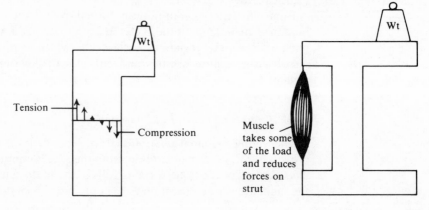

Fig. 8.2.
The forces produced in a solid strut loaded eccentrically.

Fig. 8.3.
The reduction of tension on a bone by use of muscle connections.

8.1 Bone and the skeleton

Considering that it has a density only twice that of water, bone has immense physical strength and in a young adult human can withstand forces of up to 10 tonnes to the square metre.

8.1.5 The joints

Movement takes place at the joints where bones meet. The nature of the joints varies according to the amount of movement that they permit. In the skull the sutures between bones are joined firmly by the periosteum, the connective tissue which covers all bones. Where there is relatively little movement, as between vertebrae, the joints are pads of cartilage, but where greater movement is necessary synovial joints are used.

A synovial joint (Fig. 8.4), such as that between a limb and the limb girdle, consists of two articulating bones each capped with hyaline (transparent) cartilage and enclosed by a capsular ligament. To the outside of the cavity of the joint lie the synovial membranes, whose function is to

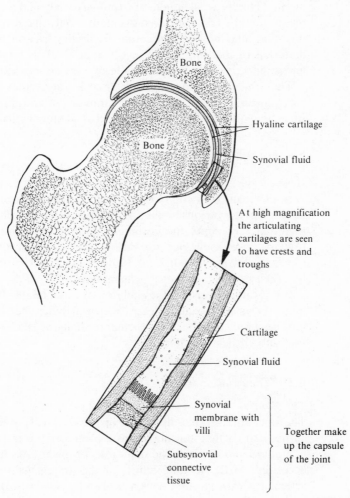

Fig. 8.4.
A synovial joint (such as the hip joint).

secrete the lubricating synovial fluid. This fluid is also derived from the cartilage itself and gives a sort of hydraulic suspension to the body. The outer layers of the capsule have fibrous thickenings which hold the bones together.

Synovial fluid has a composition basically similar to blood plasma but with less protein and the addition of the mucopolysaccharide, hyaluronic acid. By means which are not entirely understood, synovial joints have very low coefficients of friction, lower indeed than those of many bearings used by engineers!

Detailed examination of the cartilage surfaces of a synovial joint shows protuberances whose crests are some 4 micrometres high and 20 micrometres apart. As the bones of the joint move opposing crests are brought together but as the pressure increases water exchanges take place between cartilage and synovial fluid. The effect of this is to produce sheets of hyaluronic acid separating the points of highest pressure. The stresses are locally very high but the contacts shear and re-form with joint movement. (This is what engineers term hydrodynamic lubrication and it means that the resistance to movement is only that force needed to shear the molecules of the lubricating fluid; the friction between the solid materials of the joint is not involved at all. This type of superefficient lubrication can only be man-made by very sophisticated techniques.)

Thus through changes in the viscosity and other properties of the synovial fluid at different rates of movement and different shear pressures, the joints between the bones act at all times with minimum friction.

8.2 The skeleton of the mammal

It is necessary, before considering how the skeleton becomes modified for various means of locomotion, to have a clear idea of the position and function of the various bones, for the arrangement of bones in a terrestrial mammal represents the results of a long evolution and adaptation towards an efficient performance in terms of support, protection, and especially locomotion on land. In the following discussion the rabbit will be taken as the example of a mammal.

It is useful to consider the skeleton of a mammal to be made up of two major parts: the axial skeleton, which includes the skull, ribs and vertebral column, and the appendicular (or hanging) skeleton which includes the limb girdles and limb bones.

8.2.1 *The axial skeleton*

The skull
The names of the bones that make up the skull will not be described here – suffice it to say that they make a well-protected box in which the brain can be housed and in which the mandible, or lower jaw, operates efficiently according to the feeding habits of the animal (see p. 57). The skull articulates with the first neck vertebra by two large occipital condyles.

8.2 The skeleton of the mammal

The vertebral column

Around the embryonic notochord (a skeletal rod) a number of skeletal blocks or vertebrae ossify. These fall into five sections in the mammal and the numbers in each section, except for the last, are constant. Working downwards from the skull we find seven cervical or neck vertebrae, twelve thoracic or chest vertebrae, seven lumbar or back vertebrae, and a sacrum consisting of four fused vertebrae. The caudal, or tail, vertebrae are variable in number. The vertebrae play an important role in all functions of the skeleton: they protect the spinal cord by enclosing it, they support the weight of the head, tail and abdomen, and they act as the origin for powerful back muscles essential in locomotion.

(a) *The cervical vertebrae* (Fig. 8.5). These are characterised first by a large neural canal, the spinal cord having just left the medulla; secondly, by well-developed centra and small projecting processes, because they act as compression members (the weight of the head is taken by ligaments

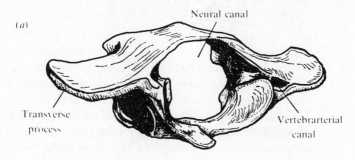

(a)

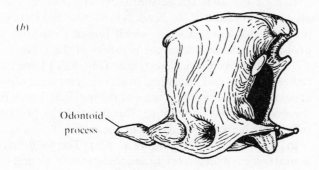

(b)

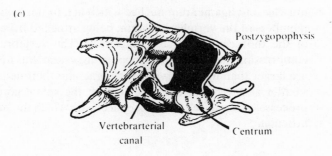

(c)

Fig. 8.5.
(a) Atlas vertebra;
(b) axis vertebra;
(c) cervical vertebra.

191

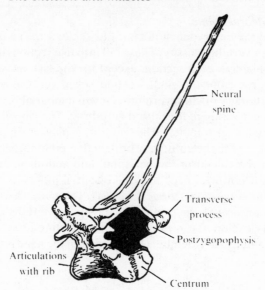

Neural
spine

Transverse
process

Postzygopophysis

Articulations
with rib

Centrum

Fig. 8.6.
Thoracic vertebra.

and muscles attached to the neural spines of the thoracic vertebrae) and need to be flexible. The first two cervical vertebrae are modified to form a universal joint for the head. The skull articulates by its occipital condyles with the concave facets of the first cervical or atlas vertebra (Fig. 8.5a). This allows the head to move backwards and forwards (or, in the rabbit, up and down) across the large flat joint that is formed. The atlas also has a wing-like transverse process for the origin of the muscles which move the head in this plane. It has no centrum.

Behind the atlas the second cervical vertebra, the axis (Fig. 8.5b), allows the rotation of the head. It is narrow and its centrum extends as a peg, the odontoid process, which projects into the atlas. In fact this process was derived from the centrum of the latter.

The other five cervical vertebrae (Fig. 8.5c) have large neural canals, reduced spines and transverse processes, and large vertebrarterial canals at each side. The zygopophyses of the individual vertebrae are flattened, the prezygopophysis pointing dorsally and the postzygopophysis ventrally. This increases flexibility of the neck.

(b) *The thoracic vertebrae* (Fig. 8.6). The long, backwardly directed neural spines of these vertebrae are characteristic and are well seen in the rabbit. They vary in length according to the size of the head and carry the muscles and ligamentum nuchae which act, in this region, as the tension members of the vertebral column, being attached from the neural spines of the thoracic vertebrae to the neck. The vertebrae themselves are compression members of the supporting system thus formed; their centra are larger than in the cervical vertebrae and their neural canals become smaller as successive nerves exit from the spinal cord. The transverse processes are reduced to a broad facet with which the tubercles of the ribs articulate.

192

8.2 The skeleton of the mammal

(*c*) *The lumbar vertebrae* (Fig. 8.7). These may be easily recognised by the long, anteriorly directed transverse processes which provide areas of attachment for the large back muscles (the sacrospinalis and longissimus dorsi) involved in bending the back and in transferring the weight carried by the posterior part of the vertebral column to the pelvic girdle. The neural spines, which are short, are also anteriorly directed and provide further points of origin for the back muscles. (Thus between the thoracic and lumbar vertebrae the stresses are coming from opposite directions and the facets and spines of the two sections therefore face in reverse directions to meet these forces. The junction is at the diaphragmatic vertebra.) The lower lumbar vertebrae that are associated with the more powerful muscles of the pelvic region have better-developed centra to take the compression forces involved in locomotion.

(*d*) *The sacrum* (Fig. 8.8). The function of the sacrum is to provide a firm attachment of the vertebral column to the pelvic girdle. Unlike the pectoral girdle, which acts partly as a shock absorber, the main force of

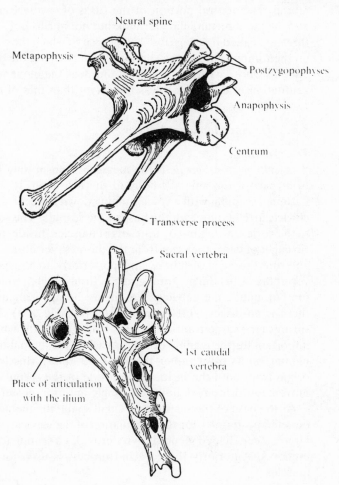

Fig. 8.7.
Lumbar vertebra.

Fig. 8.8.
The sacrum.

locomotion from the muscles of the hind limb and buttocks passes through the pelvic girdle–sacral articulation, which allows very little movement.

The sacrum consists of four fused vertebrae of which the first is modified with large articular surfaces to fit against a similar surface on the inside of each ilium. The three posterior members of the sacrum are small and lead directly into the caudal vertebrae.

(*e*) *Caudal vertebrae.* The rabbit has a short tail and the 15 caudal vertebrae which support the tail are small cylindrical bones. Their articulation is such as to allow a freer movement between them than other parts of the vertebral column.

The ribs

These long curved bones are inserted by their heads or capitula against the centra of the thoracic vertebrae, while a small dorsal spine, or tuberculum, supports them against the transverse processes of the same vertebrae. The ventral portion of the rib is of cartilage and in ribs 1–7 is attached to a rectangular sternum, but not in ribs 8–12 of which the last three are called floating ribs. The degree to which the ribs move depends on their importance in the support of the body via the pectoral girdle. In animals whose weight is taken on four legs the movement of the ribs in ventilation of the lungs is less important than that of the diaphragm.

8.2.2 *The appendicular skeleton*

The fore limb

(*a*) *The pectoral girdle* is the means of transmitting the weight of the front part of the animal's body to the fore limb. It consists of a large flattened scapula, with a spine running down the centre, and a very small clavicle articulating with the base of the scapula. The scapula is attached to the vertebral column by four sets of muscles. Inside, muscles run to the rib cage; outside, they run from the thoracic vertebrae, anteriorly to the anterior thoracic vertebrae and posteriorly to the posterior thoracic vertebrae. Thus the pectoral girdle is firmly anchored to the axial skeleton but, unlike the pelvic girdle, the anchorage is made by muscles and there is no direct skeletal connection (see Fig. 8.9). This allows more spring in the suspension of the front half of the body, which is particularly important during leaping or running when the animal takes the shock of landing on its front rather than its hind limbs. Muscles that take their origin from both the inside and outside of the scapula serve to restrict movement of the fore limb to a single anterior–posterior plane.

At the base of the scapula is a small spine, the metacromion, which is directed posteriorly for the articulation of the clavicle, while below it the large saucer-shaped glenoid cavity provides a seating for the head of the humerus. Anteriorly the coracoid process is used for the insertion of muscles.

(*b*) *The humerus* is a stout bone with the proximal end enlarged into a

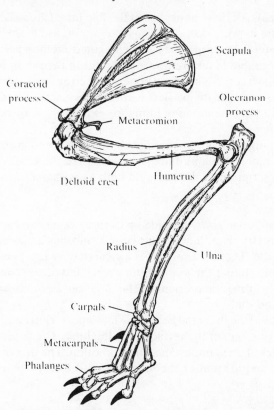

Fig. 8.9.
Left pectoral girdle and
fore limb.

head (articulating with the glenoid of the scapula). On either side of the head other processes are the sites of insertion of muscles such as the deltoid and pectoralis and others from the scapula.

At the further, or distal end of the humerus, is the condyle for insertion into the sigmoid notch of the ulna. This forms a characteristic hinge-joint which limits movement to a single plane.

(*c*) *Radius and ulna* are two long curved bones making up the shank of the limb. In the rabbit the radius is anterior to and shorter than the ulna and is fixed to each end of it. The proximal end of the radius makes up the lower part of the sigmoid notch into which the humerus fits, and the radius takes most of the weight, which it transfers to the foot via a distal articulation. The ulna is the main bone for the attachment of muscles in this segment of the limb. Its proximal end extends out beyond the sigmoid notch into the olecranon process, and on this the triceps, the main extensor of the fore limb, is inserted. The distal end of the ulna articulates, like the radius, with some small bones at the top of the foot.

(*d*) *The fore foot.* There are many bones, called carpals, at the proximal end of the foot. The number of these is very variable in mammals. In the rabbit there are two rows, the first containing the radiale, the intermedium and the ulnare, and the second containing the trapezium, the trapezoid, the centrale, the magnum and the unciform. A small extra bone called the pisiform is found on the outer carpals of mammals (and

195

reptiles). All these bones give a flexible joint (the wrist) between the arm and the hand.

Beyond the carpals are the elongated metacarpals. Again these are very variable in different mammals but in the rabbit fore limb there are five. Each consists of a long bone attached at one end to the carpals and at the other to the digits. Between these two sets of bones the ends of the metacarpals make hinge joints for extension or flexion of the digits or phalanges.

The digits are long cylindrical bones which end in claws. There are five in the rabbit fore limb and this is taken to be the ancestral number – hence the description of the tetrapod limb as *pentadactyl*.

The hind limb

(*a*) *The pelvic girdle* is made up of three bones on each side of the body (Fig. 8.10). The ilium is the largest and has a flattened wing running forwards. The inside of this is articulated by ligaments to the 1st sacral vertebra so that the joint is very strong and nearly fixed. From the outer surface of the ilium originate the large retractor muscles (see p. 214) of the hind limb.

Behind the ilium the ischium extends posteriorly and this again acts as a point of origin of the retractors of the limb. The latter are very powerful muscles of great importance in locomotion. The third bone, the pubis, lies inside and in front of the other two and forms a bony cage around the

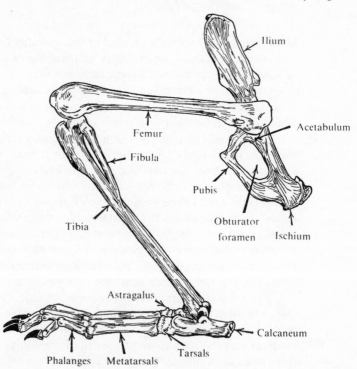

Fig. 8.10.
Left pelvic girdle and
hind limb.

genitals. Between this and the ischium is the large obturator foramen, and to the outside where the three bones meet there is a large cavity, the acetabulum, for articulation of the femur.

(*b*) *The femur*, like the humerus, is a stout cylindrical bone with enlarged ends for articulation. At the proximal end there is a round head protruding to the inside and fitting into the socket of the acetabulum. Along the shaft of the bone next to the head there are a number of wing-like processes, the trochanters, for the attachment of muscles. There is also a crest running down the dorsal length of the bone for a similar purpose. At its distal end the femur has two condyles for the articulation with the tibia. Once again there are wing-like surfaces for muscle attachments and a characteristic feature is the deep groove dorsally for the tendon of the extensor muscle of the lower part of the limb.

(*c*) *The tibia and fibula.* The tibia is large and carries all the weight down to the hind feet. It has a flattened head which articulates with the condyle of the femur forming a hinge joint and at the side of the head there is the cnemial (lower leg) crest for attachment of the patella or knee-cap. On to the latter are inserted the tendons for the extension of the tibia. The distal end of the tibia articulates with the tarsus of the hind foot. In the rabbit, as in many mammals, the fibula has become vestigial and is fused to the distal end of the tibia.

(*d*) *The hind foot.* The ankle corresponds to the wrist of the fore limb and is made up of small bones called tarsals. The first row consists of two substantial elements, the astragalus and calcaneum, with which the end of the tibia articulates. The calcaneum is prolonged posteriorly to provide a point of insertion for the calf muscle which extends the foot, and gives a degree of mechanical advantage to this muscle which is important in running and jumping.

Below the astragalus is the square centrale and beyond this a second row of tarsals – the mesocuneiform, ectocuneiform and cutoid. As with the wrist, these bones of the ankle joint are little modified from the primitive pentadactyl limb.

Following the tarsals are four elongaged metatarsals which make up the main shaft of the foot. The first digits have been lost. Finally there are the four phalanges, which end in claws.

8.3 Cartilage and connective tissue

8.3.1 Cartilage

Function
The embryonic skeleton is comprised entirely of cartilage, but in the adult it is found only at the ends of bones, the rest becoming ossified. It is also found in isolated places in the body such as the external ear, epiglottis, etc. In every case it is a supporting tissue well able to resist compression

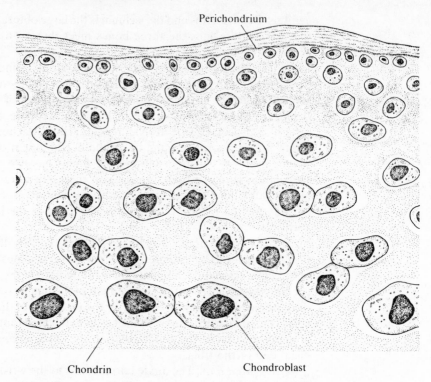

Fig. 8.11.
Cartilage.

TABLE 8.1. *The relative properties of vertebrate cartilage and bone*

	Specific gravity	Modulus of elasticity	Tensile strength*
Cartilage	1.1	130	6118
Bone	2.0	140,000	1122×10^3

* In newtons per square metre.

and intermediate in tensile strength between connective tissue and bone
(Table 8.1). Where cartilage persists as a main skeletal element in the
adult (e.g. dogfish) it is often strengthened by deposition of calcium salts.

Structure

The living cells of cartilage are chondroblasts; they arise from a surrounding perichondrium and secrete a ground mass of chondrin (Fig. 8.11). The
chondrin consists of fibres of collagen, mucopolysaccharides and water,
and its properties are due to the semi-liquid nature of its constituents.
Hyaline cartilage is the most liquid type of cartilage; fibrous and elastic
cartilage have a greater proportion of collagen, and other proteins such as
elastin.

There is no blood supply to the chondroblasts so food and oxygen must
diffuse across the matrix.

198

8.3.2 *Connective tissue*

Function

These tissues support and hold together other cells and tissue systems of the body. They also provide part of the body's defence mechanisms. Connective tissues are found under the skin, around the nerves and blood vessels, between muscle and bone and between bone and bone, as lubricating mesenteries, as synovial membranes and as fatty tissue surrounding the kidney and other organs. They also form the periosteum covering the surface of bone, providing a means of attachment for tendons and ligaments and giving rise to new bone.

Structure

The basic cells of connective tissue are fibroblasts, which are able to secrete the other structures found within the tissue. In all types of connective tissue a matrix of mucopolysaccharide is secreted in which the fibroblasts lie scattered, while what other substances are laid down depends on the position and function of the tissue. White fibres of collagen are commonly found where resistance to stretching is required. Collagen consists of protein, formed as unbranched fibres, up to 10 micrometres thick. It makes up the bulk of tendons, ligaments and the general areolar tissue between organs. Yellow fibres of elastin, also secreted by the fibroblasts, are found in the connective tissue between vertebrae and around vessels. Elastin is unlike collagen in that it will stretch freely and subsequently return to its original size. It is also present in areolar tissue.

Connective tissue also plays a part in the body's defence mechanisms and a type of white corpuscle, called a macrophage, is found scattered throughout the tissues. These are similar in appearance to the fibroblasts, from which they probably originate. They play an important role in the interaction between antigen and antibody-forming cells (see p. 151).

8.4 **Muscles**

8.4.1 *Function*

Muscles perform the work which moves parts of the skeleton relative to one another; they cannot extend actively but are pulled out usually by the action of another, antagonistic muscle. In the alimentary canal there are groups of longitudinal and circular muscles which are antagonistic to one another and operate against the hydrostatic 'skeleton' formed by the gut contents. There are two main types of muscle: smooth and striated. Smooth muscles are under autonomic or involuntary control and are particularly adapted to produce changes in length and operate in visceral activities. Skeletal muscle is striated. It is under direct central (voluntary) control and is capable of changes in tension but relatively small changes in length. Cardiac muscle is also striated.

8.4.2 *Gross structure*

A skeletal muscle is made up of many striated fibres. Each fibre may be quite long and may exceed 10 centimetres in length; they are usually between 10 and 100 micrometres in diameter. Altogether the skeletal muscles make up more than 50 % of the weight of the body. At each end of the muscle the fibres are united by their blunt ends to tendons, which are usually attached to bones but sometimes to connective tissue. The

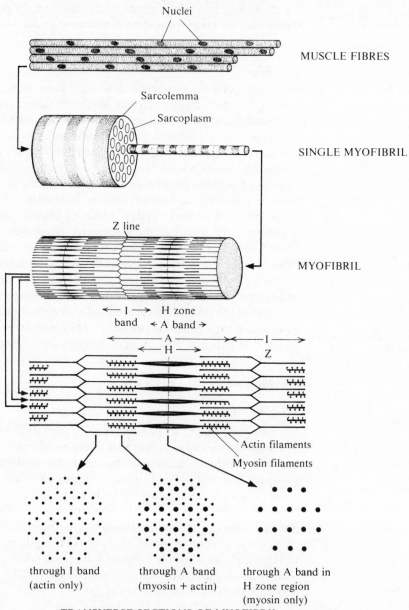

Fig. 8.12.
The structure of
striated muscle.

finer tapering ends of the fibres end in connective tissue within the muscle itself. Each muscle fibre contains hundreds of nuclei and is enclosed in a connective tissue membrane, the sarcolemma. Internally the fibre is composed of many myofibrils about 2 micrometres in diameter and the remaining protoplasm forms the sarcoplasm (Fig. 8.12). Some muscles are red in colour because of their high myoglobin content, and these are generally concerned with sustained contractions. White muscles are concerned with the more rapid or phasic movements of the organism.

8.4.3 *Detailed structure*

Development of the ultra-microtome for the cutting of very thin tissue sections suitable for electron microscopic examination enabled detailed knowledge of muscle structure to be obtained. Visual findings have been confirmed by enzymic investigations.

Besides the myofibrils the sarcoplasm of the fibre also contains a complex sarcoplasmic reticulum and mitochondria. The functional significance of these two structures becomes apparent when the mechanism of contraction and energy exchange is considered.

The individual myofibrils consist of regular alternating dark and lighter bands and lines. The broad A band (A = anisotropic) consists mainly of thick filaments of the protein myosin; at its centre is found the H zone (H = heller, German for light). The other broad band is the I (I = isotropic) and in the centre of this is a thin Z line (Z = Zwischenscheibe, German for intermediate disc). The I band consists of thin filaments of the other important muscle protein, actin.

The structure and ultrastructure of muscle at increasing magnifications is shown in Fig. 8.12.

8.4.4 *The muscle proteins and their relation to each other and to the fine structure of the muscle*

Myosin
Myosin is the main constituent of the A band and of the muscle as a whole, making up over 50 % of the total protein. Molecules of myosin have a long 'tail' which consists of two polypeptide chains wound round

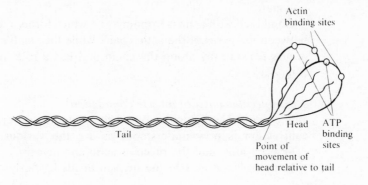

Actin
binding sites

Tail

Head   ATP
binding
sites

Point of
movement of
head relative to tail

Fig. 8.13.
A myosin molecule.

201

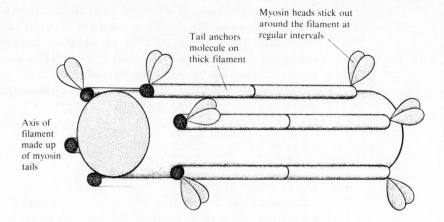

Myosin heads stick out around the filament at regular intervals

Tail anchors molecule on thick filament

Axis of filament made up of myosin tails

Fig. 8.14.
The arrangement of myosin molecules in a myofilament.

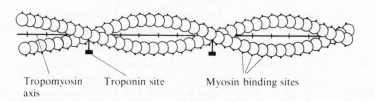

Fig. 8.15.
Actin molecules in a helically coiled polymer as found in muscle.

Tropomyosin axis   Troponin site   Myosin binding sites

each other in a helix, and a 'head' made of two identical parts and globular in shape. In muscle the head is inclined at an angle to the remainder of the molecule and it has been found to be made up of two different polypeptide chains of much shorter lengths than those of the tail. The myosin heads form cross-bridges in the muscle (see below) and each head has two binding sites for ATP and two for actin (Fig. 8.13).

In the myofibril the myosin molecules are arranged tail to tail with the heads sticking out at regular intervals (every 60° in a spiral formation) from the axis of the myosin filament. The whole arrangement (Fig. 8.14) is repeated every 43 nanometres along the length of the filament.

Actin
Actin is the main constituent of the I band. It is a much smaller molecule than myosin, found in muscle in the F (fibrous) state as a helically coiled polymer.

Combined with actin is tropomyosin, which forms a series of axial units between the gyres of the actin chain, while the small molecule troponin occurs periodically along the chain and has a role in calcium binding (Fig. 8.15).

The molecular basis of muscle contraction
While the whole picture is not yet complete, the general basis of muscle contraction is now understood, as are the various parts played by ATP, Ca^{2+} ions, and the proteins actin and myosin.

In the relaxed muscle the myosin heads lie fairly close to the main

8.4 Muscles

myosin filament and not connected to actin at all. ATP is present in the spaces between the actin and myosin. At the start of contraction ATP binds to the ATP binding site on the head of the myosin and is hydrolysed to ADP and P_i, this hydrolysis only being possible when Ca^{2+} ions are present (see below). These products of ATP hydrolysis remain bound to the myosin. Once the ATP has been hydrolysed the actin and myosin are able to bind together at their respective binding sites to form actomyosin. In doing this the myosin head rises away from the axis of the filament and forms a cross-bridge between the two proteins, being bound to the actin units in two places. The energy derived from the hydrolysis of ATP is converted into mechanical energy which causes the heads of the myosin molecules to rotate, pulling the actin filaments some 12 nanometres in the direction of the H band (the tail of each myosin molecule is serving to anchor it firmly to the body of the thick filament and thus prevents it from moving). These tiny individual movements of the myosin heads that draw

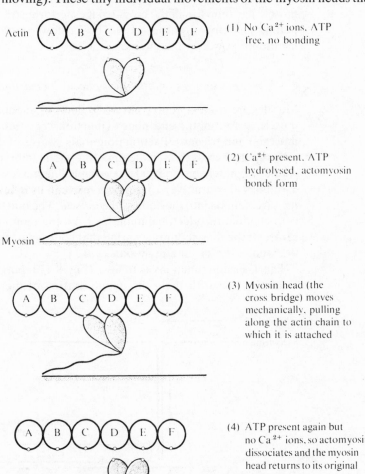

Actin

(1) No Ca^{2+} ions, ATP free, no bonding

(2) Ca^{2+} present, ATP hydrolysed, actomyosin bonds form

Myosin

(3) Myosin head (the cross bridge) moves mechanically, pulling along the actin chain to which it is attached

(4) ATP present again but no Ca^{2+} ions, so actomyosin dissociates and the myosin head returns to its original position

Fig. 8.16.
The molecular basis of muscle contraction and relaxation.

in the actin filaments lead to the contraction of the whole muscle fibre. Meanwhile ATP has been regenerated from the ADP and P_i and this, in the absence of Ca^{2+} ions, causes the actomyosin complex to break down. The myosin head falls back towards the thick filament and resumes its original orientation to the axes prior to another work stroke. Relaxation is a passive process that does not involve the actin and myosin elements; the muscle is pulled back to its resting length by the forces of gravity or of antagonistic muscles.

The work stroke, consisting of the formation of actomyosin and the oscillation of the myosin heads or cross-bridges, is repeated many times a second, and in the contraction of a whole muscle will be repeated by many millions of individual actin and myosin molecules. The sequence is shown diagrammatically in Fig. 8.16.

When a muscle contracts rapidly the minimal number of cross-bridges are made, which limits the power of the contraction. A slowly contracting muscle has time for all the cross-bridges to form and for each to exert its pulling movement on the actin. Thus a slow contraction produces the maximum power in the muscle.

8.4.5 *Nerve impulses and the part played by calcium ions*

Muscles are stimulated to contract by nervous impulses, which reach the muscle at the motor end plates (junctions between motor nerves and muscles) and trigger off action potentials (waves of depolarisation) that pass rapidly across the sarcolemma (muscle membrane). Ultrastructural studies of muscle have revealed a complex network of tubules, known as T-tubules (T = transverse), very extensive in its nature and passing from the sarcolemma into the body of the muscle. The function of this T-system is to conduct the electrical impulses from the motor end plates into the centre of the muscle, thus ensuring that contraction is simultaneous over the whole of a given muscle block.

The T-system functions as follows (Fig. 8.17). Impulses pass down the T-tubules and with them the depolarisation of the membranes of the

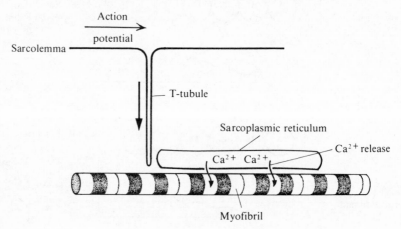

Fig. 8.17.
The role of the
T-system in muscle
contraction.

sarcoplasmic reticulum within the muscle cells. Calcium is bound in this sarcoplasmic reticulum, but on depolarisation it is released into the proximity of the actin and myosin filaments. In the presence of Ca^{2+} ions APT binds onto the ATPase sites on the myosin heads. Hydrolysis of the ATP occurs and at the same time the myosin head undergoes a molecular rotation which leads, collectively, to a shortening of the muscle.

Between the arrival of successive impulses the free Ca^{2+} ions are rapidly pumped back into the sarcoplasmic reticulum by active transport, again using ATP.

8.4.6 *Energy exchanges in contraction*

ATP is seen to provide the energy for contraction and also for the active transport of the calcium ions at the end of the contraction cycle, yet there is not very much ATP in muscle. Initially this may seem surprising, as it did to early investigators, but the explanation lies in the relationship and energy exchanges between ATP and creatine phosphate. Creatine phosphate is an energy-storing substance that exchanges energy with ATP as follows:

Creatine + ATP → Creatine phosphate(CP) + ADP

In the muscle the major store of energy for immediate use is creatine phosphate. After ATP is hydrolysed during contraction it is resynthesised from ADP and high-energy phosphate transferred from creatine phosphate. Thus at any one time the level of ATP in the muscle will not be high, although it must be available for contraction as CP cannot act as an energy source directly. As the muscle goes on contracting the CP supply becomes drained and further energy for ATP synthesis must be provided via respiratory pathways that break down the muscle glycogen. In normal working muscle high-energy phosphate is provided via anaerobic metabolism of glycogen, which has lactic acid as the end product. This is a fast but inefficient way of generating energy and the lactic acid is later transported out of the muscle for use in anabolism or complete aerobic breakdown.

8.4.7 *Innervation*

Each muscle is supplied by many nerve fibres, each of which innervates a number of muscle fibres (Fig. 8.18). When a nerve is stimulated an all-or-nothing response called an action potential may be produced. The action potential is propagated along the nerve fibre and in turn excites all the muscle fibres which it innervates. The muscle is made up of numerous motor units, each supplied by a single nerve fibre and capable of functioning independently. The total activity of the muscle being made up of the sum of the individual activities of these units. Thus the greater the number of motor units that are functioning the stronger will be the contraction of the muscle. Normal skeletal muscles have 2,000–3,000 fibres in each motor unit, but the extrinsic eye muscles have only 10–15 and hence their contractions can be more finely controlled. Where the nerve fibres terminate on the muscle fibres there is a single motor end plate. Here the nerve impulse causes a small amount of acetylcholine to be released, which changes the permeability of the muscle membrane and initates the propagation of an action potential along the muscle fibre.

After an action potential the muscle fibre takes a short time, called the refractory period, to re-establish the charge on its surface (as happens in nerve fibres), during which the muscle membrane cannot be excited. Excitation can occur, however, before the mechanical contraction has declined and when this happens at the arrival of another impulse there is a summation of the mechanical contractions which results in a state of prolonged contraction called tetanus (Fig. 8.18*c*).

An important functional distinction can be made between muscles that respond very quickly to a stimulus and those that do not. The former, called fast twitch muscles, are found in limb extensors where speed of response is critical for locomotion. The latter, which have a slower twitch but for a more sustained duration, are more important in the musculature involved in maintaining standing posture. Fast twitch muscles are used to produce the maximum speed of movement possible, bearing in mind the animal's mass and the limiting factors of the strengths of the various elements involved in locomotion. Slow twitch muscles produce the same degree of contraction, but with the expenditure of less energy, and are thus more economical for postural homeostasis.

The strength of a muscular contraction is determined by the number of motor units active at any one time. Under normal conditions it is rare for all of them to be active at once, although this condition (the maximal twitch) is often produced artificially in physiological experiments. When a mammal is in its normal posture and at rest the tension exerted by a muscle remains fairly constant, but this is not due to the continued contraction of the same muscle fibres: there is a rotation of contraction among the total muscle fibre population such that the net tension remains constant.

Fig. 8.18.
The nature of muscle contraction depends on the frequency of the stimulus. In (*a*) the interval between stimuli is greater than the refractory period of the muscle, so the muscle fibre is able to relax completely between stimuli. As the interval between stimuli decreases (*b*) it will fall within the refractory period and relaxation between successive stimuli will only be partial. An extreme case of this is shown in (*c*), where the stimuli are so close together that the muscle goes into a state of permanent contraction known as tetanus.

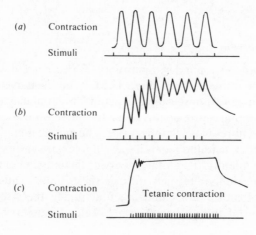

9 Locomotion

At each evolutionary level the different classes of vertebrates have shown adaptations for various means of locomotion. Thus in each class, while a majority move in a particular way, we find species adapted for contrasting modes of movement. This point can be seen in Table 9.1.

TABLE 9.1. *The types of locomotion used by different classes of vertebrates*

Class	Aquatic (swimming)	Terrestrial (creeping, walking, leaping running)	Aerial (gliding, flying)
Pisces	The majority of species	A few species such as the eel, mudskipper, climbing perch	The 'flying' fish (which in fact glides rather than actively flying)
Amphibia	Most forms, e.g. newts, salamanders frogs	Many species of frogs and toads; apodans	Gliding and flying tree frogs such as *Hyla* and *Rhacophorus*
Reptilia	Extinct dinosaurs (*Icthyosaurus* etc.), living turtles	The majority of both extinct and living species, e.g. lizards, snakes	The extinct pterodactyl and few living forms such as *Draco* the flying lizard
Aves	Marine flightless birds, e.g. penguins	Many flightless terrestrial species e.g. emu	The majority of species
Mammalia	Whales, seals, otters, etc	The majority of species	Bats and some gliding arboreal species

The modification of the body and limbs of vertebrates for different types of locomotion is an excellent illustration of the important biological principle of adaptive radiation. Through the workings of natural selection a basic organ, such as the pentadactyl limb, becomes modified for a number of specialised functions. As a consequence of adaptive radiation there is frequently convergence between the structural modifications found in animals from different groups which have evolved independently to fulfil the same sort of functions.

9.1 Swimming

Swimming is the form of locomotion found among primitive vertebrates and is best developed in fishes, but is also found as a secondary adaptation in other groups such as whales and dolphins. The main adaptations for movement in water include:

(*a*) *Streamlined shape.* This may involve reduction of the appendicular skeleton in secondarily adapted forms (for example, whales have a minute pelvic girdle).

(*b*) *Well-developed muscles* along the axial skeleton (myotomes) for producing undulatory movements, and usually a form of tail fin. In whales and dolphins the movements are in a dorso-ventral plane, whereas in fishes, including flatfishes, they are lateral. The 'tail fins' of mammalian swimmers are referred to as the flukes.

(*c*) *Stabilising elements.* Among fishes the paired and unpaired fins function as stabilising elements, but they may also play a part in active locomotion. This is also true among groups of secondarily aquatic verte-brates including seals, penguins and turtles, which make use of their paired limbs as paddles.

When a fish such as a dogfish is swimming, lateral waves of contraction are seen to pass backwards along the body. In a simple way these may be regarded as pushing against the water and hence driving the animal forwards. All parts of the body contribute to this driving force but the expanded tail is probably the dominant force. From cine films it is clear that each segment of the body, and the tail, passes through the water at an

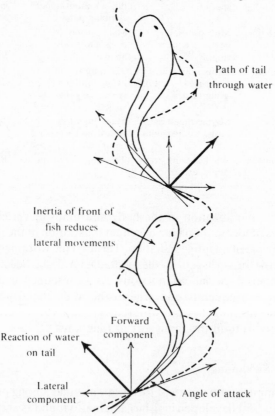

Path of tail through water

Inertia of front of fish reduces lateral movements

Reaction of water on tail

Forward component

Lateral component

Angle of attack

Fig. 9.1.
Forces developed by the tail of a fish during swimming.

angle to its surface. In this way, as they press against the water the latter reacts on the body and produces a resultant force at right angles to the body surface (Fig. 9.1). This force can be resolved into two component forces: a lateral one that pushes the body sideways and is counteracted by equal and opposite forces acting at other parts of the body, and a forward component that overcomes the resistance of the water. In most bony fishes (eels are notable exceptions) the backward passage of waves along the body is not so clear, and the most obvious movement is a sideways movement of the tail fin. This again acts as an inclined plane passing through the water at an angle to its path of motion and from it forces are developed. In these fishes the head scarcely moves to the side at all, whereas in the dogfish and eels the head is displaced laterally during the swimming movements. This lack of lateral displacement of the head in most teleost fishes is due to the greater inertia of the front end and the resistance to sideways movement provided by the fish's shape and the unpaired fins.

In mammals, such as dolphins, the propulsive thrust develops almost entirely from the flukes, which are moved at an angle to their direction of motion in an up-and-downward plane. In all these cases of undulatory propulsion the speed is determined by the frequency of tail beat and especially by the length of the animal. A rough generalisation is that the maximum speed of a small to medium-sized fish is about 10 times its body length per second. A trout 30 centimetres long, for instance, can swim at 300 centimetres per second. The highest speed recorded for a fish is about 12 metres per second for a 120-centimetre barracuda. Authentic speeds for dolphins are in the region of 30 kilometres per hour. The ability of these animals to move so rapidly is aided by the properties of their skins which, because of its structure and mucous covering, serves to reduce considerably their resistance. They may also be able to damp out turbulence, for while the dead bodies of such animals towed through water produce high drag resistance associated with turbulent flow it is likely that in life they can produce laminar flow over their body surfaces (Fig. 9.2). This would result in a very substantial decrease in the resistance of the water to their swimming. It is unlikely that the muscles of swimming vertebrates have any special adaptations for generating extra power.

The muscles which produce swimming movements in fishes are the myotomes, which have a characteristic $\gtrless$-shape when viewed from the side. This shape is the outward sign of many segmentally arranged cones of muscle which interlock with one another. The myotomes operate between sheets of connective tissue (myocommata) which connect with

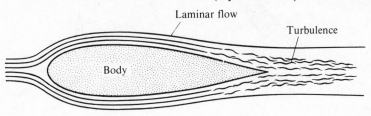

Laminar flow

Turbulence

Body

the transverse processes of the vertebral column. This complex arrangement of the muscle fibres probably enables a smoother passage of the wave of contraction along the body and also prevents the formation of excessive thickenings of the body muscles in the region where they have contracted during the passage of the wave. Each segmental myotome contracts slightly ahead of the myotome behind it and the waves pass alternately down each side of the body. The waves are co-ordinated by proprioceptive information which enters the dorsal roots of the vertebrae from the sensory nerves. If all the dorsal roots of a dogfish are severed the animal cannot swim (p. 259).

The cigar-shape of aquatic animals, though reducing their resistance as they move through the water, tends to make them roll about their longitudinal axis and to yaw and pitch about their centre of gravity. These tendencies are corrected in fishes by fins (Fig. 9.3) which, though partly passive, also operate actively and so require the presence of sense organs (semicircular canals of labyrinth) which detect movements in these

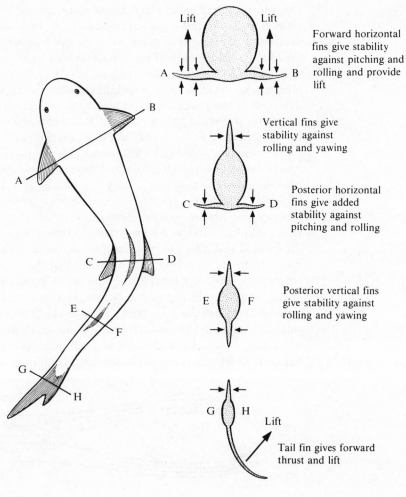

Forward horizontal fins give stability against pitching and rolling and provide lift

Vertical fins give stability against rolling and yawing

Posterior horizontal fins give added stability against pitching and rolling

Posterior vertical fins give stability against rolling and yawing

Tail fin gives forward thrust and lift

Fig. 9.3.
The role of the fins in a swimming dogfish.

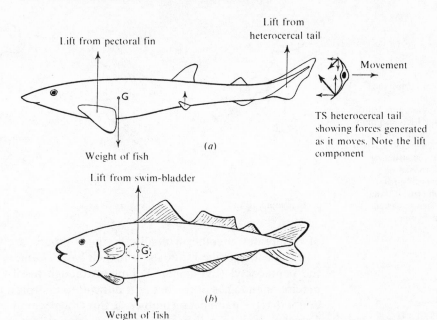

Lift from pectoral fin

Lift from heterocercal tail

Movement

G

Weight of fish

(a)

TS heterocercal tail showing forces generated as it moves. Note the lift component

Lift from swim-bladder

G

(b)

Weight of fish

Fig. 9.4.
Elasmobranch and teleost compared with respect to the vertical forces operating on their bodies.
(a) Dogfish, (b) cod.

planes. In a dogfish all the fins, both paired and unpaired, tend to hinder rotation of the body about its longitudinal axis (i.e. roll). The tail and fins behind the centre of gravity tend to maintain the fish in its direction of motion but the unpaired fins in front of the centre of gravity counteract this tendency and make the animal more manoeuvrable than if it were highly stabilised like an arrow. Although it is heavier than water a dogfish is able to swim on an even keel, partly because of the action of the pectoral fins and also because of its heterocercal tail. During its lateral movements this tail fin produces a lift force because the flexible caudal fin lags behind the stiff upper spinal column. The moment of this lift force about the centre of gravity tends to make the head tilt downwards, but this is counteracted by an equal and opposite moment which results from lift developed at the pectoral fins (Fig. 9.4).

Bony fishes do not have heterocercal tails and most of them are not heavier than water. They maintain their position free-floating in water without apparent effort by means of their swim bladder. This contains a bubble of gas whose volume can be regulated, thereby maintaining the overall density of the fish close to that of the water in which they are swimming.

9.2 Terrestrial locomotion and its evolution

In order for vertebrates to be able to live on land a number of changes were necessary in their skeletal and muscular systems. It is thought that during the Devonian period natural selection favoured these changes in a certain group of air-breathing fishes (the Crossopterygii) and that from these the first terrestrial vertebrates – the amphibians – arose. This class

211

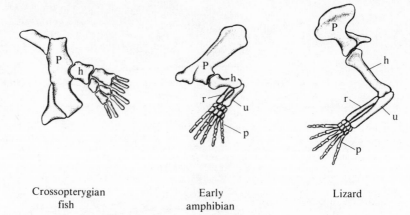

Fig. 9.5.
The paired fin and girdle of a Crossopterygian fish and the limb and limb girdle of an early amphibian and a lizard. The three structures are homologous. P, pectoral girdle; h, humerus, u, ulna; r, radius; p, carpals and phalanges.

Crossopterygian fish   Early amphibian   Lizard

of vertebrates, together with all subsequent groups, are known as tetrapods – the four-limbed animals – and their basic locomotory appendage is the pentadactyl (five-fingered) limb. Although fossil evidence of the precise homologies of the bones is difficult to interpret, there can be no doubt that the pentadactyl limb arose from the pectoral and pelvic fins of Crossopterygii (Fig. 9.5). Whereas in swimming it is the axial skeleton (i.e. the vertebral column) and segmental myotomes that are functionally important, in walking and running the bones and musculature of the limbs and the limb girdles become of increasing importance, both in support and propulsion.

From when the vertebrates first came on land it was necessary for them to support their own body weight against the effect of gravity, and this profoundly influenced the whole of their subsequent evolution. Not only did the bones of the fin change to produce the pentadactyl limb, but major changes occurred in the vertebral column and limb girdles. The all-round support of the body provided by the aquatic medium was lost and replaced by a system of struts which transmitted the weight of the body through a series of bones to the ground. For the support of the body it is clearly important that this series of bones should be fixed to the main body axis. In fishes the pelvic girdle is embedded in the myotomes and has no direct connection with the vertebral column. In the early tetrapods the primitive plate of the pelvic girdle acquired a new dorsal process (the ilium) which became attached to the transverse process of the sacral vertebrae. In this way the main propulsive forces developed from the hind limb were transmitted directly to the axial skeleton of the animal. In the fore limb the girdle became detached from the hind end of the skull, which was its position in the bony fishes. (This position was related to the region of weakness present behind the head because of the gills.) Consequently, as this girdle became reduced and the gills were lost, a neck evolved which enabled the head to have greater mobility (smooth streamlining behind the head was no longer necessary). The pectoral girdle became embedded in the musculature and acquired a shock-absorbing function. This is highly developed in mammals, where the large scapula

212

has serratus muscles on its inner face from which the body is suspended. Consequently, any shocks transmitted up the fore limbs are taken by the plate-like scapula and thence to the body through this muscular suspension. Their function is analogous to the mainspring of a car (see p. 217).

The vertebral column, which in fishes is markedly metameric (i.e. there is segmental repetition of parts) and facilitates lateral bending, now has quite different stresses acting upon the various sections along its length. As a result of this the uniform vertebrae found in the fish became differentiated into the five main types found in mammals (as described in Chapter 8).

In early tetrapods, locomotion on land was accompanied by lateral undulations of the body (Fig. 9.6). At this stage the limbs were unable to support the weight of the body, which consequently slithered along the ground. The movements of a newt or salamander are similar to those which the early vertebrates might have used. In these amphibians the limbs are spread out akimbo and their movements are coordinated with the lateral undulations of the body; as soon as the animal stops walking it tends to collapse on its belly. Despite its inefficient nature this means of progression persists in many reptiles (e.g. lizards). The amount of contact between the body and the ground can be observed by letting a newt move

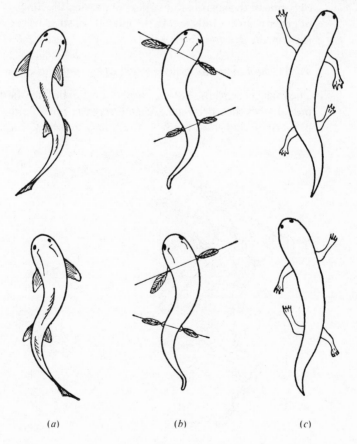

Fig. 9.6.
The evolution of terrestrial locomotion. (a) Fish; (b) fish (e.g. crossopterygian) using fins as passive extensions of body, with lateral undulations of back to assist movement on land; (c) primitive tetrapod (e.g. newt) using independent movements of limbs together with back muscles to assist locomotion.

(a)   (b)   (c)

213

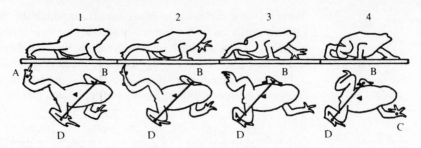

1   2   3   4

Fig. 9.7.
The way in which a
toad walks. Notice how
the centre of gravity
(indicated by the solid
triangle) always
remains within the
stable tripod formed by
the three feet that are
on the ground.

across mud and examining its tracks. During these movements each limb
is protracted and placed on the ground, and then retracted as a result of
the contraction of extrinsic muscles (retractors) and changes in the
intrinsic musculature (extensors). These actions combine to push the
distal end of the limb backwards and downwards against the ground. The
ground therefore exerts a force upwards and forwards against the limb
and provided the hand or foot does not slip, this can cause the animal to
move forwards. The rhythm of the limbs with respect to one another in
early tetrapods is again illustrated by the way in which a newt or toad
walks. The limbs are lifted in the order right fore, left hind, left fore, right
hind, right fore, etc. This, the so-called diagonal or tetrapod rhythm, is the
only one of the six possible orders of moving the limbs in which the centre
of gravity always falls within the triangle of support provided by the three
legs on the ground (Fig. 9.7).

9.2.1 *The functional arrangement of the muscles of the limbs*

The muscles working across a limb joint can be classified according to the
movements they produce. Considering the possible movements about a
ball-and-socket joint (such as the femur makes with the pelvic girdle or

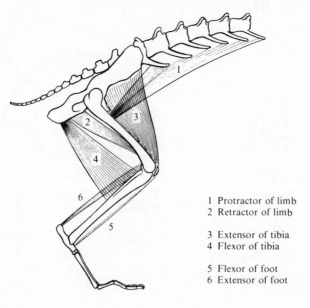

1 Protractor of limb
2 Retractor of limb

3 Extensor of tibia
4 Flexor of tibia

5 Flexor of foot
6 Extensor of foot

Fig. 9.8.
A simplified scheme of
the antagonistic
muscles in the hind
limb of a rabbit.

the humerus with the pectoral girdle) they can be subdivided as follows:

(*a*) *Protraction* – forward movement of the limb.

(*b*) *Retraction* – backward movement of the limb.

(*c*) *Elevation* – raising of the limb at right angles to the longitudinal body axis.

(*d*) *Depression* – lowering of the limb in the same plane.

(*e*) *Rotation* – movement of the limb about its longitudinal axis.

It can be seen that these fall into three major movements which can take place in opposite directions, so that (*a*) and (*b*), (*c*) and (*d*), and (*e*) may be brought about by antagonistic muscles. The action of many limb muscles combines several of these movements. Muscles which produce movements of the limb as a whole have their origin outside the limb although they are inserted on one of the limb bones, usually the most proximal (femur or humerus). Muscles with their origin outside the limb are called extrinsic limb muscles, but it must be understood that this classification is entirely a functional one and has no implications as regards the homologies of these muscles. The relative importance of the different extrinsic muscles varies from one species to another and consequently their size and arrangement will vary. When animals walk or run their propulsive power comes from the retraction of the limbs pushing the feet against the ground and thus driving the animal forwards.

Other sorts of movement are those taking place within the limb itself, and these are performed by intrinsic muscles that have both their origin and insertion within the limb itself. Once again they may be subdivided into antagonistic pairs. The main movements consist of:

(*a*) *Extension* – increasing the angle between two segments of a limb.

(*b*) *Flexion* – the decrease in the angle between two segments of the limb.

The system described above may be applied in modified form to the musculature of any pentadactyl limb. Fig. 9.8 shows the three main sets of antagonistic muscles seen in the hind limb of a rabbit. Rabbits do not

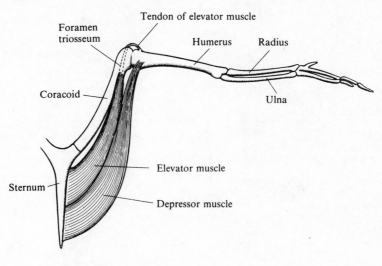

Fig. 9.9.
The wing musculature of a bird, showing the importance of the antagonistic elevator and depressor muscles.

215

rotate their limbs like climbing mammals, and so the tibia and fibula are fused. For the rabbit, and for running animals in general, the power is provided by the extension and retraction of the limb against the ground. The muscles bringing about these movements are much larger than the recovery antagonists of the flexors and protractors.

The limb and limb girdle of a tetrapod that runs do not have well-developed elevator and depressor systems as lateral movement would make the limb unstable along the axis it must travel in order to produce the driving force. The front limb of a bird, however, is powered by muscles producing a down beat against the air followed by an up beat recovery stroke, and although the situation is complicated by the thrust component and the flexions and extensions during flight, the major wing muscles are elevators and depressors (see Fig. 9.9).

9.2.2 *Adaptations for running*

The evolutionary modifications of the primitive tetrapod limb that are associated with running include a rotation and elongation of the limbs as well as changes in the limb girdles and backbone.

By rotation of the limbs from the awkward akimbo position to one where they lie directly beneath the body, the weight of the body is transferred from muscles to bones and the belly of the animal lifted permanently off the ground.

The essentials of fast locomotion on land are a long stride repeated at the smallest possible time interval. In order to increase the length of stride the distal bones of the limbs have been elongated so that the animal no

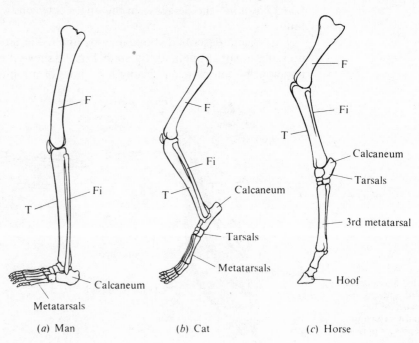

Fig. 9.10.
The hind limb modifications for walking of three mammals: (*a*) man is plantigrade, i.e. walks on a flat foot; (*b*) the cat is digitigrade, i.e. walks on its toes using extensions of the metatarsals; (*c*) the horse is also digitigrade but the foot is reduced to a single digit capped by a hoof. F, femur; T, tibia; Fi, fibula.

(*a*) Man   (*b*) Cat   (*c*) Horse

216

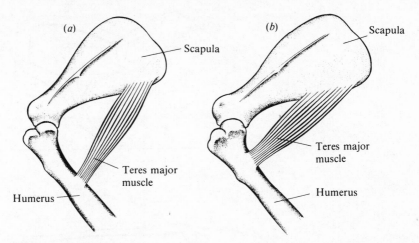

g. 9.11.
) 'Low gear' insertion
f the teres major
uscle gives power for
urrowing animals such
s badgers. (b) 'High
ear' insertion of the
res major muscle in
nning mammals gives
e humerus a greater
rc of movement.

longer rests on its flat feet, like man, but is raised up on its digits. In the case of mammals either several, two, or just one digit may be used according to the taxonomic order to which the animal belongs (see Fig. 9.10).

In contrast to the extension of the feet, the proximal elements of the limbs have gradually shortened and been taken up into the body, where they are surrounded with powerful extrinsic muscles. This has the effect of lightening the limb, which in the fleetest herbivores is little more than a chain of articulating bones and long tendons. As movement becomes more rapid and the stride lengthens, changes in the gait occur. In some cases a greater length of time is spent on the diagonal part of the cycle (the trot) or on the phase where both ipsilateral limbs are on the ground (the amble of, e.g. the camel), and at the fastest speeds there are phases when no legs are on the ground (the gallop).

In running mammals the feet, although extended and lightened, usually have a muscular (e.g. cat) or tendinous (horse) base for absorbing the shock of landing and also providing extra forward spring to the stride itself.

For speed fast muscle contraction is required, but the speed of contraction of a muscle varies inversely with its linear dimensions. For this reason running animals tend to use what physiologists describe as 'high gear' muscles which are attached very close to the point of articulation of the limb. By this means the movement of the limb bone is increased although at the same time power is lost. In fact movement through the air does not require a great deal of force. Generally a compromise is made between the use of high gear muscles for rapid limb movements and low gear muscles, which have a more powerful sustained contraction, in supporting the animal (Fig. 9.11).

In many of the fastest-running animals there are dorsal and ventral movements of the backbone which effectively increase the stride (Fig. 9.12); these are found especially in the largest cats and deer. The cheetah is the fastest land mammal and at one phase of the cycle the hind

217

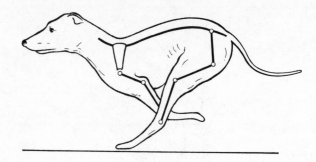

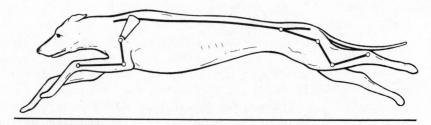

Fig. 9.12.
A dog running shows
how extension of the
spine and the
movements of the ends
of the limb girdles
increase stride length.

limbs are in front of the fore limbs preparatory to their rapid retraction, which provides the main propulsive thrust. Aquatic mammals such as the whale reveal their mammalian descent by the up and down movements of their tails (the tails of fish move from side to side).

As already stated the speed of the movement of the limb is a critical factor in fast movement and this is increased by moving many muscle and joint systems at the same time to produce a synergistic effect. The push forwards is produced by retraction of the limb girdles, extension of the spine, and extension of the limbs themselves. All these combine to exert a backwards and downwards force upon the ground, a force whose result-ants serve to drive the animal forwards and upwards.

Speed depends on the stride length and the rate of stride. For example a horse and a cheetah both have a stride length of approximately 7 metres, but a cheetah can attain a speed of 110 kilometres an hour, compared with a horse's maximum speed of 80 kilometres an hour because it can make up to 3.5 strides per second compared with the horse's 2.5. The big cats are the fastest mammals for short sprints but the long-legged herbi-vores such as the deer and horse family can maintain high speeds for long distances.

9.2.3 *Metabolic cost of running in mammals*

Running is a metabolically expensive activity and the smaller the mam-mal concerned the more rapidly does its oxygen consumption increase with its speed of running. From this it follows that, in terms of energy

ig. 9.13.
he relative metabolic osts of running for ree mammals of ifferent mass. (The oints of intersection ith the vertical axis epresent the resting asal metabolic rate of ach animal.)

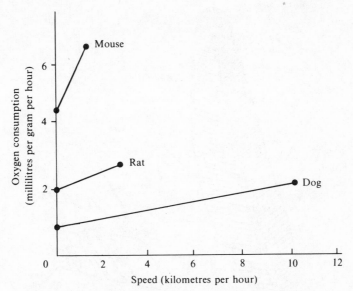

consumption, it pays to be a large rather than a small running mammal. The relationship between energy expenditure and speed for three mammals of different mass is shown in Fig. 9.13.

9.3 **Flying**

It is possible to distinguish two types of flight – passive or gliding flight and active or flapping flight. Many groups of vertebrates have become adapted for gliding but only birds and bats have evolved active flight. The essential structural requirement for flight is the possession of a membrane or patagium, which is an extension of the body that can support the body in the air. Birds are the only vertebrates in which the patagium is limited to the fore limb and has no connection with the hind limb. This release of the hind limb has enabled it to become modified for a variety of functions, for example swimming, climbing or running. The patagial membranes of the bird fore limb are extended by the presence of feathers, which are a diagnostic character of these creatures. The whole body is clothed with them and besides their function in creating a smooth contour to reduce air resistance in flight, they provide important insulation of the body against heat loss.

9.3.1 *Feathers*

The feather is largely composed of the protein keratin, which has the properties of being both light and strong; it develops as a secretion from an epidermal papilla. The basic structure is best described by considering one of the main plume feathers from the wing (Fig. 9.14). Each of these is organised about a central stem or rachis which forms the quill of the

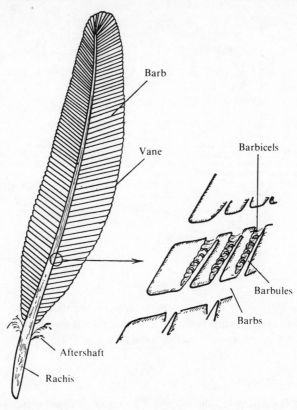

Fig. 9.14.
Bird plume feather.

Barb

Vane

Barbicels

Barbules

Barbs

Aftershaft

Rachis

feather, upon which are inserted a number of bars on each side. The barbs bear still smaller branches (the barbules) on both their proximal and distal faces. The distal barbules interlock with the proximal ones of the neighbouring barbs by means of microscopic barbicels, so that an almost airtight membrane (the vane) is formed on either side of the rachis. Such plume feathers are found attached to the wings and to the tail. Other feathers which form the general surface of the body (contour feathers) do not have such complex interlocking of the barbs, for the small hooks are absent and the vanes are more fluffy. Down feathers, found between the contour feathers, are the simplest of the adult feathers; they are loose and soft and may have a short rachis but are usually sessile. The smallest types of feathers are known as filoplumes, which are very inconspicuous except when a dead bird is plucked. They are reduced feathers with a few barbs at the tip of a slender hair-like rachis.

Much of the bird's preening activity is directed towards sorting out these feathers and arranging them correctly.

9.3.2 *Structure of the bird's wing*

The way in which the basic pentadactyl limb is modified to form the wing of a bird is one of its most outstanding adaptations (Fig. 9.15). All the

220

main segments are present – the humerus is large, the radius and ulna elongated, but the carpal and metacarpal regions have undergone fusion, reduction, and marked elongation. Only three digits are represented and these are now thought to represent the second, third and fourth rather than the first, second and third as is often stated. The anterior one (i.e. second) is short and bears the bastard wing. The middle digit (third) is the longest and has quite an elongated fused metacarpal region. From the functional point of view the wing is made up of the arm and the hand; the former bears the secondary feathers and the hand has the primaries attached to it. The structure of these two types of plume feathers may be differentiated by the relative sizes of the vanes on the two sides of the rachis: in the secondary feathers they are equal but in the primaries they become more and more unequal as the feathers become nearer to the front, or leading, edge of the wing. The spacing of the feathers with respect to one another is maintained by a fenestrated membrane (vinculum) through which they are attached to the bone. Two important structural features of the skeleton are that when the elbow joint is extended the wrist joint automatically extends. Secondly, in the extended position the wrist cannot be rotated, but this can be done very easily in the flexed position.

Between the head of the humerus and the wrist and from the elbow to the hand are stretched the thin sheets of tissue which form the patagial membranes. That on the anterior edge is the most important and it is maintained in a stretched condition by the presence of special muscles (the tensor patagii muscles). The rest of the wing is very muscular and has in its skin a large number of contour and down feathers which fill in between the bases of the large plume feathers. In the extended condition the whole structure forms an excellent aerofoil when seen in profile.

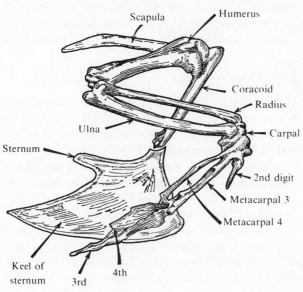

g. 9.15.
:ctoral girdle and
ng of a bird.

221

9.3.3 *Modifications of the limb girdles*

Although all parts of the skeleton of a bird are modified for flying, notably by lightening of the bones, it is mainly the fore limb and the pectoral girdle which have become especially changed. The pectoral girdle has become closely attached to the large keeled sternum by a stout coracoid bone. Dorsally this bone forms one of the constituents of the foramen triosseum (the other two being the scapula and furcula); these same three bones form the glenoid cavity with which the humerus articulates. The large keeled sternum forms the point of origin of the important muscles for lowering and raising the wing. The pectoralis (P. major) forms up to 25 % of the body weight in a pigeon and is inserted on the crest of the humerus. The coracoid forms a compression member between the sternum and the wing; its strength resists the forces exerted during the downstroke of the wing beat. Also with its origin on the sternum is the supracoracoideus (P. minor), which is very much smaller than the pectoralis (about one-fiftieth in a flapping flyer) and it sends a tendon through the foramen triosseum to be inserted on the dorsal side of the humerus. Thus two muscles with their origin and insertion on the same bones have opposite effects. Raising of the wing is normally accomplished by the pressure of the air and it is only during take-off and landing that the supracoracoideus muscle is involved to any extent.

The pelvic girdle of birds is greatly modified because of their bipedal habit. The ilium is largely fused with the sacral vertebrae which form the so-called synsacrum. Ventrally the ischium and pubis do not join with the corresponding members on the opposite side, so that there is no complete girdle in a bird. The tail is very much reduced in modern birds although it was elongated in early flyers such as *Archaeopteryx*. The hind limb of most birds is well provided with muscles which move it forwards and backwards, and there are tendinous arrangements which ensure that the claws clasp a branch when the weight of the body is taken on the legs.

9.3.4 *Types of flight*

The principles of flight are the same for both gliding and flapping flight. When the wing is inclined at an angle up to 15° to the flow of air, it develops a lift force mainly because the more rapid flow across its upper surface produces a region of reduced pressure (Fig. 9.16). The relative movement between the air and the wing may result from either movements of the air, as in rising air currents, or the loss of height of the bird, or by the active movements of the wing through the air. Under all these conditions there is a net resultant force on the wing which can be resolved into a component at right angles to the air flow (lift) and one in the same direction as the air flow (drag). When flying at uniform velocity the lift component must be equal and opposite to the weight of the body and the drag component must be opposed by another force driving the animal through the air. When a bird glides from the top of a tree to the ground in

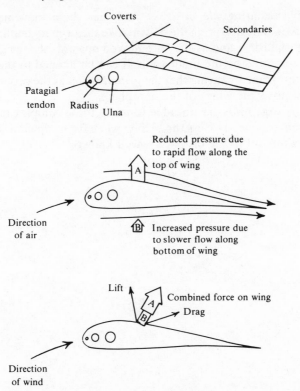

Fig. 9.16.
Forces acting on a
bird's wing during
flight.

still air, the weight may be resolved into two components, one acting in a direction opposite to the lift component of the aerodynamic force, and the other acting in the direction of movement of the animal, which is a thrust component equal and opposite to the drag component of the aerodynamic force (Fig. 9.16).

Birds make use of differences in air velocity either at different heights above the sea (for example the albatross) or different wind velocities at the same horizontal height, or by the use of ascending air currents over deserts (for example vultures). These conditions for gliding were first appreciated by Lord Rayleigh in 1883, who said that if a bird was not flapping its wings and yet kept on its course, then one of the following three conditions must apply: either (*a*) the course was not horizontal, (*b*) the wind was not horizontal, or (*c*) the wind was not uniform in velocity.

Flapping flight may either be that which takes place at take-off or the hovering of a species such as a humming bird. In humming birds the wing is very flexible about its base and it moves in a figure of eight so that lift is developed throughout the whole cycle. During the upstroke (relative to the bird) the forces are developed from the back of the wing. In medium-sized birds like pigeons, the take-off and landing flight involves complex movements in which the upper surface of the primary feathers strike against the air on the upstroke and so develop lift. During the down-stroke, of course, lift is produced by the movement of the wing at an angle to the air. In fast flight the complex movements of the wrist no longer take

223

place during the upstroke and the whole movement is much more economical. During the downstroke the primary feathers produce lift and propulsion, and as they are moved upwards the weight of the animal is supported by the inner wing. It can be likened to the mechanism of an aeroplane by saying that the primaries act as the propellers and the inner wing as the wings of the aeroplane.

Large birds are not able to make these complex take-off movements and they need to flap their wings when facing upwind. The noise of a swan taking off in the distance is well known.

10 Nervous coordination

Each of the physiological processes described so far has its coordinating mechanisms which regulate its actions so that they are appropriate to the conditions prevailing at a given time. For this to be achieved communication between different parts of the body is essential and vertebrates have two main means by which this is done: the first is by nervous impulses (electrical signals) and the second by hormones (chemical signals). These two types of signal are associated with the nervous and endocrine systems respectively.

In general the body needs to respond to stimuli from both the external and internal environments in order to produce the adaptive changes necessary for it to survive. Although the distinction is not exact, the nervous system tends to operate in the former case and the endocrine system in the latter. It should be realised, however, that the two are very closely linked both in origin and function and that the normal functioning of the body depends on the integrated action of the two. For example, the nervous system depends for transmission between individual cells on the liberation of chemicals at their junctions with one another and also with effector organs, and nerve impulses can cause the release of hormones from endocrine organs (e.g. adrenals). In other words the two systems are not even distinct in the type of signal used. However, during the course of evolution the hormone or chemical signal has been selected to bring about slow and general changes in the metabolism of individual cells (such as growth), while the nervous system tends to be involved in rapid local responses to stimuli (such as movement).

The nervous system, with which we are concerned here, consists of specialised cells or neurones, linked together via the central nervous system (see p. 255) to form networks that connect the organs which receive stimuli (receptors) and those which carry out actions or responses (effectors) (Fig. 10.1). It is an exceedingly elaborate system in mammals,

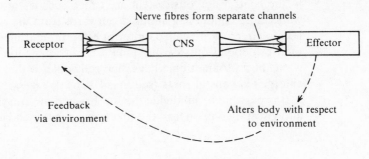

Fig. 10.1.
The basic pattern of the nervous system of a mammal, showing the role of the central nervous system (CNS) as a link between the receptor and effector systems.

with channels capable of rapid conduction (up to 120 metres per second in some nerve fibres) and a great specialisation and multiplicity of pathways between receptor and effector whereby a whole variety of responses is possible.

Before considering how the pathways of nervous communication are organised it will be as well to deal with the basic components of the system – that is, the neurone and its capacity to generate and conduct impulses, and the synapse across which the impulses must pass – as well as to discuss the origin of this physiological system from the embryonic tissues of the organism.

10.1 Structure and origins of the nervous system

It is a characteristic of chordates (the phylum to which the vertebrates belong) that they have a hollow, dorsal nervous system; this originates from the ectoderm which overlies the skeletal notochord in the embryo. Other sources of nervous tissue in the embryo are the crests of the invaginating ectoderm (the neural crests), which give rise to collections of cells (the dorsal root and autonomic ganglia) outside the neural tube. In addition, local ectodermal thickenings associated with cranial nerves form placodes which sink in and give rise to other nervous structures, which include the labyrinth and lateral line system as well as the olfactory organ.

The chief functional units of the nervous system are neurones, but there are other cells, called neuroglia, which make up as much as half the nervous system. They are generally thought of as supporting cells but they also play a vital role in the nutrition of the neurones. In all, the nervous system of man comprises some thousands of millions of cells, a system which makes the most complex electronic computers appear simple in comparison.

10.2 Units of nervous function

10.2.1 *Neurone*

Most neurones are characterised by a long protoplasmic process or axon, which is specialised for the conduction of impulses from one part of the cell to another. Neurones which conduct impulses from receptor organs to the central nervous system are called sensory or afferent neurones, while those conducting impulses outwards from the central nervous system to effectors are called motor or efferent neurones. The former have their cell bodies outside the spinal cord in a ganglion of the dorsal root, while the motor neurones have their cell bodies within the spinal cord and emerge via ventral roots (see Fig. 10.26).

It is possible to subdivide the sensory and motor systems into somatic and visceral components, the first being concerned with coordination of

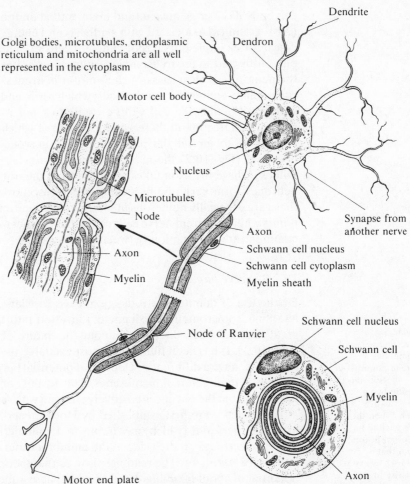

Golgi bodies, microtubules, endoplasmic reticulum and mitochondria are all well represented in the cytoplasm

Dendrite

Dendron

Motor cell body

Nucleus

Microtubules

Node

Axon

Myelin

Axon

Schwann cell nucleus

Schwann cell cytoplasm

Myelin sheath

Synapse from another nerve

Node of Ranvier

Schwann cell nucleus

Schwann cell

Myelin

Axon

Motor end plate

Fig. 10.2.
The structure of a motor neurone. A longitudinal section through a node (left) and a transverse section through the axon (right) are also shown.

the skeletal or striated muscle systems and the second with the control of visceral activities such as the smooth muscle of the gut. The visceral system, both motor and sensory, mainly coordinates functions of the body below the level of consciousness, and forms the autonomic system. It will be discussed more fully later in this chapter.

Despite the large number of afferent and efferent neurones, the vast bulk of the CNS is made up of intermediate (or connecting) neurones which lie entirely within the brain and spinal cord. These tend to have branched axons; some may be very long and form extensive tracts which pass up and down the spinal cord, but others have very short axons and only communicate over short distances. Their function is to link afferent and efferent neurones.

In structure the neurone has a cell body or soma which is the main nutritional part of the cell and is concerned with the biosynthesis of materials necessary for the growth and maintenance of the neurone. The

227

cell body has a large nucleus and many mitochondria and other granules (Nissl granules) associated with synthesis and energy exchange. In some cases, for example dorsal root ganglion cells of mammals, the cell body plays little part in the transmission of impulses as it is on a side branch of the neurone. In most cases, such as motor neurones, the cell soma is the site of many endings from other cells which form 'end buttons' all over its surface. Whether the cell will be excited or not is determined by the summated effect of all these endings, some of which may tend to excite but others to inhibit the production of a nervous impulse. In motor neurones (Fig. 10.2) the nervous impulse arises at the point (the axon hillock) where the axon takes its origin from the soma. Other regions of neurones form very fine branching processes known as dendrites, and these are especially numerous in intermediate neurones. These are again points where different nerve cells come into contact with one another and communicate information between their surfaces.

10.2.2 *Nerve impulse*

By the use of minute electrodes (e.g. glass capillary microelectrodes of less than 1 micrometre tip diameter) inserted into the soma or axon a great deal has been discovered about the nature of the nerve impulse (Fig. 10.3). It has been found that most excitable tissues, whether nerve or muscle, have a difference of electrical potential between the inside and outside of their external membranes. This resting, or membrane, potential is such that the outside is positively charged with respect to the inside.

These facts were first established by Hodgkin and Huxley using giant axons of the squid (which may be up to 1 millimetre in diameter), by placing electrodes on each side of the membrane and connecting them to a sensitive voltmeter. The readings showed that there was a difference of potential of about 60 millivolts (0.06 volts) across the membrane; this is largely due to differences in the concentration of K^+ ions inside and outside the cell.

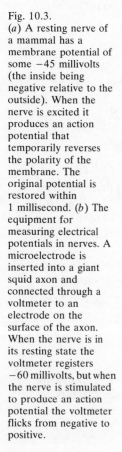

Fig. 10.3.
(*a*) A resting nerve of a mammal has a membrane potential of some −45 millivolts (the inside being negative relative to the outside). When the nerve is excited it produces an action potential that temporarily reverses the polarity of the membrane. The original potential is restored within 1 millisecond. (*b*) The equipment for measuring electrical potentials in nerves. A microelectrode is inserted into a giant squid axon and connected through a voltmeter to an electrode on the surface of the axon. When the nerve is in its resting state the voltmeter registers −60 millivolts, but when the nerve is stimulated to produce an action potential the voltmeter flicks from negative to positive.

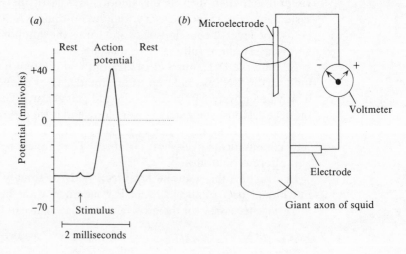

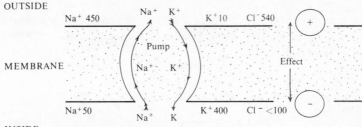

ig. 10.4.
he differences in ionic
oncentrations across
he membrane of a
esting squid axon. (All
alues are in millimoles
er litre.)

The actual ionic concentrations obtained by Hodgkin and Huxley in their investigations on squid axon* are shown in Fig. 10.4. The concentration of Na^+ was some 9 times greater outside than inside and the concentration of K^+ 40 times greater inside, as well as there being a very much higher concentration of Cl^- on the outside. These differences were explained as being due to the activity of sodium pumps in the nerve membranes which by pumping positively charged sodium ions from inside to outside cause a passive efflux of negatively charged chloride ions in their wake and a tendency for positively charged potassium ions to move passively into the cell to replace the Na^+ ions.

The axon membrane is also a great deal more permeable to Na^+ than K^+ ions and the resting potential reflects this fact as well as the activity of the 'pump'.

On stimulation of the neurone, a change takes place in the potential recorded by the voltmeter whereby the charge across the membrane is reversed (see Fig. 10.3). As the outside now becomes negative relative to the inside, current flows from the resting part of the nerve into this region and consequently sets up small local circuits which themselves excite the neighbouring regions of the axon and hence the self-propagating impulse travels away from the region that was initially excited (Fig. 10.5).

This changing potential which passes along the nerve fibre and thus corresponds to the passage of an electric impulse has been shown to be due to a change in the permeability of the membrane that causes an inrush of Na^+ ions across the membrane followed by an outrush of K^+ ions. The sodium pump mechanism removes the Na^+ ions that have entered and soon repolarises the surface of the membrane to its resting potential. It is not surprising to learn that the repolarising of the membrane requires a lot more energy and releases more heat than the more passive change in its permeability that leads to the depolarisation.

The impulse travels in one direction only in the nerve because once a section of the membrane has been depolarised it goes through a refractory period, during which it cannot be excited again, while in front of it excitation is possible. By the time the original section of membrane is

* This basic mechanism established for the squid giant fibre has now been accepted as being the means of nerve impulse conduction in other animals including vertebrates. There are great difficulties in applying the techniques used in squid axons to the fine nerves of the vertebrates, but all the evidence we have points to a very similar mechanism.

229

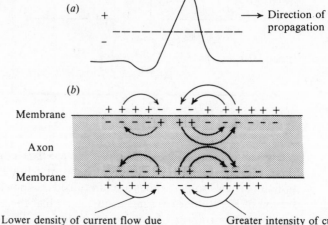

Fig. 10.5.
The propagation of a nerve impulse. (*a*) An action potential (showing changes in potential inside the axon.) (*b*) The local currents set up by the action potential that cause propagation of the impulse. (Note that local currents pass into the axon from all directions as its permeability changes.)

Lower density of current flow due to refractory period prevents propagation backwards

Greater intensity of current flow in forward direction produces excitation

once again capable of excitation (about 1 millisecond later) the impulse has itself moved on. (Artificial stimulation of a neurone will lead to impulses passing in both directions from the point of stimulation, but only those going in the 'correct' direction will be able to pass across the polarised synapse (see p. 231).) The existence of the refractory period also means that there is a limit to the frequency of impulses that a nerve fibre can transmit, this is usually 500–1,000 impulses per second.

Under normal conditions the nerve impulse originates in the cell body or the sensory endings, although, as mentioned above, neurones can be caused to generate impulses anywhere along their lengths by appropriate stimuli. A generator potential precedes the action potential, and this varies in amplitude according to the strength of the stimulus. When the generator potential reaches a critical or threshold value the action potential, which has a fixed value, is triggered and passes down the neurone. The nerve impulse is thus an all-or-nothing response, for weak stimuli that produce a generator potential below the threshold do not trigger an action potential. The fixed value of the action potential also means that an increase in the stimulation at a sensory ending when it is above threshold leads to an increase in the frequency of the impulses that are generated but not a change in their individual nature.

To summarise, the nerve impulse seems to be a wave of potential change generated in special parts of the cell and passing down the whole length of the axon; it is always the same size from a given neurone and varies only in the frequency at which it is propagated. The rate of transmission depends on the cross-sectional area of the fibre and the nature of its external covering. In vertebrates the conduction velocity of the impulses is increased because of a specialisation of the outer sheath of the axon called the myelin sheath. Nerve fibres which have this sheath, called medullated fibres, conduct more rapidly than those without (non-medullated fibres). Fig. 10.2 shows the structure of a medullated fibre and it can be seen that the sheath is interrupted at various places, called the nodes of Ranvier, where the axonal membrane is exposed to the tissue fluid. The sheath cells are rich in lipid and effectively insulate the fibre so that it is only excitable at the nodes, and it has been shown that the impulse is in fact conducted from node to node. This saltatory, or jumping, conduction enables the velocity of conduction to be much increased without enlarging the diameter of the fibre.

10.2.3 *The synapse*

We must now consider what happens when the nerve impulse reaches the end of the axon along which it has been conducted. The endings may either be associated with an effector organ such as a muscle fibre or gland cell, or, if the axon belongs to a sensory neurone or intermediate neurone, will make connections with other neurones. Regions where two neurones are functionally connected are known as synapses. They represent discontinuities of both structure and function. In many ways the endings of motor axons on muscle fibres are similar and have been shown to have properties in common with those of true nerve-to-nerve synapses. In all of these cases electron microscope studies have shown a very close approximation of the two membranes of the excitable cells (10 nanometres).

Despite this very short distance for transmission of the impulse, there is undoubtedly a relatively long delay (0.5–0.9 milliseconds) between the arrival of the nerve impulse at the end of one fibre (pre-synaptic fibre) and the setting up of an impulse in the post-synaptic fibre. Part of this time is taken up by processes whereby small amounts of specific chemical transmitters are liberated which depolarise the post-synaptic membrane and thus set up of an impulse in that neurone. Acetylcholine is the best known of these transmitters and there is no doubt that it operates at the neuromuscular junctions of vertebrate skeletal muscle (Fig. 10.6). One simple experiment that shows this is that poisons known to render the post-synaptic membrane insensitive to acetylcholine block transmission of the nerve impulse. Such a poison is curare, which has long been known as a poison by South American Indians, who used it on their arrowheads. After the acetylcholine has been liberated and has produced its effect at specific sites on the post-synaptic membrane, it is destroyed by enzymes called cholinesterases. Many drugs are known which specifically prevent the activity of these enzymes and consequently have a marked effect on transmission at the neuromuscular junction. Well known amongst these anticholinesterases are eserine and some of the nerve gases such as DFP (diisopropyl fluorophosphonate).

Besides acetylcholine a number of other transmitter chemicals have been isolated; thus adrenaline and noradrenaline (also called epinephrine and norepinephrine) act as transmitters in parts of the sympathetic system.

The synapses dealt with so far have been of the excitatory type, but it should be appreciated that inhibitory synapses are equally important in

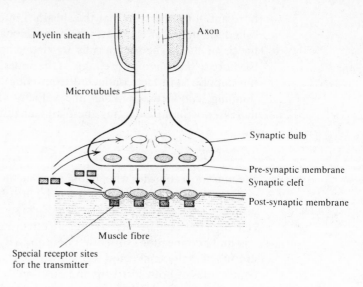

Myelin sheath

Axon

Microtubules

Synaptic bulb

Pre-synaptic membrane

Synaptic cleft

Post-synaptic membrane

Muscle fibre

Special receptor sites
for the transmitter

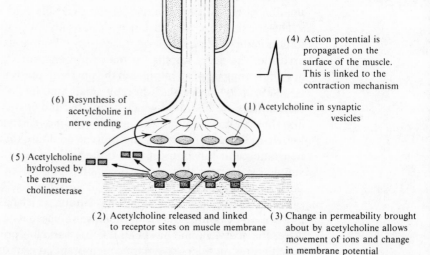

(4) Action potential is
propagated on the
surface of the muscle.
This is linked to the
contraction mechanism

(6) Resynthesis of
acetylcholine in
nerve ending

(1) Acetylcholine in synaptic
vesicles

(5) Acetylcholine
hydrolysed by
the enzyme
cholinesterase

Fig. 10.6.
(*a*) A neuromuscular
junction. (*b*) The way
in which acetylcholine
acts as a transmitter at
a neuromuscular
junction.

(2) Acetylcholine released and linked
to receptor sites on muscle membrane

(3) Change in permeability brought
about by acetylcholine allows
movement of ions and change
in membrane potential

the nervous system. Inhibitory synapses have transmitters that reduce the
permeability of their membranes to the passage of ions; thus instead of
setting going an action potential on the far side of the synaptic cleft these
transmitters actually inhibit it. There is some evidence that excitatory
synapses have thicker pre-synaptic membranes and wider synaptic clefts
than the inhibitory synapses. In the former the cleft is of the order of 0.2
nanometres. Detailed ultrastructure of the synaptic bulb reveals the
presence of microtubules that appear to run in organised arrangements. It
is possible that these provide a means of rapid and direct transport for the
synaptic vesicles loaded with transmitter substance, bringing them from
the inner part of the bulb to the pre-synaptic membrane for discharge.

232

10.2 Units of nervous function

Whether or not a given nerve cell will fire depends on an integration of stimuli from both positive and negative synapses that connect with its dendrites. A further aspect of integrative function is found in the next section.

Transmission at the neuromuscular junction has been investigated in great detail because of its ready accessibility in physiological experiments. But similar studies, such as the use of drugs, have been applied to many parts of the central nervous system (CNS) and to ganglia of the autonomic system. For instance, if the pre-synaptic fibres of a sympathetic ganglion are stimulated, while it is being perfused, acetylcholine can be detected in the fluid leaving the ganglion during stimulation. This is most clearly seen if the perfusing fluid also contains an anticholinesterase such as eserine, that prevents the breakdown of acetylcholine. Notable among the properties of synapses is their polarity. That is, the impulse can only cross the synapse in one direction. Hence, although it is possible under certain conditions for a nerve impulse to pass both ways along an axon, when it reaches a synapse it can only produce an effect on another neurone or motor cell in one direction. This is due to the properties of the two membranes on either side of the synapse.

10.2.4 Integrative mechanisms

The nature of the synapses and the detailed connections between neurones provides one of the most important ways in which integration can take place. Thus only in relatively few cases does any nerve axon make a synapse with only one other axon. Such 1:1 synapses are found (for example between the primary sensory neurones from the fovea of the retina and first-order intermediate neurones), but in such cases the synapses are simply relay stations, each impulse in the pre-synaptic fibre producing one in the post-synaptic element with no integration. More usually, several pre-synaptic fibres have their endings on a single post-synaptic neurone, and in these instances the arrival of a single impulse in one axon is insufficient to excite the post-synaptic cell, presumably because it does not liberate a sufficient quantity of transmitter. However, another impulse arrives very soon after, along either the same axon or, more usually, along other axons ending on the cell, then the post-synaptic element may be excited. This type of connection is illustrated in Fig. 10.7(a) and is called convergence. The effects of impulses arriving along different pathways or the same pathway are said to summate with one

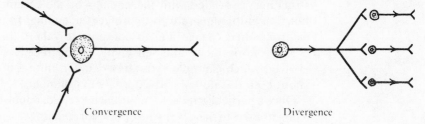

Fig. 10.7.
Two types of connections found in the central nervous system.

Convergence   Divergence

233

another. It is also common to find in the CNS that impulses passing along a single axon may produce effects on a large number of post-synaptic cells. This is called divergence (Fig. 10.7*b*), and well-known examples are found among the giant fibre systems of invertebrates and in the Mauthner neurones found in fishes and some larval amphibians in the spinal cord, where they are effective in initiating sudden swimming by the animal. But such divergent connections have been established in many functional tracts descending from the brain of mammals to the spinal cord.

In considering the CNS in this way it is inevitable that a picture is built up of nerve cells responding only to the arrival of impulses in preceding members of the chain, that may be presumed to have originated from the receptors. That this is not necessarily true is now well established, because many nerve cells throughout the animal kingdom are known to exhibit activity in the absence of any preceding input in the way of nervous impulses. We have seen a clear example of this 'spontaneity' when considering the respiratory centre of mammals and fishes, for even when the medulla is completely isolated some cells continue to discharge with frequencies and patterns almost identical to those which acompany the normal respiratory movements of the intact animal. And the fore brain of vertebrates shows clearly defined electrical activity. In man there is the alpha rhythm at 8–13 cycles per second, the beta rhythm at 14–30 cycles per second, the theta rhythm at 4–7 cycles per second and the slow delta rhythm at 0.5–3 cycles per second.

10.3 Receptor organs

10.3.1 *Basic mechanisms*

In experimental work with the nervous system it is usual to employ small electric shocks for stimulating the axons. This is because such stimuli are easily both measured and varied in frequency and intensity. But nerve fibres and cell bodies can also be excited by mechanical, chemical, light, and temperature stimuli. This applies particularly to some of the branching endings found in the skin of primitive organisms. Such receptors play a role in detecting changes in the external environment of the animal, but the information they communicate to the CNS is not very specific because it may have arisen from any of these types of stimulation. During evolution there has been a division of labour among receptor organs so that they became more and more specialised in the type of stimulus or modality to which they respond. This is usually due to changes in accessory structures associated with the endings of the receptor neurones. The function of this apparatus is to protect the endings from certain types of environmental change and to concentrate others. In this way a high degree of specificity and sensitivity of the receptor mechanisms has evolved which can give extremely detailed information to the animal about both its internal and external environments.

Once a particular receptor neurone is excited, impulses are transmitted along its axon to the CNS. All impulses originating in a given receptor cell

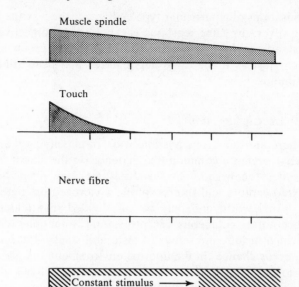

ig. 10.8.
hanges in the
equency of nerve
npulses in a muscle
pindle, touch receptor
nd medullated nerve
bre in response to a
onstant stimulus.

are identical but they vary in the frequency with which they are generated. Frequency is therefore the code, and the relationship between stimulus and frequency is a characteristic property of each receptor. When a sense organ is suddenly stimulated, for example by shining a light or suddenly exerting a pull on a muscle, the initial frequency of sensory impulses is very high but falls away with time, although the stimulus intensity remains constant. This property is fundamental to all sense organs and is called adaptation; it is often due to the accessory structures that are associated with the receptor neurones rather than to the neurones themselves. In some cases the frequency falls very rapidly (Fig. 10.8) and these are said to be rapidly-adapting sense organs. Examples are those concerned with touch and the detection of sudden movements. Other receptors are characterised by a relatively slow decline in the discharge frequency and are classified as slowly-adapting. Such receptors play a vital role in sensing the more or less steady conditions of the organism with respect to its internal and external environments. Examples are the receptors in muscles which detect their tension, or those in a joint which signal the angle at which it is held. In all sense organs the intensity of the stimulus, for example the brightness of a light or the amount of pull exerted on a muscle, determines both the maximum initial discharge frequency in the sensory fibres and its level at all stages in the adaptation curve. In sense organs where there are more than one receptor cell, which is not uncommon, different cells may become excited at different intensities; hence the CNS receives an input that is variable not only in frequency of impulses along a given axon but also in the number of axons transmitting the impulses. Each sensory neurone has connections within the CNS which, because of their nature, give rise to

235

sensations of a particular type. Whatever the way the fibres are excited, however, the same sensation is experienced subjectively. A well-known example is the way pressure on the eye-ball produces the sensation of light just as if the optic nerve fibres had been stimulated by a light stimulus.

10.3.2 *Classification*

There are numerous possible ways of classifying the different types of sense organ; a common one depends on the site of the stimulus which excites the organ. On this basis we may divide sense organs into exteroceptors and interoceptors. Exteroceptors detect changes in the external world and may be sub-divided into teloreceptors (distance receptors), cutaneous receptors, and sometimes chemical receptors (which include the senses of taste and smell). Interoceptors are stimulated by changes in the internal environment and are divisible into proprioceptors stimulated by the position or activity of the body itself), visceroceptors (stimulated by part of the viscera), and chemical interoceptors, such as those detecting the oxygen tension in the carotid bodies.

In another system of classification the nomenclature is based upon the type of environmental change or energy to which the receptor is particularly responsive. On this basis we may distinguish chemoreceptors (smell and taste), mechanoreceptors (touch, pressure, hearing, etc.), photoreceptors (light, e.g. eyes), thermoreceptors (temperature) and undifferentiated endings, which produce the sensation of pain (see p. 276).

In certain fishes it is apparent that there are also receptors which are sensitive to the deformation of small electric fields produced by the fish itself or by other fish. This is true of sharks and dogfish as well as of so-called 'electric fish'. The ampullae of Lorenzini in cartilaginous species may function as sense organs for such electrical stimuli.

Whatever the type and classification of a receptor the basic mechanism is the same. The environmental change is concentrated in some specific way upon nerve endings which produce small changes in potential which in turn generate nervous impulses that are transmitted to the CNS. In all cases the essential change which precedes the generator potential is either chemical or mechanical. Thus, in the eye there is a lens which, together

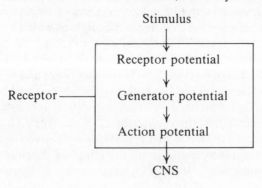

Fig. 10.9.
The sequence of events in the response of a receptor. (In some receptors the receptor and generator potential are one and the same.)

with other devices, focuses an image on the receptive retinal cells where photochemical changes take place. The chemical events excite the receptor cells to produce a slow change of electrical potential which in turn generates nervous impulses. Thus the sequence of events in the receptor may be summarised as shown in Fig. 10.9.

The actual sensations received by the brain seem to be determined largely by the pattern of stimulation of the peripheral receptors, so that a very complex system of sensory input is found to exist.

10.3.3 *Mechanoreceptors*

Touch

Recordings made from small nerves supplying mammalian skin show that both rapidly- and slowly- adapting receptors are present. The former are suitable for the detection of touch and vibration, whereas the latter give information about continuous pressure on the skin. There are at least three different types of sensory ending involved. The first are associated with the base of hairs, particularly on the 'windward side'. These receptors especially those in the vibrissae of some mammals, are extremely sensitive to small movements. Some can detect movements produced by air currents and in this way act as teloreceptors. A single nerve fibre may serve endings at the base of hairs over an area as great as 5 square centimetres. A second type of ending is found on relatively non-hairy regions such as the fingers, where the main cutaneous sense is due to Meissner's corpuscles which lie in papillae which extend into the ridges of the fingertips. The corpuscles consist of spiral and much-twisted endings each of which ends in a knob. A third type of end-organ is the very sensitive and large Pacinian corpuscles which are usually situated quite deep in the body and in the limbs may lie close to the tibia; they are also well known from the gut mesenteries. They have been very well investigated physiologically and those in the limbs probably form the basis of a vibration sense, for they are very rapidly-adapting as well as being highly sensitive; for example, a deformation of 0.5 micrometres if applied for 100 milliseconds will excite them. Their rapid adaptation has been shown to be due to the properties of the complex accessory corpuscle arrangement which protects the end-organs.

The precise relationship between end-organs observed in sections of skin and the different sorts of response recorded from cutaneous nerve fibres has not been established in all cases. As indicated above, a single sensory fibre may be stimulated by movements of hairs over a relatively large area and the responses are rapidly-adapting, but touch receptors are generally more slowly adapting and their receptive areas may be limited to one or more very discrete spots on the skin. Pressure receptors have a much higher threshold and have no sharply localised receptive spots. That pressure reception is due to a deformation of the receptor by a pressure gradient on the skin, and not pressure as such, is indicated by a classical experiment of Meissner. If a finger is thrust into a vessel of

237

mercury the sensation of pressure is only obtained at the interface of the mercury and air, where the skin is deformed by the pressure gradient, there being no sensation of pressure within the mercury.

A great deal of work has been done on the so-called 'two-point threshold', that is, the determination of the minimum distance at which two mechanical stimuli can be distinguished from one another. This distinction can be very fine (one or a few millimetres) in certain parts of the body such as the tips of the fingers, the tongue, and the lips, but in other parts it is relatively poor; for example, on the thigh, arms and neck it may not be possible to distinguish single points applied to the skin at distances of 5 centimetres or more apart. At first it might be thought that the basis for this discrimination would depend upon the size of the areas of skin supplied by the sensory endings. This is certainly involved but the mechanism must be more complex, because the areas which may be discriminated are smaller than the minimum areas innervated by a given sensory fibre. The fields of these sensory endings overlap one another and the pattern emerging from their stimulation is one factor involved in the integration mechanism.

Proprioceptors

These often somewhat neglected receptors are generally slowly-adapting and are represented in the muscles of amphibians and mammals by the muscle spindles and tendon organs. Each spindle consists of a few muscle fibres (intrafusal fibres) with contractile striated portions at their two ends and between which are the sensory endings. The spindles are therefore in parallel with the main mass of the contracting muscle fibres. On the other hand, tendon organs are in series with these fibres. When a muscle contracts and its tension increases, the tendon organs will be excited, but only at a relatively high intensity or threshold and they do not play such an important part in the regulation of body movements as do the muscle spindles. The tendon organs are stimulated during the 'clasp-knife' reflex, which results in the inhibition of an extensor muscle when it is stretched more than a certain amount and protects the muscle from damage due to overloading. The afferent fibres from the muscle spindles are large (20 micrometres) and have a series of spiral turns round the intrafusal fibre. By recording from their nerves it has been shown that these endings are excited either when the muscle is stretched passively, or if the intrafusal fibres themselves contract. The discharge is slowly-adapting and increases in frequency the greater the stretch of the muscles. As will be discussed later, these endings form a vital part of the stretch reflex (see p. 257), which in turn plays an important role in regulating the posture and movements of mammals.

If the motor nerve fibres to the ordinary muscle fibres are excited electrically, the muscle shortens and the discharge of impulses from a muscle spindle is interrupted; in this way reflex excitation of the motor neurones is reduced during the contraction. The contractile regions of the intrafusal fibres are also innervated and have end-plates at both ends.

10.3 Receptor organs

These motor nerve fibres are relatively thin (gamma (γ) fibres, 3–8 micrometres in diameter) and when excited they increase the tension within the muscle spindle and hence augment its discharge frequency. This motor system is therefore able to bias the receptors in definite directions, for example during slow postural movements when the γ efferents are excited by descending pathways from the brain (Fig. 10.10). Contraction of the intrafusal muscle fibres will then alter the input from the spindles in relation to the load that the muscle is working against. Consequently, the ensuing reflex excitation of the main motor neurones of the muscle will be greater or less according to this load and the resulting contraction will be graded to overcome the load itself. This example illustrates the importance of proprioceptors as parts of feedback loops in the control of muscular activities. The fact that the muscle spindle afferent fibres are large in diameter and rapidly conducting ensures that delays in this control loop will be brief. It is useful in reflex experiments because electrical stimulation of the sensory nerves excites these large fibres at the lowest threshold.

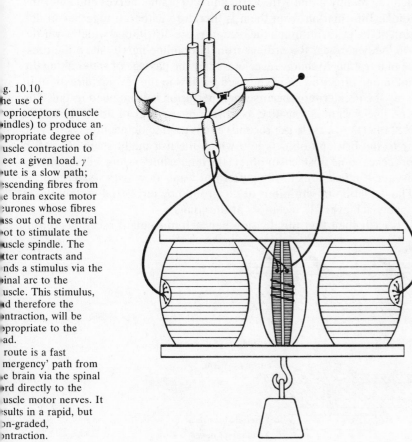

Fig. 10.10.
The use of proprioceptors (muscle spindles) to produce an appropriate degree of muscle contraction to meet a given load. γ route is a slow path; descending fibres from the brain excite motor neurones whose fibres pass out of the ventral root to stimulate the muscle spindle. The latter contracts and sends a stimulus via the spinal arc to the muscle. This stimulus, and therefore the contraction, will be appropriate to the load.
α route is a fast 'emergency' path from the brain via the spinal cord directly to the muscle motor nerves. It results in a rapid, but non-graded, contraction.

γ route

α route

Other interoceptors

Within the vertebrate body there are many receptors which respond to the mechanical conditions of the internal organs. The input from these receptors often plays an important part not only in the control of particular physiological mechanisms, but may also influence certain behaviour patterns of the whole animal. For example, the receptors of the stomach wall may be concerned in the arousal of 'hunger'. Stretch receptors in the carotid and aortic sinuses of tetrapods have important roles in the regulation of blood pressure; endings with similar properties are found in the branchial vessels of fishes.

Vibration sense and lateral line organs

These receptors are intermediate between mechanoreceptors which detect movements of the body and those described as hearing organs which detect higher frequencies of air- or waterborne vibrations. The lateral lines of fishes and some amphibians comprise canals just beneath the surface of the skin which open at intervals to the exterior. Intermittently within each lateral line canal there are groups of sensory hairs called neuromast organs. These sensory hairs are innervated by fibres which run mainly in the Xth, VIIth and Vth cranial nerves and enter the medulla. Information from them is therefore gathered together in one region of the CNS, although the receptors are distributed widely over the whole body surface. Recordings from lateral line nerves show the presence of a resting discharge, accelerated by the passage of water along the canal in one direction and decreased by flow in the reverse direction. In the ray the acceleration occurs when perfusion is from head to tail. The lateral line system is sensitive to vibrations in water of up to about 100 hertz (1 hertz = 1 cycle per second) and is used to detect objects moving near to the fish and objects into which the fish might swim. As a fish approaches some obstacle or object the lateral line system will detect any consequent alterations in the pattern of water flow over its surface.

The detection of vibrations of the ground by terrestrial vertebrates is probably achieved by receptors at the joints.

It is clear then that the division between movement receptors, prop-

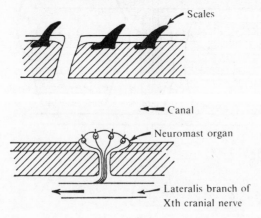

Fig. 10.11.
Longitudinal section of the lateral line organ of a dogfish.

240

10.3 Receptor organs

rioceptors and other mechanoreceptors involved in hearing etc., is not a sharp one and that they form a continuous spectrum. However because of its separation as an organ of special sense it is convenient to discuss the ear separately.

10.3.4 *Sound and equilibrium receptors*

The ear: equilibrium reception and hearing
The inner ear of vertebrates develops from a placode (epidermal thickening) to form a hollow otocyst which subsequently differentiates into two parts – a lower region (pars inferior) which forms the sacculus and lagena (cochlea in higher tetrapods) and an upper region (pars superior) composed of the utriculus and three semicircular canals (only two in some cyclostomes). The whole structure (Fig. 10.12) represents a portion of the lateral line system which has become closed off from the exterior except for the ductus endolymphaticus, which remains open in fishes. The semicircular canals have small groups of hairs, corresponding to the neuromast organs which detect movement of the fluid (endolymph) within the labyrinth (the system of canals of the inner ear). Each canal is oriented at right angles to the others and its neuromasts respond to angular accelerations in the plane of the canal. The neuromast organs are contained within the ampullae at one end of each of the canals (Fig. 10.13a). Their sensory hairs are embedded in a mucilaginous cupula which functions as a pendulum that is swayed by movements of the endolymph.

Ultrastructure investigations on the sensory cells have revealed a complicated mechanism for converting the lateral forces from the movement of the cupula to vertical deformations of the receptor cell. Fig. 10.13(b) shows a single hair cell. From the top of the cell project a number of stereocilia attached to a cuticular plate which is itself embedded in the surface of the hair cell. A single, longer kinocilium takes its origin from the cell cytoplasm rather than from the cuticular plate. When lateral

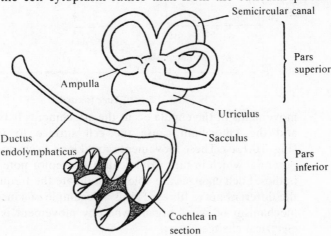

Fig. 10.12.
The inner ear of a mammal.

241

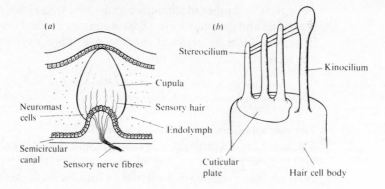

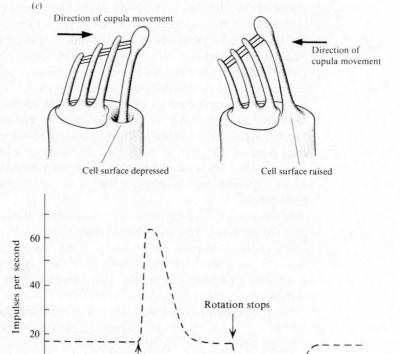

Fig. 10.13.
(*a*) Section through an
ampulla. (*b*) A sensory
hair. (*c*) The response
of sensory hairs to
movements of the
cupula.

Fig. 10.14.
The responses of a
sensory cell in the
semicircular canals to
rotation.

movements of the cupula occur the attachments between the stereocilia
and the kinocilium cause the cell surface to move up or down
(Fig. 10.13*c*). These movements could lead to changes in the membrane
potential which in turn would cause a receptor potential in the hair cell
bodies. Such changes are able to modulate the frequency of impulses in
the different nerve fibres. This is an example of a mechano-transduction
mechanism of sense organs whereby movement is converted into an
electrical change.

242

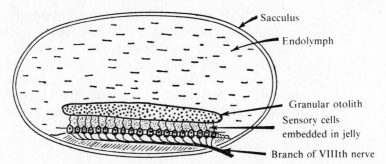

Sacculus

Endolymph

Granular otolith

Sensory cells embedded in jelly

Branch of VIIIth nerve

g. 10.15.
ransverse section of a
acula.

The sensory cells of the ampullae fire off a continuous resting discharge. When the receptors in a given plane are stimulated by movement that causes their rotation the discharge increases, though as constant velocity is reached it returns to its original level. With the cessation of movement the resting discharge immediately dies away, but then returns to its resting level. These changes are shown in Fig. 10.14.

Neuromasts are also modified to form maculae (Fig. 10.15). Their sensory hairs are embedded in mucilage which contains calcareous bodies or otoliths, which may be quite large in some fishes. The degree of stimulation of the sensory hairs of the maculae depends on the position of the labyrinth and they therefore provide tonic information essential for orientation of the body with respect to gravity. They will also respond to linear acceleration of the body because of the inertia of the otoliths. The difference in density between the calcareous bodies and the surrounding medium means that the otoliths will also vibrate when subjected to sounds. Like the ampullae the maculae also show a pattern of resting discharge which is increased or decreased according to the nature of the stimulus.

In fishes the maculae are known to detect vibrations of higher frequency than those which stimulate the lateral line organs. Some fish can detect even higher frequencies (10,000 per second) by the use of the swim-bladder wall, which vibrates when sound waves are passed through the body and, because of its resonant properties, acts as an amplifier. In fish such as roach, tench, and minnows there is a series of bones (Weberian ossicles) which transmits these vibrations from the swim bladder to the inner ear. These small bones are derived from vertebrae and the sounds are transmitted to the sacculus and lagena. Fish with such connections between the swim bladder and the inner ear may be trained to respond to sounds of higher frequencies than can be detected using the ear alone.

In tetrapods the lagena is specialised to detect airborne vibrations, which are usually transmitted to it from the outside by means of a bone, the stapes (columella auris), which is homologous with the hyomandibula of fishes. The external tympanic membrane vibrates in response to the original sound and then the stapes transmits the vibrations across the middle ear to the fenestra ovalis (oval window), which communicates with the fluid (perilymph) surrounding the inner ear.

243

Nervous coordination

Frogs have been conditioned to respond to sounds of 50–10,000 hertz but they cannot distinguish different pitches or frequencies within this range. In the reptiles an extension of the sacculus forms the beginnings of a cochlea but there is little evidence for pitch discrimination. Despite the small increase in complexity of the labyrinth, birds have a considerable range of pitch discrimination which may extend as high as 25,000 hertz in the pigeon. It is in the mammals, however, that the labyrinth has become modified to such a great extent by the elongation of the cochlea and its coiled structure (Fig. 10.12).

The mammalian ear

Essentially, the mammalian ear consists of (*a*) an outer portion which functions as a sort of trumpet concerned with the collection of the airborne vibrations, (*b*) the middle ear in which lie the three auditory ossicles serving to transmit these vibrations to the fenestra ovalis, (*c*) finally the internal labyrinth with its elongated cochlea, which is the site of the nervous discrimination of the different sounds that have been transmitted to it (Fig. 10.16).

The external ear or pinna, with its direction-finding capacity, is well developed in some mammals such as bats which have very acute hearing; the pinna is sometimes moved reflexly in order to facilitate the collection of sound from a particular direction. The three ear ossicles lie in the middle ear, which communicates with the pharynx by the Eustachian tube (a structure derived from the hyoid gill slit). This tube is normally closed by a muscle but opens during swallowing to allow the pressure on the two sides of the tympanic membrane or ear drum to be equalised. The three ossicles function not only to transmit vibrations of the tympanic membrane but also to concentrate the pressure change on the fenestra ovalis,

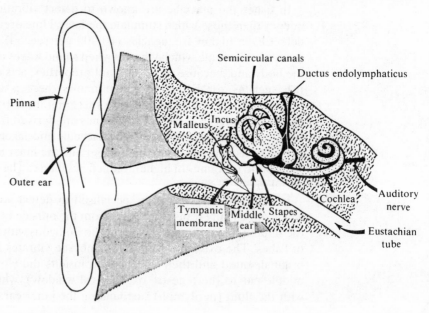

Fig. 10.16.
The human ear.

244

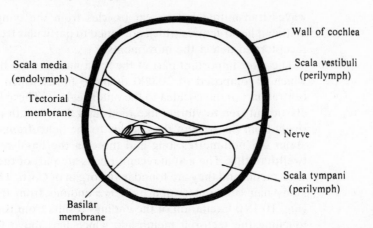

Labels: Wall of cochlea; Scala media (endolymph); Scala vestibuli (perilymph); Tectorial membrane; Nerve; Scala tympani (perilymph); Basilar membrane

Fig. 10.17. Transverse section through part of the cochlea.

with which the inner ossicle (stapes) communicates. The stapes only weighs 1.2 milligrams and the fenestra ovalis on which it acts like a piston has an area of 3.2 square millimetres, which is about $\frac{1}{22}$ the area of the tympanic membrane. Consequently the ossicles act like a hydraulic press which steps up the force of the vibrations twenty-two fold, although their amplitude is correspondingly reduced. In this way the ear absorbs the major part of the sound energy which impinges upon the tympanic membrane and transmits it to the inner ear. There are two muscles in the middle ear which may contract by means of reflexes when very intense sounds fall upon the ear. By increasing the stiffness of the transmission system they decrease its sensitivity and thus protect the internal mechanisms from overstimulation just as the iris protects the eye.

We must now consider in detail the structure and function of the cochlea. The cochlear duct (or scala media) is an extension of the internal labyrinth which is attached to one wall of the bony tube in which it lies. As with other parts of the labyrinth it contains endolymph, whereas the remainder of the tube contains perilymph. This perilymphatic space is divided by a partition (the basilar membrane) into a lower scala tympani and an upper scala vestibuli (Fig. 10.17). Vibration of the fenestra ovalis is transmitted not to the cochlear duct or scala media itself but to the scala vestibuli. Consequently, when sounds are transmitted from the stapes the whole of the fluid within the scala vestibuli is set into vibration. Close to the fenestra ovalis is the fenestra rotunda (round window) with which the scala tympani communicates. The scala tympani and scala vestibuli only communicate through a very fine pore (helicotrema) through which vibrations in the auditory range cannot be transmitted. Consequently, changes in pressure of the scala vestibuli produced by inward movements of the stapes result in pressure increases which can only be transmitted to the scala tympani by movement of the spiral lamina. The scala tympani connects with the fenestra rotunda, which will move outwards. The basic mechanism is illustrated in Fig. 10.18, which indicates the condition of the cochlea when it is drawn out from a spiral condition. The importance of the scala media is clearly that it will vibrate at the same frequency as the

245

waves transmitted by the ear ossicles from the tympanic membrane. Within it lie structures which are tuned to particular frequencies and also receptors to detect the movements.

The most important part of the scala media is the basilar membrane, which is composed of 20,000 or more fibres which project from the central part of the cochlea to the outer regions. These fibres are stiff and elastic but free to vibrate like reeds and their length increases progressively from the base of the cochlea to the helicotrema. The longest are about 0.5 millimetres long and those in the basal regions about one-twelfth of this. The actual receptor cells are part of the cochlear duct or scala media and they are found in the organ of Corti. This is composed of many hair cells innervated by nerve endings from the cochlear nerve (Fig. 10.19). Excitation of the endings results from the hair projections touching the tectorial membrane which lies above them in the scala media. When the basilar membrane vibrates, the hair cells are excited

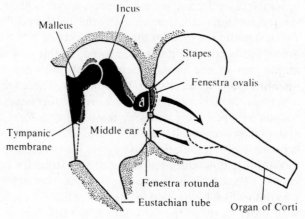

Fig. 10.18.
The route by which sound impulses are transmitted through the middle and inner ear.

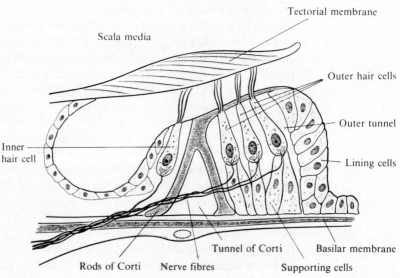

Fig. 10.19.
The ultrastructure of the organ of Corti.

246

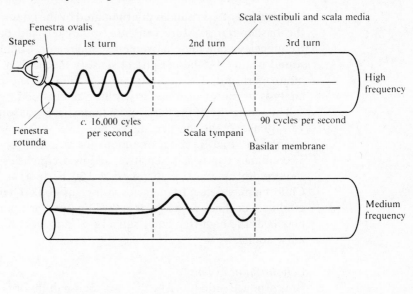

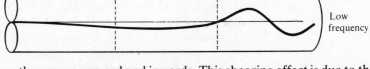

Fig. 10.20. The amplitude of displacement in different regions of the basilar membrane in response to sound of different frequencies.

because they move upward and inwards. This shearing effect is due to the presence of a supporting structure called the rods of Corti. This arrangement leads to an amplification of the pressure changes in the fluids and a greater shearing force on the sensory cells.

Vibrations of the stapes transmitted to the scala vestibuli produce deformations of the fibres of the basilar membrane. Because of their elasticity they bend towards the fenestra rotunda and initiate the passage of a wave down the basilar membrane. This is similar to the way in which a wave travels down a water hose when the end is moved up and down very rapidly. The pattern of these waves varies according to their frequency. At low frequencies (up to 60 hertz) this vibration produces volleys of nervous impulses in the auditory nerve fibres that are synchronous with each wave. As the intensity of the sound is increased the number of spikes in each group increases and hence the pattern of impulses gives information about both the loudness and pitch of the sound. Above 60 hertz, however, vibrations of the basilar membrane are unequal along its length and are maximal in certain regions because of the elastic properties of the basilar fibres. Because of the difference in tension in the basilar fibres and the mass of fluid which needs to be moved, the high-frequency sounds resonate at the base of the cochlea where the fibres are short and the mass of fluid to be moved is slight. Low-frequency sounds travel a longer distance in the cochlea, almost to the opposite end, before moving the basilar fibres, and these longer fibres resonate at lower frequencies (Fig. 10.20).

247

Nervous coordination

Loudness of the sound is discriminated by the intensity of movement of the basilar fibres. More cells are stimulated and discharge at higher frequencies when these movements are greater. Not all of the analysis of sound is done by the peripheral sensory mechanism, although this is a major part of it. The auditory parts of the CNS sharpen the peripheral analysis in ways which are as yet little understood.

The sensitivity of the ear is quite fantastic, for at some frequencies the vibrations of the basilar membrane are only one-hundredth of a nanometre, that is, about one-tenth the diameter of a hydrogen atom. The human ear is less sensitive at low frequencies, being only one-thousandth as sensitive to tones of 100 hertz as it is at 1,000 hertz. Children may detect frequencies as high as 40,000 hertz, but this ability gradually declines with age. It is well known, for example, that children may hear the high-frequency squeaks of bats, but these are inaudible to most adults.

Echolocation in bats

Bats produce sounds in a beam from their mouths or noses of frequencies between 30 and 100 kilohertz. The pattern of the sonar is probably characteristic for each species of bat. In the Vespertilionidae family, to which British Pipistrelle bats belong, the sounds are emitted in very short bursts of between 5 and 10 milliseconds duration. There are some 50 separate pulses per second and the frequencies start high and fall off towards the end of each pulse.

It seems that the pattern of sonar emission may be very complicated and that both a long distance location focus and a fine focus mechanism are employed. By having fast pulses of different frequencies imposition of one set of sound waves on another can occur.

In the family Rhinolophoidea, to which the British horseshoe species belong, the sonar system is different. Pulses are of about 100 milliseconds length and are emitted some six times a second. No frequency variation takes place and these bats make use of the Doppler effect for their echolocation. By this effect the pitch of the sound it emits appears to the bat to increase as it approaches an object and to decrease as it moves away.

Both families of bats described have very large ears to receive the sounds reflected from their own sonar emissions. The cochlea of the inner ear is very large compared with other mammals of comparable size.

Echolocation in marine mammals

It is now known that whales and dolphins produce ultrasonic sounds of between 20 and 200 kilohertz and that these allow them to detect objects around them. These mammals have no vocal chords and the sounds are made by release of air from the lungs into the sinuses of the upper respiratory tract.

The air-filled middle ear allows interruption of the returning sound and it is clear that the hearing of these animals is very acute. More recently it

248

has been shown that seals also use sonar echolocation, especially in darkness.

This extra sense is of obvious biological advantage to these large and fast moving marine mammals that live in an environment often in total darkness.

10.3.5 *Light receptors*

Light forms the part of the electromagnetic spectrum that has wavelengths between 200 and 10,000 nanometres, but within this range human eyes are only sensitive between 400 and 750 nanometres (visible light). Nearly all living organisms can detect and respond to light stimuli and their receptors are able to detect one or more of the following: (*a*) the intensity of the light, (*b*) its direction, (*c*) the pattern of stimulation and the formation of images, (*d*) the frequency of the light wavelengths (i.e. colour), (*e*) the plane of polarisation of the light.

Among the vertebrates the form of the eye is relatively constant (Fig. 10.21), although it is modified a great deal in relation to the particular habitat and mode of life of the organism. In the majority of vertebrates it is the principal sense organ of the whole body. Each eye arises in the embryo as an outpushing of the 'tween-brain which, as it reaches the surface ectoderm, induces a local thickening which sinks in to form the lens of the adult eye. The stalk of the original outpushing forms the optic nerve. Because of its mode of development, the retina is inverted, that is, the light must pass through the nerve fibres before it impinges on the photosensitive elements. As with other receptors the stimulus (i.e. light) results in the passage of nervous impulses along nerve fibres to the CNS. These impulses are generated in the nerve endings of the retina at the back of the eye, and it is their frequency which conveys information on the light intensity. The change of light energy into electrical energy requires the presence of mechanisms for the absorption of the light, and this is the function of specific pigments and highly specialised

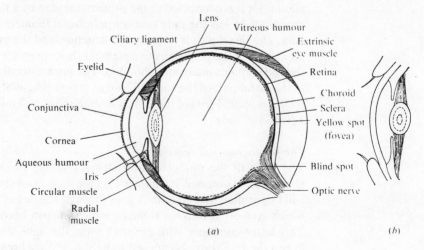

Fig. 10.21. The eye. (*a*) Eye in normal distant focus. The sclera pulls out the lens, giving long focal length. (*b*) Eye in near focus. The ciliary muscles draw the sides of the sclera together taking the strain off the lens, which becomes more spherical.

249

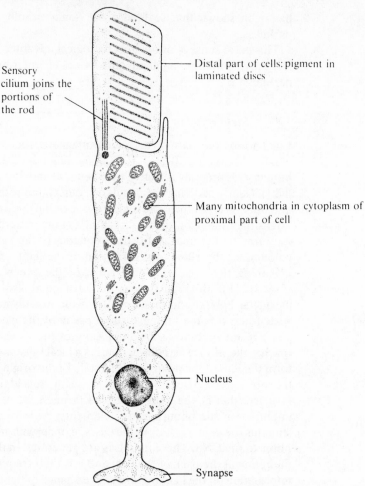

Sensory cilium joins the portions of the rod

Distal part of cells: pigment in laminated discs

Many mitochondria in cytoplasm of proximal part of cell

Nucleus

Synapse

Fig. 10.22.
Ultrastructure of a rod.

rods (Fig. 10.22) and cones of the retina. Electron microscope studies have revealed that the distal part of the rod appears to be a very complex cilium, for it is connected to the proximal region by a thin stalk, which in cross-section has the nine pairs of peripheral filaments characteristic of cilia. The distal part is a laminated structure and the many discs contain molecules of the photosensitive pigment rhodopsin. The proximal part of the rod contains many mitochondria. The photochemical reactions occur in the distal part of the rod and lead to action potentials being generated by the proximal part of the cell. A human retina will have a total of some 120×10^6 rods.

The photochemical reactions
The basis of the photochemical changes taking place in the rods has already been outlined in Chapter 1, in the section on vitamin A (p. 13). It will be recalled that vitamin A goes through a series of chemical changes which convert it to *trans*-retinene which in turn becomes *cis*-retinene. This latter combines with opsin to form the light-sensitive rhodopsin. When the rhodopsin is exposed to light it at once breaks down (via two

250

intermediates) to retinene and opsin once again, and it is thought that this breakdown changes, for a brief instant, the ionic environment in the cytoplasm of the rod. This change is communicated as a stimulus via the cilium that joins the two portions of the rod and leads to generator and action potentials being produced. The biochemical events following a sharp and instantaneous flash of light always take about one-tenth of a second, regardless of the duration of the flash.

There are some 6×10^6 cones in the human retina, mainly concentrated at the fovea, or yellow spot, on the retina at which light is normally focused. Cones require a greater degree of light intensity to stimulate them and they respond to light of different wavelengths.

Three types of cone are known to exist, each containing a different pigment with a light absorption peak in one particular band of the spectrum. Thus the blue-detecting cones contain cyanolabe, which has an absorption maximum at 430 nanometres, the red-detecting cones contain erythrolabe with peak absorption at 575 nanometres, and the green-detecting cones contain chlorolabe with its peak absorption at 540 nanometres (Fig. 10.23). The intermediate colours are sensed by the brain according to the relative strengths of stimuli of one, two or all three types of cone. Thus light of wavelength around 580 nanometres will stimulate the red cones strongly and the green cones much less. The brain interprets such a combination of stimuli as orange. Colour blindness is caused by the absence of one of the types of cones from the retina.

The presence of both rods and cones in different proportions enables vertebrates to see efficiently over a wide range of light intensities in their natural habitat. Diurnal animals generally have a larger number of cones relative to rods, whereas the number of rods is higher in nocturnal animals. Apart from such differences in proportion there are also adaptations whereby the rods and cones can be moved into a protective pigment layer in intense light and they may move relative to one another. In other cases the pigment layers may be mobile.

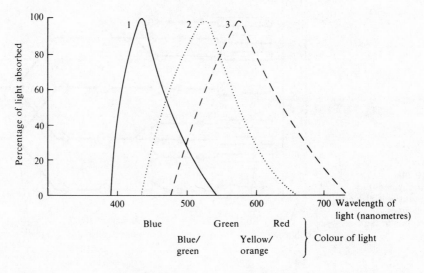

Fig. 10.23.
The absorption spectra
of the three types
of cones:
1 blue-detecting;
2 green-detecting;
3 red-detecting.

251

Acuity of vision

When the rods and cones are excited the nervous impulses they generate pass through a network of fibres and across several synapses before they reach the nerve fibres contained in the optic nerve entering the brain (Fig. 10.24). At the fovea each photosensitive element makes a connection with a single chain of neurones along a sort of 'private' line. The information transmitted is therefore very detailed and confers a high degree of acuity on the visual system. From other parts of the retina, however, several visual elements make contact with a given intermediate neurone and consequently the total acuity is reduced. Moreover the excitation of a single rod or cone may result in impulses passing up several nerve fibres and these same nerve fibres may be excited by other rods. By and large it is only the cones that have 'private lines' to the brain, and the rods have more diffuse connections. It must be remembered that before the light can reach the rods and cones it must pass through all the other layers of the retina.

It now seems likely that all this complexity acts essentially as a filter system which is able to sort out instantaneously changes in the incoming picture which are of survival interest to the animal. A great deal may well be happening in the visual field at any one time which is quite irrelevant to the animal, and it makes sense that a peripherally situated sorting mechanism should exist. A very good example of this is the frog's eye, which is heavily biassed towards detecting flies (or objects of similar size and shape).

The efficiency of the acuity mechanism is increased by the changing size of the pupil and by the presence of the pigmented choroid. The aperture of the pupil is controlled by receptors in the retina that link up via an autonomic reflex arc to the radial and circular muscles of the iris. Changes in the light intensity thus lead to a reflex adjustment of the

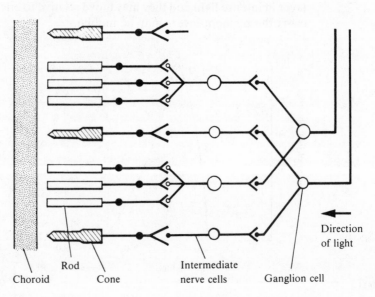

Fig. 10.24.
The arrangement of light-sensitive cells and their nervous connections in the retina.

Choroid   Rod   Cone   Intermediate nerve cells   Ganglion cell   Direction of light

tensions of these antagonistic muscles that maintains an optimum aperture of the pupil. The brain also compensates for changes in light intensity by recording the same intensity of image of the degree of dilation of the pupil. The pigmented choroid prevents the reflection of any of the light back into the eye after it has passed through the retinal sensory system. The absence of this dark pigment (which corresponds to the black inner surface of a camera) leads to a marked fall in visual acuity such as occurs in albinos who hereditarily have no melanin pigment in any part of their bodies. In nocturnal animals this black pigment layer is replaced by a reflecting layer on the choroid called the tapetum. This increases the sensitivity of the eye to light because the rods and cones will be stimulated both as the light passes through to the tapetum and when it is reflected back again. As we have stated above, however, there will certainly be a loss of acuity. Light reflected out through the pupil produces the bright shining eyes of mammals such as cats. In some dogfishes and sharks the tapetum is formed of a layer of cells filled with guanin granules which may be covered or uncovered by a movement of the pigment cells, according to the light intensity. This device has been evolved many times independently; for instance tapeta are also found in the eyes of seals, whales, and various species of fish, carnivores, etc.

The eye as an optical instrument and its aberrations
As an optical instrument the vertebrate eye may be compared to a camera. Light enters the eye via the pupil, is refracted by the lens and is focused on the sensitive retina, where the photochemical reactions take place and the sensory fibres are stimulated. The greatest change in refractive index occurs at the air/cornea interface (the refractive index of the cornea is 0.34), so it is here that the light rays are refracted most (though there is also a considerable refraction at the lens itself). For purposes of discussion it is convenient to think of all these refractive regions as one, and in this case the eye could be represented schematically as a reduced eye. The single lens of such an eye is about 17 millimetres in front of the retina, on which it produces an inverted image. Nevertheless the brain perceives objects with their correct orientation because it can recognise an inverted image as the normal one. The shape of the lens itself may be changed from moderately convex to very convex in order that light rays from distant and near objects respectively can be brought to a focus on the retina. This is known as accommodation.

In the normal eye with all parts relaxed the parallel rays from a distant object are focused on the retina, but in certain conditions these rays may become focused either behind the retina (long-sightedness) or in front of the retina (near-sightedness or myopia). In long-sightedness when the eye accommodates and makes the lens more convex, a distant object can be brought to focus on the retina. So long as the muscles concerned in accommodation do not have to contract too strongly, the lens is still able to become even more convex and so enable the individual to focus on nearer objects. Where the far-sightedness is extreme, however, the parallel rays may only be brought to focus on the retina by the insertion of another convex lens in a pair of glasses. In myopia the eye is unable, even when completely relaxed, to focus distant objects at the retina. This may be because the eyeball is too long or the lens system is convex. The eye cannot compensate for these deficiencies and it is only by the insertion of a concave lens in front of the eye that distant objects can be brought into focus. In myopia there is definitely a 'far point' at which vision is no longer acute because the rays are not brought into focus at the retina, just as there is a 'near point' when the accommodating mechanism can no longer focus the object. Of course the eye always has a near point, even if the person is long-sighted. In general, though, the near point of long-sighted people is farther from the eye than it is in people with normal eyes, and in old age this becomes even more accentuated.

Astigmatism is due to refraction errors in the lens system of the eye, usually caused by an oblong shape of the cornea or the lens. Thus the refracting surface may be the shape of the bowl of a spoon, with two planes at right angles to each other having different curvatures. Under these conditions it is possible that, for example, objects in one of these planes may be brought to focus on the retina whereas objects in the other are focused in front of the retina. In fact all combinations are possible; one of them is called mixastigmatism, where distant objects in one plane are focused behind the retina and in the plane at right angles they are focused in front of the retina. Correction for astigmatism of this common type can be achieved by the use of a cylindrical lens so that the refractive power will be altered in one direction and not in the direction at right angles to it. Spherical lenses in front of an astigmatic eye can bring the rays that pass through one plane into focus on the retina, but can never bring all the light rays into focus at the same time.

The role of the brain in vision

The information from the retina is brought by the optic nerves to the visual cortex at the back of the brain, the right eye sending information to the left part of the cortex and the left eye to its right part. The brain interprets this information as vision and has various tricks for picking out meaningful patterns from a limited primary realm of information. It produces a consistency of image size so that objects half as near to the eye do not double in size but are discerned as having their actual dimensions. The information derived from the retina and unscrambled in the cortex will then be interpreted by the frontal lobes of the brain.

Variations from the typical mammalian eye

A number of mammalian herbivores, as well as some species of fishes, have evolved a non-spherical eye with a 'ramp retina'. Because of the shape of the retina the lens is able to focus simultaneously on different parts of the retina the light from distant objects and from objects near to the animal (Fig. 10.25). This is obviously a useful adaptation for herbivores, that are thus able to see both the grass they are eating and the

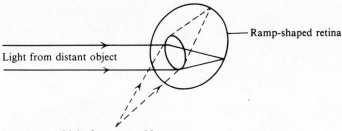

Light from distant object

Ramp-shaped retina

Light from near object

Fig. 10.25.
The herbivore eye,
showing how the ramp
retina allows light from
distant and near
objects to be focused
simultaneously.

environment around them. They also have the eyes in the side of the head, which gives them panoramic vision.

It should be noted that in frogs and some species of fish focusing of the eye is achieved not by changing the focal length of the lens, as has been described for the mammal, but by changing the distance between the retina and a fixed lens. This, of course, is the way in which a camera works. Some fish which swim at the water surface may even have two lenses, one suitable for vision in air and the other for vision in water.

10.4 The central nervous system

As described above, the central nervous system (CNS) consists of a great mass of nervous tissue lying between the receptor and effector organs – which constitute, respectively, the input and output elements of the system. The CNS serves to modify the relationship between input and output but also initiates a great deal of nervous activity on its own. In all higher animals behaviour is largely determined by such central control rather than by activities in the peripheral nervous system.

The CNS develops as a dorsal tube which is originally undifferentiated along its length. However, because of the modification of the head region for feeding and the presence of the major sense organs, the anterior region became differentiated as a brain; the remainder forms the spinal cord.

The arrangement of the neuronal pathways within the spinal cord is less complex than that of the brain region and it retains more of the segmented condition of ancestral chordates. In the brain many complex 'supra-segmental' systems are superimposed on the basic pattern. It is thus easier to describe the arrangement of the neurones and their pathways within the spinal cord before considering the complex brain.

10.4.1 *The spinal cord*

Functional components and the anatomical basis of the reflex
In the centre of the spinal cord is a small canal which contains cerebro-spinal fluid. The central part of the cord itself is the grey matter, composed of cell bodies of neurones and their dendrites, while the outer region is the white matter that consists of axons passing up and down the cord. Some of these axons run for short distances but others form recog-

255

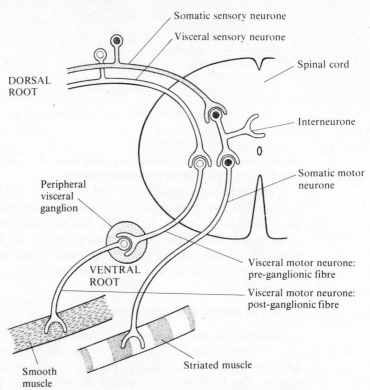

Somatic sensory neurone

Visceral sensory neurone

Spinal cord

DORSAL
ROOT

Interneurone

Somatic motor
neurone

Peripheral
visceral
ganglion

VENTRAL
ROOT

Visceral motor neurone:
pre-ganglionic fibre

Visceral motor neurone:
post-ganglionic fibre

Striated muscle

Smooth
muscle

Fig. 10.26.
The connections of the
somatic and visceral
neurones through the
spinal cord.

nisable tracts descending from the brain and carrying information to the spinal nerve centres. There are also tracts of ascending axons which transmit impulses to the appropriate regions of the brain.

From each segment of the cord a ventral root leaves, made up of motor fibres carrying impulses to the skeletal musculature (the somatic-motor component) and to the gut and other derivatives of the embryonic lateral plate (the visceral-motor component). In the nerve cord and ventral root the visceral fibres are placed dorsally to the somatic ones. In the dorsal region of the cord the segmented dorsal roots carry sensory fibres which bring information from the skin and proprioceptive organs of the skeletal muscles (the somatic-sensory component) and others which transmit sensory impulses from the smooth muscles and gut (the visceral-sensory component). (The dorsal roots of some lower vertebrates also contain visceral-motor fibres, but in the higher vertebrates these are entirely in the ventral roots.) The arrangement is shown diagrammatically in Fig. 10.26.

In all higher vertebrates the dorsal and ventral roots join soon after their exit from the cord to become a single spinal nerve carrying both motor and sensory fibres. In the body region of some lower chordates (e.g. lampreys and amphioxus) and in the head of all chordates, the two roots remain separate – either completely or for long distances.

There are thus two reflex pathways running within the spinal cord – that concerned with the skeletal muscles and their coordination, called

256

the somatic arc, and that concerned with coordination of smooth muscles etc., called the visceral arc. The latter forms part of the autonomic system (p. 260).

Spinal reflexes

The main functions of the spinal cord can be illustrated by reference to the reflexes which exist when the cord is severed from higher centres by cutting it behind the medulla of the hind brain (i.e. in a spinal preparation). A simple example is the so-called stretch reflex (Fig. 10.27) by which muscles that have been stretched contract more strongly because of

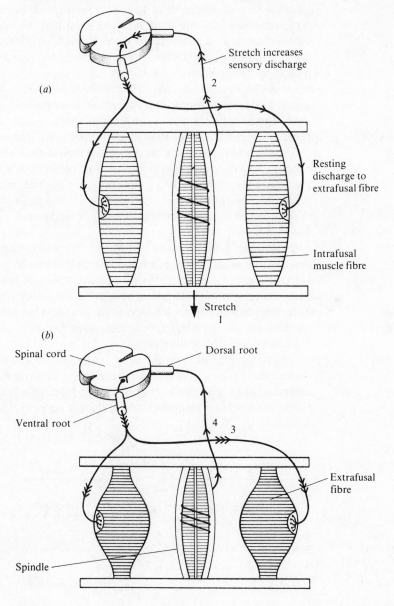

Fig. 10.27.
The stretch reflex:
(a) stretch, (b) result.
The numbers indicate
the sequence of events
and the frequency of
nerve impulses is
indicated by the
arrowheads.

excitation through a spinal path. An everyday example is the reflex producing the knee jerk, produced when the tendons of the knee cap are stretched and thus cause excitation of the muscle spindles, resulting in mechanical contraction of those muscles to which they are attached. In such a reflex only two neurones need be involved: the afferent (sensory) neurone conveying impulses from the stretch receptors (which are the muscle spindles in the body of the muscle) and the efferent (motor) neurone which produces the muscular contraction.

Another spinal reflex is the flexion reflex. For example, if the toe of a spinal cat, that is one with the spinal cord severed behind the medulla, is pinched, its limb will be flexed away from the point of stimulation. At the same time as it lifts one limb, it can be seen that the opposite one of the pair is extended – a natural reflex which would stop it falling over. This is the crossed extension reflex. From detailed investigations of the flexion reflex by microelectrode techniques, it appears that at least three neurones are involved – the afferent, intermediate neurone (within the spinal cord) and efferent. This is a simple example and it must be realised that normal reflex pathways are often more complex, especially with regard to the number of intermediate neurones that are involved.

Classical work on the reflexes of the spinal cord and the integrative properties of the nervous system was done by Sir Charles Sherrington and his associates in the early part of this century. They established certain principles about reflex mechanisms that are still valid in spite of the later application of far more refined techniques, and some may be summarised as follows:

(*a*) A single muscle can be the effector organ for many different reflex arcs (i.e. its motor neurones form the final common path).

(*b*) If a given sensory pathway exciting a muscle is stimulated then a certain strength of contraction results; if another pathway is excited a different strength of contraction is produced. This is because different proportions of the total motor neurones innervating the muscle are excited by the two pathways.

(*c*) Stimulation of both pathways simultaneously either gives a total contraction which is greater than the sum of separate contractions (spatial summation) or it gives one that is smaller (occlusion). The former is due to the excitation of neurones which do not receive threshold excitation

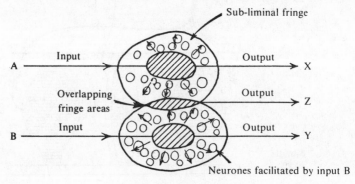

Fig. 10.28.
Spatial summation.
Input A gives output
X; input B gives output
Y; input A+B gives
output X+Y+Z.

from either pathway alone (Fig. 10.28), whereas the latter results from an overlap of the neurones which were excited to supra-threshold by both pathways and therefore summation is not possible.

The above shows in itself how complicated can be the performance of a given effector system following stimulation through spinal reflex arcs.

Spinal reflexes, posture, and locomotion

(*a*) *Tetrapod locomotion.* By analysing the movements of a four-legged animal, such as a dog, when it is walking slowly, it is possible to see how a combination of the spinal reflexes is involved in producing the normal walking rhythm. In this rhythm the legs move in the order right fore, left hind, left fore, right hind, right fore, etc. Thus when one leg is lifted the opposite leg of the pair must support a greater proportion of the body weight, as must also the other limb on the same side. In order to achieve this the stretch reflex is called into action, so that the extensor muscles of the three legs on which the dog is standing increase their activity in proportion to the weight they support. In addition to the stretch reflex the crossed extensor reflex also operates, increasing the extensor action of the contracted limb, and a further type of reflex, called the ipsilateral extensor reflex, increases the tension in the extensor muscles of the leg on the same side. All three reflexes operate in the normal walking of the dog.

Of vital importance in all of these reflex responses is the inhibition by one muscle of the antagonistic muscle in the same joint. Thus stimuli which result in the excitation of a given flexor muscle have the effect of inhibiting activity in the extensor of the corresponding extensor muscle. The basis of this reciprocal inhibition of antagonistic muscles depends on the different effects on the motor neurones of the excitatory and inhibitory inputs to the CNS. The former tend to depolarise the soma membrane and thus to produce excitation, whereas inhibitory pathways increase the polarisation and therefore hinder the propagation of an impulse.

The mechanisms for controlling walking described above may be compared with those involved in the control of respiration (see p. 74). In both cases there is centrally directed spontaneity of action dependent on built-in patterns of coordination. These innate patterns excite appropriate descending pathways in the brain and so cause the activity.

(*b*) *Swimming.* Just as the locomotory rhythms of a tetrapod can be interpreted in terms of spinal reflexes, so the swimming movements of fishes involve reflexes of both segmental and intersegmental types. If the spinal cord of a dogfish is cut behind the medulla, the fish shows swimming movements which may persist for many days without stopping. These movements may be inhibited if the fish touches the bottom of the tank, and are increased in amplitude by putting a clip on the tail – clearly showing that sensory input from the skin affects these movements. The receptors involved in the coordination of normal rhythm are mainly situated in connective tissue between the skin and the muscle, and if all sensory input to the cord is removed by cutting the sensory nerves (i.e.

cutting all the dorsal roots) then the spontaneous swimming movements cease.

From experiments of the above type on vertebrate locomotion we are led to the conclusion that the spinal cord contains all the reflex pathways that, when they interact with each other, produce normal locomotion. Of course these reflex paths are not independent of the higher centres in the intact animal and these higher centres are necessary for the finer gradations of movement, as well as being needed to assess the appropriateness of responses and in many cases to initiate them.

The autonomic nervous system and visceral reflexes

The motor component of visceral reflex arcs, whose path through the spinal cord has already been described, makes up the autonomic system. This system coordinates the activities of the viscera which take place below the conscious level and it can be subdivided into the sympathetic and parasympathetic systems (Fig. 10.29).

In the mammals the distinction between the two systems is clear both structurally and functionally. In both there are ganglia outside the CNS. At each ganglion a pre-ganglionic neurone (i.e. one leading from CNS to the ganglion) synapses with a non-medullated post-ganglionic neurone (that is, one leading from the ganglion to the organ (e.g. heart, intestine) innervated).* There is, however, an anatomical distinction between the sympathetic and parasympathetic systems in the relative lengths of the pre- and post-ganglionic fibres. In the sympathetic system the ganglion is near the CNS but in the parasympathetic – for example, the vagus – the ganglion is located near the organ innervated. Some of the ganglia of the sympathetic system are in an approximately half-way position. One of the results of this arrangement is that the effects of parasympathetic stimulation tend to be very localised, while sympathetic excitation spreads more widely over the area innervated. This spread is also a feature of the grosser structural arrangements, as the outflow of the sympathetic system extends throughout the thoracic and lumbar regions of the spinal cord, whereas the parasympathetic fibres have their exits only in the cranial and sacral regions of the CNS. The spread of sympathetic stimulation is further encouraged by the passage of some pre-ganglionic neurones up and down the sympathetic chain which joins some of the sympathetic ganglia.

Functionally the two systems differ in the nature of the transmitter substance released at the endings of the post-ganglionic neurone (Fig. 10.30). In both parasympathetic and sympathetic ganglia the preganglionic fibre are cholinergic, i.e. liberate acetylcholine to excite the post-ganglionic neurone. But while parasympathetic neurones produce acetylcholine at their endings on smooth muscle or glands, in the sympathetic system the post-ganglionic neurone is adrenergic, i.e. liberates

---

* The visceral-motor components of the cranial nerves which supply the muscles of lateral plate origin operating the visceral arches have no ganglia, do not form part of the autonomic system and are sometimes distinguished as the special visceral-motor component.

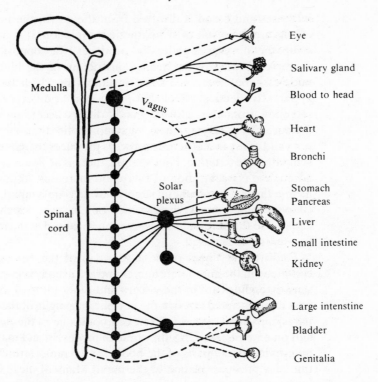

Eye

Salivary gland

Blood to head

Heart

Bronchi

Stomach
Pancreas

Liver

Small intestine

Kidney

Large intenstine

Bladder

Genitalia

Medulla

Vagus

Spinal
cord

Solar
plexus

Fig. 10.29.
The autonomic nervous
system:
— sympathetic;
, sympathetic ganglia;
−− parasympathetic;
−< parasympathetic
ganglia.

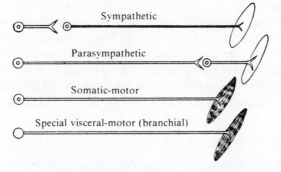

Sympathetic

Parasympathetic

Somatic-motor

Special visceral-motor (branchial)

Fig. 10.30.
The various types of
nerve–muscle systems.
The black
post-ganglionic
sympathetic neurone is
adrenergic; all the
other neurones are
cholinergic.

the adrenaline-like noradrenaline. However sympathetic fibres to sweat glands in man are cholinergic. The adrenal gland itself is a sort of giant post-ganglionic neuronal system liberating vast quantities of adrenaline into the blood system, and it generally acts in conjunction with sympathetic stimulation.

The most important difference between the two systems, which is clearly related to their structure, is that the parasympathetic innervation is more localised and produces more discrete responses than does the sympathetic system. The sympathetic generally functions as a whole and its effects may be summarised by considering the reactions of the body which take place in a state of emergency. Under these conditions the heart rate is increased, the blood pressure is raised, the respiratory rate increases and blood is diverted from the gut to the muscles. All these effects are the result of sympathetic stimulation. It is true that the parasympathetic system generally produces the opposite effects – for example, slowing of the heart, constriction of the blood vessels to the muscles – but there are some instances in which both systems have apparently identical effects but play a part at different times. Thus, the secretion from the salivary glands differs according to whether it is produced by parasympathetic or sympathetic stimulation. Similarly, it is not valid to say that adrenaline always produces the same reaction as does sympathetic excitation. For example, there is no sweat secretion when the adrenaline concentration of the blood increases, although these glands have a distinct sympathetic supply but no parasympathetic innervation. Blood contains marked quantities of enzymes which destroy acetylcholine, so that its effects are inevitably localised to the regions where this transmitter is released.

The afferent fibres (visceral-sensory) of the autonomic system are contained in the autonomic nerves and are usually non-myelinated. Their sensory endings are in the viscera and other parts of the body and they pass uninterrupted through the peripheral ganglia of the sympathetic and parasympathetic systems. Their cell bodies lie in the dorsal root ganglia and on entering the CNS they branch profusely and may have effects on many widespread parts of the body; for example, stretching of the intestine may produce dilation of the pupil! Many of these visceral afferents perform vital roles in the functioning of other physiological systems such as respiration and circulation. Notable examples are the afferent fibres which arise from the carotid sinus and carotid body and enter the brain stem in the IXth cranial nerve. These receptors detect, respectively, dilation of the carotid artery and changes in the oxygen and carbon dioxide content of the blood. The integration of autonomic functions is maintained by specific paths of the CNS, and in this the hypothalamus is of very great importance, as will be discussed later (p. 278).

10.4.2 *The brain*

The brain forms as a series of enlargements at the anterior end of the neural tube. These enlargements develop to process information gathered by the main sense organs which, as in all bilaterally symmetrical animals, become concentrated at the front of the animal: it is interesting to note that in the primitive chordate amphioxus, for example, which has relatively few sense organs anteriorly, there is only a slight enlargement of the neural tube to form a brain. The enlargement is mainly due to an increase in the number of nerve cells but also to the greater number of nerve fibres which enter this part of the spinal cord from the teloreceptors (distance receptors). The regions where these nerve endings are to be found form the primary sensory centres of the brain, where the initial sorting out of the information takes place.

During the evolution of the vertebrate brain we find, however, that new association, or correlation, areas appear which are quite distinct from the primary sensory areas (although it is possible these primary sensory areas may also have some correlation functions). In the correlation centres information from several different sense organs is gathered together and analysed. As a consequence of these processes, which as yet are little understood, descending motor pathways are excited and set into action movements which are related to the pattern of sensory stimulation to which the body is subjected. The association centres also function as storage elements and form the sites of memory, and it is here that the possibilities of learning are mainly found.

As mentioned previously the brain is in a state of constant electrical activity and most of the effector activity taking place is actually initiated from the brain. The brain is thus much more than a set of intermediate nerve circuits between input and output systems.

Development

In early vertebrate embryos there are usually three swellings of the anterior part of the neural tube, forming the fore, mid and hind brains. A little later the fore brain becomes constricted into two regions: an anterior end brain or telencephalon, and a 'tween brain or diencephalon. The mid brain or mesencephalon does not subdivide, but the hind brain forms the anterior metencephalon and posterior myelencephalon (Fig. 10.31).

A further process which takes place in higher vertebrates is that these regions do not retain their primitive arrangement but the axis of the brain becomes bent, typically in three places. This results in a more compact arrangement, particularly in mammals, where the cervical flexure results in the axis of the brain being almost at right angles to the axis of the spinal cord. In the later development of most mammals this flexure becomes reduced but in man it persists in the adult.

The central canal of the spinal cord continues into the brain where it forms the ventricles, which contain cerebrospinal fluid as does the spinal canal. This fluid is produced by the vascular choroid plexi which develop in the thin roof of the 'tween brain and medulla. The whole CNS is covered by an inner membrane or meninge (the pia mater) which is very vascular and helps to form the choroid plexi. Outside the pia mater is the arachnoid layer which forms a delicate web-like meshwork (hence its name). The cerebrospinal fluid fills in the space between the arachnoid

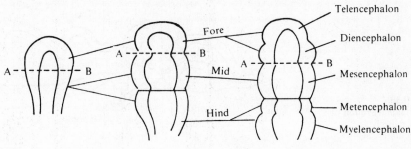

Fig. 10.31.
Development of the vertebrate brain.

263

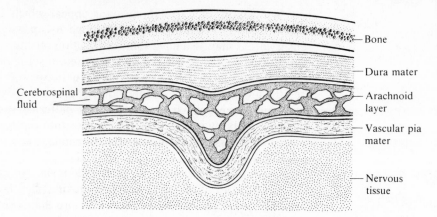

Fig. 10.32.
The membranes
covering the central
nervous system.

Bone

Dura mater

Cerebrospinal
fluid

Arachnoid
layer

Vascular pia
mater

Nervous
tissue

layer and pia mater and forms a cushion about the CNS as well as serving
to convey oxygen and nutrients to the inside when it is secreted at the
choroid plexi. The outer meninge is the tough dura mater which protects
the entire CNS. The arrangement of membranes covering the CNS is
shown in Fig. 10.32.

Evolution of the brain

As we have seen, the brain of amphioxus is scarcely visible, but in
cyclostomes and fishes the brain is well-marked and has the same divisions as in man. In these lower forms the main subdivisions of the brain
retain their primitive association with the important sense organs. Thus
the telencephalon may be referred to as a 'smell brain', the mid brain as a
'sight brain', and the myelencephalon as the 'ear brain'. The relative sizes
of these different portions of the brain are correlated with the degree of
development of these teloreceptors and their importance in the life of the
animal. A longitudinal section of a generalised vertebrate brain is shown
in Fig. 10.33.

As we ascend the vertebrate series there are ndications of increasing
cephalic dominance (from the Greek *kephalē*, head) as shown, for
example, by the effects of severing the brain from the rest of the CNS. If
the spinal cord is cut just behind the medulla in man or a mammal there is

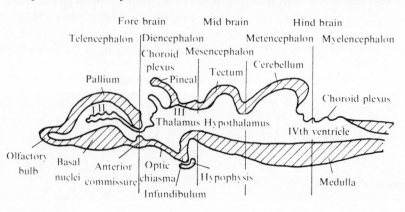

Fig. 10.33.
Longitudinal section of
a generalised
vertebrate brain.

Fore brain   Mid brain   Hind brain

Telencephalon   Diencephalon   Metencephalon Myelencephalon
Choroid   Mesencephalon
plexus   Cerebellum
Pallium   Pineal   Tectum

Choroid plexus

III
III
Thalamus Hypothalamus

IVth ventricle

Olfactory
bulb   Basal
nuclei   Anterior
commissure   Optic
chiasma   Hypophysis   Medulla

Infundibulum

complete loss of movement, whereas, as we have already mentioned, in a dogfish there persists a rhythmic swimming which may continue for several days. The hind brain remains relatively unchanged in the vertebrate series for it is mainly concerned with the control of visceral functions such as respiration and circulation. It has evolved in relation to the changes in these systems but there is no fundamental change in function which might give rise to alterations in the external form. The cerebellum, which forms part of the metencephalon, does increase in size in birds and mammals, where it plays a large part in the control of delicate movements. It is also quite large in some fishes, where movement must be oriented accurately in all three planes. The mid brain remains as a centre for optic stimuli throughout the vertebrates, and in fishes the roof or tectum forms the most important correlation centre, where tracts converge from the olfactory and vestibular parts of the brain. In fishes, amphibians and reptiles there are usually two optic lobes but in mammals these are replaced by the corpora quadrigemina.

It is the fore brain, however, which changes most markedly in the evolution of vertebrate brains (Fig. 10.34). The fore brain of cyclostomes shows a relatively slight enlargement at the front end of the neural tube but in fishes and amphibians there is an enlargement of the roof and/or floor of this region which increases in mammals and birds. Enlargements of the roof or pallium lead to the great expansion of this region in the

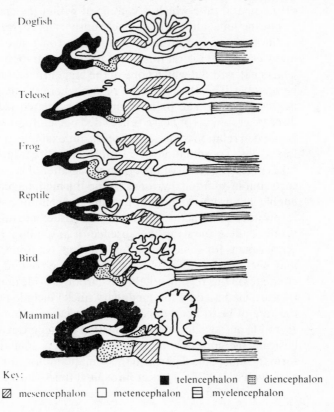

Dogfish

Teleost

Frog

Reptile

Bird

Mammal

Key:

■ telencephalon   ▦ diencephalon

▨ mesencephalon   ☐ metencephalon   ☰ myelencephalon

Fig. 10.34.
Regions of the
vertebrate brain.

mammals so that it covers the whole of the mid brain and forms a major part of the brain (80 % by weight). The floor of the telencephalon becomes enlarged in the birds and forms the corpus striatum. In both cases this great development of regions within the fore brain is correlated with its increasing importance as an association centre. In the lower forms it is concerned with olfactory sense alone but in the reptiles are found the first signs of pathways from the mid and hind brains entering the roof of the fore brain. These and other projection areas are enormously developed and have been mapped in great detail in man and other mammals. Regions concerned with motor functions have also been mapped, by the application of small electric currents through fine electrodes. Another general feature of fore brain evolution is the development of the cerebral cortex. The primitive position of the cell bodies in the CNS is near the central canal, but in the brains of higher vertebrates many are also found in the surface of the brain. The number of cells in these regions is enormous and the possible ways in which the information entering through the sense organs can be sorted appear to be almost infinite.

The main regions of the brain and their functions

The precise position at which the brain begins and the spinal cord ends is not clearly defined, for the medulla oblongata passes imperceptibly into the spinal cord. In general it may be taken that the brain includes that part of the neural tube which is contained within the cranium.

The medulla, although relatively unspecialised, forms a most vital part of the brain of vertebrates because it is concerned with the control of many essential visceral functions such as respiration, circulation and the heart beat, without which none of the higher functions of the brain would be possible. The medulla is the region where the majority of the cranial nerves (V–XII) take their origin and associated with these nerves are discrete groupings of nerve cells called nuclei. In some cases the medulla is lobed in relation to these, as occurs, for example, in the vagal and facial lobes. The coordination of respiratory and cardio-vascular responses takes place as a result of the bringing together of inputs of many different sorts, but in addition neurones within the medulla have properties which enable it to produce rhythmic outputs along motor fibres that result in the respiratory movements. The medulla of fishes and higher vertebrates can produce these bursts in the absence of any inflow from the peripheral sense organs for some time after isolation. Neurones within the medulla may also respond directly to changes in the content of the blood which circulates in this region, notably its carbon dioxide tension. Other reflexes that are mediated by the medullary nuclei include those which regulate the rate of heart beat, degree of constriction of the capillaries, salivary secretion and swallowing. The position of the nuclei and nerve endings is basically the same as in the spinal cord: that is, somatic-sensory neurones form the most dorsal part and the somatic-motor components the ventral regions (Fig. 10.35). The cerebrospinal fluid which forms in the anterior

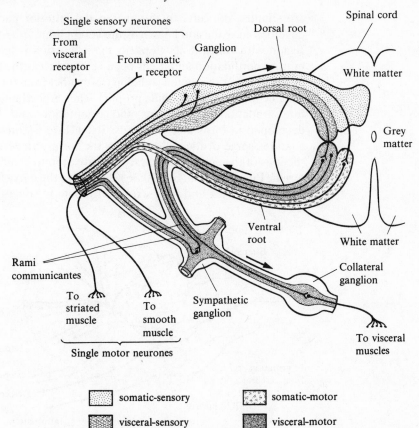

Single sensory neurones

From visceral receptor

From somatic receptor

Ganglion

Dorsal root

Spinal cord

White matter

Grey matter

Ventral root

White matter

Rami communicantes

To striated muscle

To smooth muscle

Sympathetic ganglion

Collateral ganglion

Single motor neurones

To visceral muscles

somatic-sensory   somatic-motor

visceral-sensory   visceral-motor

Fig. 10.35.
The spinal arc,
indicating the major
nerve tracts. These
carry many individual
sensory and motor
neurones.

choroid plexus passes posteriorly and leaves the cord through three foramina in the roof of the medulla.

The VIIIth nerve, which terminates in the vestibular nucleus of the medulla, contains important afferent fibres which play a large part in the control of posture and balance in vertebrates. These afferents are able to affect directly some of the descending motor pathways responsible for producing the animal's movements. Drastic lack of input on one side results in rolling of a fish or tetrapod, though some compensation may occur. A frog, for example may be able to compensate for labyrinthectomy within six to eight weeks. In fish and amphibians, section of the nerve cord in front of the medulla leaves the animal still capable of relatively normal locomotion. In the dogfish, even if the cord is sectioned behind the medulla the resulting spinal preparation continues to make spontaneous swimming movements for many days. In birds and mammals, however, locomotion is not possible even in a medullary animal (i.e. CNS sectioned in front of the medulla) but it continues to make its basic respiratory and cardio-vascular responses.

As we have seen, the acoustico-lateralis centres (i.e. those concerned with hearing and balance) form a very important part of the medulla and are situated dorsally. In this region the anterior part of the medulla develops an extension known as the cerebellum. In accordance with the region with which it develops, its main function is that of balance and motor coordination and it has been referred to as the 'gyroscope of the body'. It forms the metencephalon (as distinct from the myelencephalon which comprises the medulla proper). The cerebellum is concerned with fine gradations of posture and orientation and becomes greatly developed in birds and mammals. Its surface is complexly folded and contains some of the cell bodies in the form of a cerebellar cortex. In sections of the cerebellum these folds are seen to be lined by the characteristic Purkinje cells (Fig. 10.36). These are the only efferent cells of the cerebellar cortex and their axons terminate in some of the subcortical

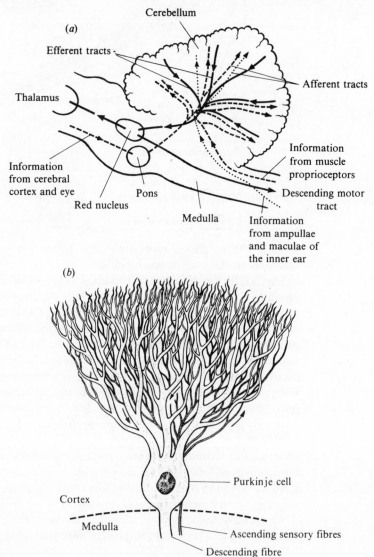

Fig. 10.36. (*a*) The main connections of the mammalian cerebellum. (*b*) A Purkinje cell from the cortex of the cerebellum.

268

nuclei. The close connection between the cerebellum and cerebral hemispheres is further indicated because the latter develop both ontogenetically (life-history of the individual) and phylogenetically (race-history of the species) at about the same time as the so-called neocerebellum. Correlated with the development of this interrelationship in mammals is the pons, which contains nuclei that relay cerebral impulses to the cerebellum. The pons also contains transverse fibres interconnecting the two sides of the cerebellum. There appears to be some functional localisation within the cerebellum, for stimulation of a given region either inhibits or facilitates the effects produced by stimulation of corresponding areas of the motor cortex. The cerebellum is characterised by electrical activity of high frequency (150–250 hertz).

Removal of the cerebellum in a dog, cat or monkey gives rise to complex symptoms which are essentially the same in each animal. They become incapable of making voluntary movements and cannot stand. They make periodic, spasmodic movements of the leg, neck and tail and these extremities become rigidly extended. After a period of four or five weeks some sort of voluntary movements return in a dog and the animal is able to move about on its four legs, but very unsteadily. The cerebellum is highly developed in birds and is very important in their sense of balance. Birds cannot walk or peck, let alone fly, several days after its removal. In man indications of damage to the cerebellum are shown by the tendency to produce oscillatory movements of the hand (i.e. a tremor) when the patient tries to follow a given object through space. It should be noted that the cerebellum is mainly concerned with the regulation of the anti-gravity postural muscles. Following removal of the cerebellum the animal becomes very uncoordinated on land but is much less disturbed in water, where it is able to make effective swimming movements.

By and large, then, the cerebellum is concerned with the more detailed regulation of the posture and voluntary movements of the animal and in doing this it is closely interconnected with the cerebral hemispheres. Its precise mode of function remains uncertain. It is particularly well developed in primates and man, which rely on their ability to manipulate objects in their environment.

In many ways the mid brain is the most conservative region and changes relatively little in the vertebrate series. Its relative importance as a centre correlating different types of sensory input decreases, however, from being dominant in fish to a relatively subsidiary role in birds and mammals. The roof or tectum receives a point-to-point projection of the retina in fishes and it also receives olfactory, acoustico-lateralis and proprioceptive inputs. In lower vertebrates it is also the region where the main descending somatic motor tracts have their origin. Its roof is usually formed into a pair of optic lobes but in mammals there are four such lobes, referred to as the superior colliculi and inferior colliculi. The former retain their optic function and are mainly concerned with the reflex control of eye movements; control of visual responses requires the visual cortex. Localised stimulation of the superior colliculi can lead to

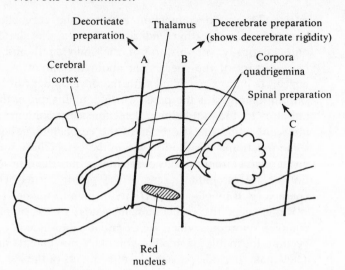

Decorticate preparation

Thalamus

Decerebrate preparation (shows decerebrate rigidity)

Cerebral cortex

A   B

Corpora quadrigemina

Spinal preparation

C

Red nucleus

Fig. 10.37.
The three main levels at which the brain can be sectioned for experimental purposes.

discrete movements of the eyes. The inferior colliculi are concerned with auditory functions and reflexes. For example a cat without its auditory cortex can discriminate tones and detect sounds of low intensity, but this ability is lost when the inferior colliculi are also removed. The floor of the mid brain is important because it contains a large group of fibre tracts or crura cerebri which connect the spinal cord with the anterior cerebral regions. In addition to these there are many nuclei, primarily of the IIIrd and IVth cranial nerves, which control the eye muscles. Another part of this basal region, or tegmentum, forms a mass of grey matter and the so-called reticular formation. This is a tangled complex of branching dendrites that also contains groups of cell bodies. It appears to be important in affecting the level of excitability of other parts of the brain in a non-specific way. The reticular formation is present in all vertebrates but there is a progressive differentiation of this structure into more well-defined nuclei in the higher forms. The red nucleus, which first appears clearly defined in the reptiles, is an important part of the reticular formation playing a major role in the regulation of posture and movement in these and all higher vertebrates. This nucleus is a part of the extrapyramidal motor pathway (see p. 274).

Transection of the mid brain behind the red nucleus (Fig. 10.37) leads to exaggerated contraction of the anti-gravity muscles and produces the condition known as decerebrate rigidity. The limbs are held stiffly and point downwards and backwards and the tail and head are held high up in spite of their weight. Section of the brain in front of the mid brain of a dog or cat produces a condition in which the animal is able to right itself and stand awkwardly, but in primates standing is not possible. Decerebrate rigidity is lost when the brain stem is cut behind the vestibular nucleus and it can also be abolished in a limb by severing the dorsal roots, showing that the input from the proprioceptors is important in its maintenance. The control of the posture of the limbs is produced from the inputs derived

270

from the muscle receptors and from the eyes. All these inputs come together in the medulla and mid brain but if the latter is removed then the postural limb reflexes take on an exaggerated activity. As we have seen in at least the cat and dog, sectioning in front of the mid brain leads to a condition in which the animal can maintain its standing position. But for the proper control and voluntary direction of these movements the fore brain is essential. In primates this part of the brain is important even for the proper adjustment of the muscles and the maintenance of posture and righting reflexes.

After considering the control of movement it is apparent that the fore brain plays an increasingly important part and is the dominant motor region of the brain in the mammals. Not only is it the region from which the motor pathways take their origin (e.g. the pyramidal tract) but it is also the region where a vast amount of information is assembled from inputs derived from all parts of the body. The diencephalon forms a smaller and more posterior part of the fore brain but contains many important regions. The walls of this region are thickened and form the thalamus above and the hypothalamus below. The thalamus is of great importance as a relay station between other regions of the brain and the

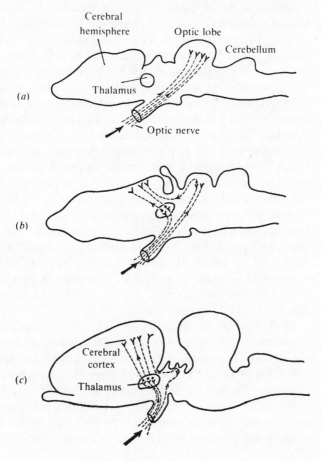

Fig. 10.38.
The centralisation of information reception in the fore brain.
(a) Amphibian.
(b) Reptile. The optic lobe is reduced and information from the eyes passes to the cerebral cortex via the thalamus. (c) Mammal. The optic lobes are very small (represented by the corpora quadrigemina) and most of the sensory information from the optic nerves passes to the cerebral cortex (neopallium) via the thalamus.

271

cerebral cortex. Notably, for example, all optical pathways to the cerebral hemispheres pass through the thalami. In fact it is the presence of such pathways which causes the formation of the neopallium, which first appears in reptiles and then becomes so elaborated in mammals that it covers over almost the rest of the brain (Fig. 10.38).

Pathways from the cerebral cortex descending to the lower regions of the CNS also pass through the thalami, which are not simply a relay station on the upward and downward pathways but are involved in some complex integration with the cerebral cortex. There are several types of nuclei within the thalami and these are concerned with the afferent projection systems from the cerebellum and various other parts of the fore brain. There are scarcely any regions of the brain which are not connected with the thalami.

The ventral region of the 'tween-brain is expanded to form the hypothalamus. As with the thalami, the importance of this region of the brain has become increasingly recognised by neurophysiologists. The hypothalamus is mainly concerned with the regulation of visceral functions, for many visceral-sensory fibres end there. Its action is not only through the autonomic system but also via the endocrine system. This is emphasised by the presence on the base of the hypothalamus of the pituitary, or hypophysis, an endocrine gland that secretes several hormones. The pituitary consists of two lobes. The posterior lobe secretes two hormones actually produced in the hypothalamus but stored in the pituitary; one of these is ADH, the hormone that regulates urine flow. The anterior lobe secretes at least six hormones, whose production is triggered by releasing factors synthesised by neurosecretory cells in the hypothalamus and transmitted to the pituitary via the hypophysial portal system. (The hypothalamus–pituitary system is discussed in more detail in Chapter 11.) The hypothalamus also contains cells which detect the body temperature and it is the site of the thermostat of endothermic animals. There are also osmoreceptors in the hypothalamus which form the sensory side of a neurohumoral reflex controlling the body fluid osmotic pressure. Injection of small quantities of sodium chloride into the hypothalamus results in a goat drinking profusely. Following the work of Hess, the hypothalmus has been studied a great deal in recent years using implanted electrodes. He first showed that by stimulation of local regions a cat could be made to sleep and to awaken. Other centres within the hypothalamus when excited may lead to vomiting, salivation, sniffing, licking, defence responses, etc.

In the roof of the diencephalon is found the anterior choroid plexus, where the cerebrospinal fluid is secreted and passes into the ventricles of the brain. The cavity of the diencephalon is the third ventricle, and anteriorly this communicates with paired lateral ventricles which are the cavities of the most anterior part of the brain (the telencephalon).

As we have indicated earlier, it is the telencephalon which changes most during the evolution of the vertebrates. In the lower forms it is entirely concerned with the receipt and sorting out of olfactory inputs. In

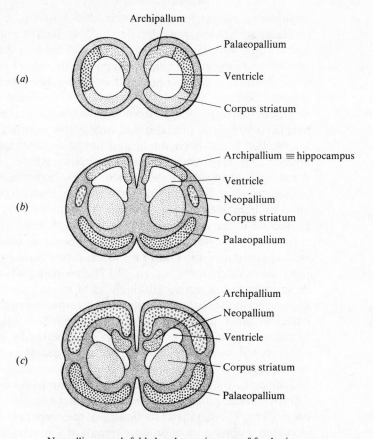

Archipallum

Palaeopallium

(a)

Ventricle

Corpus striatum

Archipallium ≡ hippocampus

Ventricle

Neopallium

(b)

Corpus striatum

Palaeopallium

Archipallium

Neopallium

Ventricle

(c)

Corpus striatum

Palaeopallium

Neopallium much folded and covering rest of forebrain

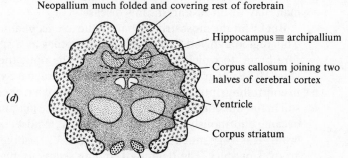

Hippocampus ≡ archipallium

Corpus callosum joining two halves of cerebral cortex

(d)

Ventricle

Corpus striatum

Pyriform lobes≡ palaeopallium

Fig. 10.39.
Changes in the
telencephalon of
different vertebrates as
seen in transverse
section.
(a) Amphibian;
(b) reptile and bird;
(c) primitive mammal
(monotreme); (d)
advanced mammal
(primate).

the mammals this function is still performed by the older parts of the fore brain, namely the rhinencephalon. This is divisible into the archipallium and the palaeopallium. The former gives rise to the hippocampus and the latter to the pyriform lobes (Fig. 10.39). The cell layering on the outside of these regions of the brain is simpler than that of the rest of the telencephalon. It has fewer layers and is referred to as the allocortex (a specific region of the cerebral cortex), as distinct from the isocortex which covers the neopallium. The pyriform lobes certainly retain an olfactory function, for after the synapses in the olfactory lobes, pathways transmit

273

impulses to the pyriform lobes. Stimulation of this region of the brain produces actions that are related to feeding, such as retraction of the lips and sniffing. Removal of the pyriform lobes leads to the loss of olfactory conditioned reflexes. The hippocampus, on the other hand, seems to have more doubtful relationships to olfactory function, and stimulation and removal experiments suggest that it is more important in the production of emotional responses concerned with fear, anger and defence. It is particularly large in primates and whales; the significance of this is not known but it has been implicated in the establishment of long-term memory. The large size of the hippocampus in whales also contrasts with the absence of any olfactory cortex or olfactory organ. The primitiveness of these two regions of the brain is further emphasised by the fact that the input to them does not pass through the thalamus as is the case with the projection to the other parts of the telencephalon.

In birds the region of the brain which becomes most enlarged is the ventral portion of the end brain which forms the large corpora striata (sing. corpus striatum) (Fig. 10.39). The neopallium is also developed in these animals but not so strikingly as in the mammals. The mammals likewise have the corpus striatum, which forms one of the so-called basal nuclei in the ventral parts of the telencephalon. The corpora striata function as motor centres which may be activated by impulses from the olfactory bulbs and they seem to play an important part in the control of motor activity. Together with the reticular formations of the mid brain and medulla they form an accessory motor pathway (known as the extrapyramidal pathway) distinct from that coming directly from the neopallium. The detailed functioning of the corpora striata is not understood but lesions in them produce tremor of voluntary movements and in man may result in Parkinson's disease.

But by far the most striking part of the telencephalon is the neopallium. This region evolved first of all in the reptiles as a special region of the fore brain, to which fibres passed from the optic input via the thalami (Fig. 10.38). It gradually replaced the ancestral pathway to the mid brain in controlling optic perception, although as we have seen the optic tectum still maintains control of eye movements. In the mammals the neopallium became enormously enlarged and in man it almost covers the rest of the brain. It is very much folded on its surface, forming gyri separated by grooves or sulci. The degree of folding seems to increase in the higher mammals, reaching its most complex form in man. The cortex contains an enormous number of cells (10^{10} have been estimated). As indicated above, it is a layered structure with at least six distinct layers being clearly recognisable. Most of these cells are relatively small, although in the motor area are found the largest cells – the Betz cells – which give rise to axons passing down the spinal cord as the pyramidal tract, which plays an important part in the control of voluntary movements. In addition to the many dendritic connections of the cortical cells, there are also many axons passing to and from the surface. Some idea of their density is known from studies on the visual cortex of a cat. The density of afferent fibres was 25,000 per square millimetre and the efferent fibres were three times as dense. Some cortical cells have dendritic endings (apical dendrites) which reach to the surface and extend several millimetres along it. The functioning of this enormously complex structure is only scantily understood and presents one of the greatest challenges of modern biology. Considerable evidence indicates some localisation of function within the cortex (Fig. 10.40) and, as indicated above, one area (the motor cortex) is intimately concerned with movements of different parts of the body. For example, stimulation of localised regions in this area may produce very discrete muscular movements and all parts of the body seem to be represented in it. Just posterior to this voluntary motor area is a somatic sensory region, where electrical activity may be recorded when sensory stimulation is applied to different parts of the body. The representation in this area is not directly proportional to the size of the area concerned, and regions such as the hands and the face, notably around the mouth and lips, are especially well represented for their size (which correlates with the observations derived from two-point stimulation mentioned earlier: p. 237). The visual cortex is the projection area for visual pathways after they have passed through the thalami; others are concerned with auditory functioning and there are specific olfactory cortical projections. The relative sizes of these different sensory projection areas (Fig. 10.41) are closely related to the importance of the particular sense in the life of the animal. For example, in a pig the region of the cortex concerned with the tactile sensations of the snout is extremely large; the auditory area is well represented in the dolphin. The relative sizes of the visual area in

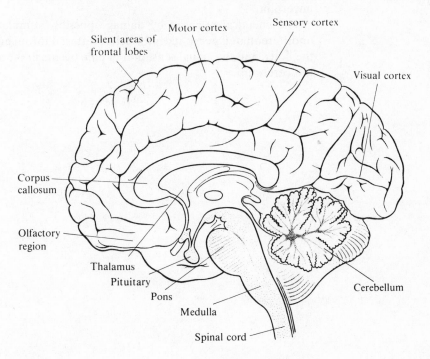

Fig. 10.40.
Localisation of function in the cerebral cortex of man. (The auditory cortex is laterally situated and cannot be indicated on a longitudinal section.)

275

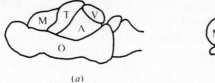

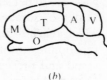

(*a*)

(*b*)

Fig. 10.41.
Relative sizes of
sensory projection
areas in various
mammals. (*a*) Shrew;
(*b*) pig; (*c*) monkey.
Pr, pre-frontal;
M, motor; T, tactile;
A, auditory; V, visual;
O, olfactory.

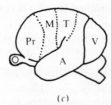

(*c*)

different types of shrew is correlated with whether they are nocturnal or diurnal.

A note on the nature of pain

A lot of research has been done recently on this topic and it now seems that pain results from actual damage to the tissues. This damage may be the result of heat, cold, lack of blood flow, mechanical or chemical stimuli and it is thought that the stimuli cause the release of special chemicals from the damaged tissues. Histamine is one possible such substance and bradykinin another, but the exact nature of the 'pain substance' is as yet uncertain.

The chemical released by damage possibly stimulates anatomically undifferentiated nerve receptor cells scattered throughout the tissues of the body, and the actual sensation of pain occurs in the sensory cortex of the cerebrum.

11 The endocrine system

11.1 **The nature of hormones**

An endocrine gland is one specialised to produce hormones, which travel to target cells in the body and produce an effect there.

Hormone action is particularly concerned with metabolic activities – both anabolic (building-up) and catabolic (breaking down) – in the cytoplasm. By altering the balance of these, hormones are able to coordinate such long-term changes as growth and maturation in a way that nervous control could not. Thus during the lifetime of the animal the balance of hormones, sometimes called the hormone spectrum, is continuously changing and controlling the direction of the animal's activities. (Certain prolonged conditions, such as exposure to heat or cold, may actually lead to an adaptive increase in the size of the endocrine organ concerned.)

Both nervous and endocrine mechanisms make use of basic properties of the cell such as secretion and the propagation of impulses. It is not possible to decide whether nervous or hormonal mechanisms arose first in evolution and the two are closely connected. As many substances produced in the cell affect the rates of nature of the reactions taking place in the cytoplasm (an obvious example being messenger RNA), it is not difficult to see how, during evolution, certain cells may have become specialised to produce secretions that can affect cells at a distance from themselves, and thus how an endocrine system of coordination arose.

Some hormones, such as neurotransmitters (e.g. acetylcholine), act at a distance of micrometres from their place of secretion; others, such as the gastric hormones, act at a distance of centimetres; and many hormones, including those from the pituitary and other major endocrine glands, have target cells far from their sites of secretion. We can in fact distinguish three ways in which the secretion of a cell can be effective elsewhere (Fig. 11.1). In the first case there are cells, such as those in the hypothalamus, which produce substances in their cell body that then travel to distal processes of the cell where they are released. The second type are those, such as the neurones, which have long cell extensions and use electrical transmission to cause release of secretions from their ends. Finally there are the more common type of endocrine cells which release hormones in their immediate vicinity and rely on the blood to transport them. In this context it is interesting to note that the endocrine glands are supplied with a copious and steady supply of blood at all times, quite

277

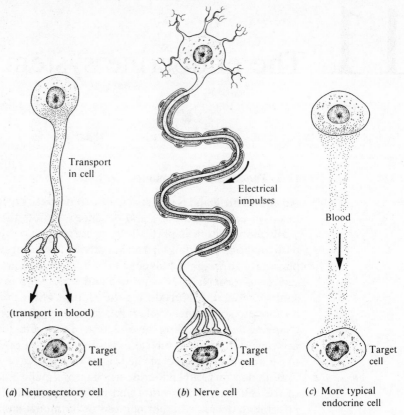

Fig. 11.1.
The three ways in which a hormone can reach its target cells. (*a*) Neurosecretory cell. Chemical synthesised in the cell travels to the distal processes of the cell, where it is released. It may then act locally or on distant cells after transport in the blood. (*b*) Nerve cell. A local hormone (equivalent to a synaptic transmitter) is made and released at the end of the cell process. The stimulus for its release is transmitted through the cell as an electrical impulse. (*c*) More typical endocrine cell. The hormone synthesised in the cell is transported to distant target cells in the bloodstream.

(*a*) Neurosecretory cell   (*b*) Nerve cell   (*c*) More typical endocrine cell

unlike the situation in the muscles and the gut where the supply fluctuates according to the immediate needs of the body.

So far as it is possible to generalise, hormones produce more gradual changes than do nerve impulses. There are, however, a number of instances of rapid endocrine coordination, such as the effects of adrenaline or the response of the stomach to gastrin. Sometimes hormones may act antagonistically to nerve impulses. Both systems make use of feedback mechanisms, and it is generally true that an increase in the concentration of a specific hormone in the blood triggers off a negative feedback response in the hypothalamus–pituitary complex that reduces its further secretion. By such a balance of stimulation and suppression a dynamic equilibrium of hormone levels is maintained.

Hormones are physiologically extremely active molecules and act in the body in very low concentrations (around 10^{-10} molar).

11.2 The hypothalamus–pituitary complex

11.2.1 *Anatomical relations of the complex*

For many years the pituitary gland, situated on the floor of the fore brain was thought to be the master endocrine organ, whose secretions, termed trophic hormones, coordinated the activity of all other endocrine glands. The pituitary was visualised as a conductor, directing in concert a subordinate endocrine orchestra.

In recent years this has been seen to be an oversimplification of the situation, and it is now customary to consider the pituitary and the hypothalamus, that lies just above it and from which part of it is derived, as working together as an integrated complex. The basic structure of this complex is illustrated in Fig. 11.2.

The anterior pituitary (also called the adenohypophysis) is derived from an upgrowth from the buccal cavity roof. Besides its own blood supply it receives blood via a portal system which comes from the capillaries of the median eminence of the hypothalamus and which, as the hypothalamico-hypophysial portal veins, run down the front of the hypophysial tract before entering the anterior pituitary.

The posterior pituitary (also called the neurohypophysis) and the median pituitary are derived in the embryo from the hypothalamus, and in the adult the posterior pituitary preserves a close neuronal connection with this part of the brain. Connecting the hypothalamus with the posterior pituitary are the two large hypothalamico-hypophysial neurosecretory tracts. Sectioning the hypophysial tract in such a way as to interrupt these two nerves shows that secretory products accumulate on the proximal side of the cut. Clearly then, there are neurosecretory cells in the hypothalamus whose secretions travel along the cell processes to the posterior pituitary.

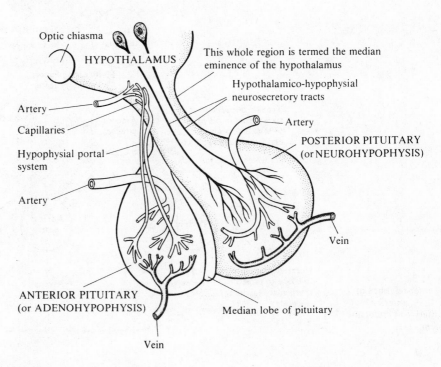

Fig. 11.2.
The hypothalamus–pituitary complex.

The endocrine system

11.2.2 *Functional integration between the hypothalamus and the pituitary*

Anterior pituitary

The role of the hypothalamus has already been described (p. 272) and it will be remembered that it is the regulating centre of many autonomic functions including blood pressure and the osmoregulation and temperature control of the body. It is also the seat of emotional states and of hunger and thirst, as well as being adjacent to the higher centres.

In our present context the hypothalamus also turns out to be the link between the CNS in general and the endocrine glands, through the influences that it has on the pituitary. It has been shown that the hypothalamus produces a number of polypeptide releasing factors, that are carried via the hypophysial portal vein to the anterior pituitary, where they bring about release of particular trophic hormones. Most releasing factors are associated with one trophic hormone, but follicle stimulating hormone (FSH) and luteinising hormone (LH) share a single releasing factor and prolactin is regulated by an inhibitory factor (PIF) not a releasing factor.

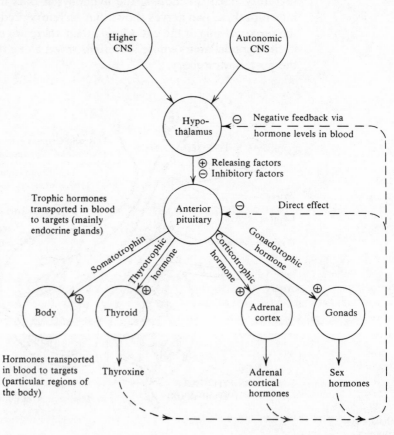

Fig. 11.3.
The inter-relations of the CNS, anterior pituitary and endocrine system.

280

A more detailed picture of the inter-relations of the CNS, the anterior pituitary and the endocrine system is given in Fig. 11.3.

Posterior pituitary
The posterior pituitary secretes the peptide hormones oxytocin, which causes contractions of the uterus during childbirth, and vasopressin (or antidiuretic hormone, ADH), which increases water reabsorption in the kidney. Their primary action is on smooth muscle and in many cases that of the vascular system.

The production of vasopressins and oxytocins is recognised from the neurohypophysis of all vertebrates, and there appears to have been some evolution in their chemical structure.

About 1940 it was noted that neurohypophysial extracts from reptiles, birds, amphibians and bony fish caused more uptake of water by frogs than would have been expected if they had only contained the mammalian neurohypophysial substances. Heller recognised that these non-mammalian neurohypophysial extracts contained a unique 'water-balance principle' which was later recognised clinically as arginine vasotocin (AVT). It is possible that AVT represents the most primitive vasotocin. A number of lines of evidence, both morphological and physiological, suggest that the evolution of the posterior pituitary has been important during adaptation of the land habit by vertebrates.

While certain selected topics related to coordination of the body by hormones will be considered in more detail below, the general information on each gland and its secretions is given in tabular form in Table 11.1. (Hormones and the coordination of the alimentary canal are considered on p. 37, their role in the regulation of water and ionic balance on p. 177, and their role in reproduction on pp. 302–20 in the chapter following.)

11.3 **Role of hormones in the coordination of growth and metabolism**

The growth of a mammal represents a balance between protoplasm manufacture and its destruction; that is, between anabolic and catabolic processes within its cells. During early life the former are predominant and material is rapidly incorporated into the body from the environment. The major influence during this period is the growth hormone somatotrophin from the anterior lobe of the pituitary, which seems to encourage the removal of amino acids from the body fluids and their condensation into proteins in the body cells. At the same time the metabolism of fats is also regulated so that the incorporation of fat and its laying down in the connective tissues is kept at a minimum.

Once the juvenile period is passed, the adult phase establishes itself and the hormone spectrum maintains the body in a state of metabolic equilibrium, with anabolic and catabolic processes approximately balanced. Meanwhile the influence of the gonadal or sex hormones causes

TABLE 11.1. *The endocrine organs of the mammal and their secretions*

Origin and structure	Secretion(s)	Control of secretion

Pituitary
Situated below the hypothalamus of the fore brain and made up of three parts. The anterior lobe (adeno-hypophysis) is derived from the roof of the buccal cavity, while the posterior lobe (neurohypophysis) is a downgrowth of the fore brain. Blood vessels pass from the hypothalamic region of the brain and form a plexus in the anterior lobe while the posterior lobe receives nerve tracts from hypothalamic and supraoptic parts of the brain. Between anterior and posterior lobes is a median lobe.
 	
    Posterior lobe hormones are synthesised in the hypothalamus and pass down the hypothalamico-hypophysial tract to the pituitary. Anterior lobe hormones are synthesised *in situ* but only secreted into the blood in response to specific releasing factors from the hypothalamus.
 	
    Histologically the anterior lobe is composed of several different types of cells; thus the chromophobes appear to be early stages in the development of the acidophilic and basophilic cells that actually synthesise the hormones. There are also types of neutrophils present		

anterior lobe		
	Somatotrophin (STH: also called growth hormone)	Somatotrophin releasing factor (SRF) is secreted from a centre in the lower part of the hypothalamus close to the feeding and hunger areas; deficiency of food derivatives in the blood can cause increase in SRF production
	Thyroid stimulating hormone (TSH: also called thyrotrophic hormone)	Release of thyrotrophin releasing factor from the hypothalamus is, in turn, controlled by the levels of thyroxine in the blood, low levels increasing production and vice versa
	Corticotrophic hormone or adreno-corticotrophic hormone (ACTH)	Release of corticotrophin releasing factor from the hypothalamus is controlled by steroid levels in the blood and by direct nervous stimulation of the hypothalamus as a result of stress

attern of secretion	Mode of action	Excess and deficiency
ecreted throughout life. When owth has mostly ceased after dolescence, the hormone continues promote protein synthesis roughout the body	It is thought that the hormone produces its effects in two ways: (1) by promoting the transport of amino acids from the body fluids into cells where they can be mobilised into proteins; (2) by activation of specific genes in cells which are involved in synthesis instructions. STH also acts through the pancreas to cause an increase in the level of blood sugar, possibly by inhibiting the cells of the islets of Langerhans to decrease insulin production. Current views indicate this hormone to be the major controlling factor for the pancreas	If produced in excess during early life leads to gigantism of whole body, if later to abnormal development of hands, feet, jaw, etc. (known as acromegaly). If there is under-secretion dwarfism results, as well as other symptoms associated with lack of thyroid and adrenal hormone
ecreted throughout life but at articularly high levels during the eriods of rapid growth and development	Acts directly on the cells of the thyroid gland increasing both their numbers and their secretory activity	Results as for disturbance of normal functions of thyroid (see below)
teady level of secretion is rapidly creased by all types of stress: e.g. old, heat, pain, fright, infections	ACTH appears to be vital to the early stages of the synthesis pathways of the adrenal steroid hormones. It is now thought also to act with STH (see above) in control of pancreas secretion	Results as for disturbance of normal adrenal functions (see below)

TABLE 11.1. *continued*

Origin and structure	Secretion(s)	Control of secretion
anterior lobe (contd)	Gonadotrophic hormones: follicle stimulating hormone (FSH), luteinising hormone (LH: called interstitial cell stimulating hormone ICSH, in the male), prolactin (sometimes inappropriately called luteotrophic hormone, LTH)	FSH and LH/ICSH share a common hypothalamic releasing factor. Prolactin is continuously produced from the pituitary and is *inhibited* by prolactin inhibiting factor (PIF) from the hypothalamus
	(Pancreotrophic hormone: no such hormone is now thought to be produced, control of the pancreas being brought about by STH and ACTH)	
posterior lobe	Antidiuretic hormone (ADH: also called vasopressin)	Secretion caused by decreases in blood pressure, blood volume, and osmotic pressure of the blood detected by osmoreceptors in hypothalamus. External sensory stimuli also influence hypothalamic neurosecretory cells
	Oxytocin	Release stimulated by distension of cervix, decrease in progesterone/suckling, and neural stimuli during parturition and suckling
median lobe	Melanocyte stimulating hormone (MSH)	*Inhibition* of secretion controlled by hypothalamus
Thyroid Originated in primitive chordates from the endostyle, a feeding groove in the floor of the pharynx which also had the function of iodine uptake and regulation. In mammals it consists of two lobes situated about the larynx, the lobes being composed of many alveoli with a secretory endothelium producing a fluid called the colloid which contains the thyroid hormone	Thyroxine (or tetraiodo–thyronine: T4) and tri-iodothyronine or T3 (which has a structure similar to thyroxine but with 3 iodine atoms rather than 4)	The thyroid is stimulated to release its hormones according to the levels of these already present in the bloodstream and via a feedback mechanism involving the hypothalamus (see above). The hormones are bound to a protein (thyroglobulin) in the gland and the action of TSH releases them from this protein and into the bloodstream

ttern of secretion	Mode of action	Excess and deficiency
st secreted in quantity at puberty, d thereafter during the repro- ctive phase of life. Blood hormone els control secretion via feedback echanisms	FSH in females stimulates follicle development and secretion of oestrogens from the ovaries; in males it stimulates development of the germinal epithelium of the testis and sperm production. LH works with FSH to stimulate oestrogen secretion and rupture of mature follicles. It also causes the luteinisation (lit. 'turning yellow') of the latter and acts synergistically with prolactin to maintain the corpus luteum (and hence the progesterone it secretes). ICSH in the male stim- ulates the interstitial cells of the testis to secrete testosterone. Prolactin stim- ulates milk production (and acts with LH as described above)	Hypophysectomy (surgical removal of the anterior lobe) causes atrophy of the testis and it can be assumed that failure to secrete the gonadotrophic hormones would lead to sterility
anges in the physiological features the blood that control secretion n result from many causes but pecially malfunction of osmoregulatory stems	Increased levels cause increased water reabsorption in distal parts of kidney	A lack of this hormone produces dia- betes insipidus, characterised by pro- duction of large quantities of dilute urine and great thirst
pid increase in production during our is maintained during nursing young	Primary action is on smooth muscle, particularly in the uterus during child- birth, and milk ejection from mammary glands	
ternal light intensity is a govern- factor in MSH levels. More MSH secreted in pregnancy	Stimulates melanocytes in skin to produce the brown pigment melanin, which darkens the skin	Excess MSH is secreted in Addison's disease, one of the symptoms of which is darkening of the skin
e thyroid is active continuously but oduces higher levels of secretions ring periods of rapid growth and xual maturation and in stress uations such as cold and hunger	The two hormones act in essentially the same way except that metabolic responses to thyroxine start about 48 hours after a stimulus and may persist for several weeks while the responses to tri-iodothyronine commence within 2 hours and last for a shorter period (at most 2 weeks). Thyroxine and tri-iodothyronine act on the basal metabolic rate by stimulating the breakdown of glucose and release of heat and generation of ATP. They possibly do this by direct stimulation of enzymes involved in oxidative phosphorylation in the res- piratory pathways. They also act in conjunction with somatotrophin in bringing about growth, and act directly on brain cells causing them to differentiate	Excess thyroxine produces a condition called Graves' disease, with exophthalmic goitre and increase in the basal metabolic rate. This can lead to cardiac failure if prolonged. The cause of Graves' disease is the production of an abnormal body protein which continuously stimulates the thyroid to excessive secretion. If congenitally deficient the lack of thyroxine causes cretinism, where the individual fails to develop normally and the mind is retarded. There is also failure to develop sexually. Deficiency later in life, perhaps due to iodine shortage in the diet, produces a swelling of the neck (goitre) and may lead to a slowing down of the metabolism and laying down of excess fat. The condition is known as myxoedema

285

TABLE 11.1. *continued*

Origin and structure	Secretion(s)	Control of secretion
Thyroid (contd)	Calcitonin	High Ca²⁺ concentration in the blood causes stimulation of the synthesis and release of calcitonin; low levels of Ca²⁺ suppress its manufacture
Parathyroid Originates, like the thyroids, from pouches that bud off from the embryonic pharynx. In man the glands are found embedded in the posterior part of the lateral lobes of the thyroid	Parathormone	Low levels of blood Ca²⁺ stimulate the parathyroid directly to increase parathormone production whereas high levels of Ca²⁺ suppress its release
Pancreas In the mammal the pancreas is found as rather diffuse glandular tissue in the first loop of the duodenum. However, in primitive chordates, e.g. lamprey, the pancreatic rudiments are found as cells in the wall of the gut and the regulation of sugar assimilation at this point probably evolved to its control throughout the animal. The pancreas has important digestive functions in mammals but the endocrine cells are clearly demarcated from the digestive and are grouped together as the islets of Langerhans (some 1–3% of organ). The islets contain large numbers of β cells with alcohol-soluble granules and are associated with insulin production. The smaller number of α cells with alcohol-insoluble granules probably secrete glucagon	Insulin and glucagon	The pancreas is under control of the pituitary trophic hormones STH and ACTH and responds directly to the level of blood glucose. Where this is high, secretion of insulin is stepped up and that of glucagon depressed, while low blood sugar causes inhibition of insulin and an increase in glucagon secretion
Adrenals These consist of two different type of endocrine organ which in fishes (e.g. dogfish) are found as long chains of secretory cells along the aorta. In the mammal the outer layer, or cortex, originates from mesodermal cells near the embryonic gonads, while the inner layer, or medulla, comes from cells associated with the formation of the sympathetic ganglia. The inner layer is characterised by taking up the stain chromic acid and for this reason its cells are called chromaffin cells. The medulla produces the hormones adrenaline and noradrenaline. The adrenal cortex has three distinct zones: the outer or zona glomerulosa, secretes mineralocorticoids such as aldosterone; the middle, or zona fasciculata, is full of lipid-containing cells and secretes glucocorticoids; and the inner reticulans secretes androgenic hormones	Adrenaline and noradrenaline	Levels are normally maintained in the blood system by a feedback system involving the medulla itself: low levels of hormone lead to synthesis and secretion and vice versa

attern of secretion	Mode of action	Excess and deficiency
	Is antagonistic to parathormone from the parathyroid glands and prevents removal of Ca^{2+} from the bones	Excess or deficiency leads to a disturbance of calcium metabolism with its associated effects on nerve, skeleton, muscle, blood etc.
	Causes a release of PO_4^{3-} from the kidney tubules which leads to its release from the bones and teeth. As the PO_4^{3-} ions are balanced by a simultaneous release of Ca^{2+} the level of this ion is increased in the blood. It also increases Ca^{2+} uptake by the walls of the gut and is involved in the release of Ca^{2+} from the store in the bones. In the last respect it is antagonistic to calcitonin (see above)	Under-activity causes a drop in blood Ca^{2+} which in turn leads to muscular tetany. Over-activity would lead to a progressive demineralisation of the bones similar to rickets, as well as to the formation of massive kidney stones. Both conditions are liable to be fatal
	The action of insulin at the cellular level is considered on p. 295. In general it depresses blood glucose levels, in a variety of ways which include increasing glycogen synthesis and increasing cell utilisation of glucose. It also stimulates both lipid and protein synthesis, which in turn reduce glucose levels. Insulin inhibits the hydrolysis of glycogen in the liver and the muscles. Glucagon is essentially antagonistic to insulin and causes an increase in blood glucose levels. It does this mainly by promoting breakdown of glycogen to glucose in the liver and muscles. It also increases the rate of breakdown of fats	Failure to produce insulin leads to a condition called diabetes mellitus. The symptoms of this are high level of blood sugar, sugar in the urine, a disturbance of the body's osmotic equilibrium and derangement of the nervous system. Toxic metabolites from fat (which need 'glucose energy' for their oxidation) also accumulate and are only lost from the kidney with valuable metal cations. The body becomes dehydrated. If excess insulin is produced the utilisation of sugar is too great and its level falls in the blood which upsets nerve and muscle functioning. Glucagon abnormalities seem rare as endocrine disorders. Tumours on the β cells will cause excess glucagon and consequent high blood glucose levels; this in turn damages the α cells with the results described above
oth adrenaline and noradrenaline re secreted in stress situations by imulus from the sympathetic nerves hich serve the medulla.	Action at the cellular level is considered on p. 296. Essentially adrenaline dilates blood vessels in certain parts of the body such as the skeletal muscles and increases the heart's output. Noradrenaline constricts blood vessels but again only in certain areas, such as the gut, so the effect of the two hormones are synergistic in raising blood pressure. Adrenaline and noradrenaline promote the release of glucose from liver glycogen and reinforce the effects of the sympathetic system	Rarely found, but in excess these hormones lead to abnormally high blood pressures. In rats whose adrenal medulla has been removed surgically the ability to withstand any stress situation – such as cold – is markedly diminished

TABLE 11.1. *continued*

Origin and structure	Secretion(s)	Control of secretion
Adrenals (contd)	Cortical hormones: cortisol, corticosterone, aldosterone	ACTH from the pituitary stimulates secretion. Aldosterone is secreted also in response to angiotensin produced by the kidneys and liver in response to low Na^+ levels in the blood. It is also produced by the lungs in response to a decrease in the volume of blood circulating
	Androgen	
Gut	The hormones and their actions are discussed on p. 46	
Gonads **ovary** The ovaries consist of a germinal epithelium and a large number of follicles in various stages of development	Oestrogens (several hormones all with very similar effects, of which the most important is oestradiol)	
	Progesterone	Gonadotrophic hormone from the placenta causes secretion in the foetus, LH from the pituitary synthesis and release from puberty onwards
** testes** The testes consist of many coiled seminiferous tubules where the spermatozoa develop and, between the tubules, regions of interstitial cells which produce gonadal hormones (e.g. testosterone)	Testosterone	
Placenta	Progesterone and oestrogens	
	Gonadotrophic hormone (in man called human chorionic gonadotrophin, HCG)	

288

ttern of secretion	Mode of action	Excess and deficiency
ιe adrenal cortex is active at all nes but especially so following ock or stress situations and in-ctions	Cortisol is a glucocorticoid hormone which brings about an increase in blood glucose level mainly by its pro-duction from protein and by an-tagonising the action of insulin. Corticosterone is both a glucocorticoid and a mineralocorticoid; it increases blood glucose levels and regulates mineral ion balance. Both cortisol and corticosterone act to allow the body to withstand long-term situations, but the exact mechanisms involved are not clear. Aldosterone is a mineral-ocorticoid which conserves the level of Na^+ ions in the body by prevent-ing their loss from the kidney tubules	The destruction of the adrenal cortex, such as occurs in Addison's disease, will lead to general metabolic disturbance, in particular weakness of muscle action and loss of salts. Stress situations, such as cold, which would normally be overcome, lead to collapse and death. The reverse of this is found in Cushing's disease where too much cortical hormone is produced. Symptoms are an excessive protein break-down into glucose and glucocorticoids, and resulting muscular and bone weak-ness. The high blood sugar disturbs the metabolism as in diabetes
	Androgens cause development of the secondary male characters. Very small amounts of androgens are secreted in both the male and female adrenal glands	A tumour on the inner part of the adrenal cortex in a female can cause excess of androgens to be produced and thus the development of certain male charac-teristics. Such cases are very rare
ιestrogens are secreted by ripening llicles (and, in many species, by in-rstitial cells of the ovary) whose evelopment has been initiated by SH from the pituitary (see pituitary ιnadotrophins above)	Oestrogens bring about the develop-ment of the secondary sexual characters in the female and, at a point during the oestrous or men-strual cycle, exert a positive feed-back which results in a sharp rise in LH output by the pituitary	Deficiency of the sex hormones, for one reason or another, leads, in the young to failure to mature sexually and sterility in the adult
·oduced by the ruptured follicle in sponse to LH from the pituitary	Progesterone inhibits further FSH secretion from the pituitary, thus pre-venting any more follicles from ripening. It also affects the uterus and other areas of the female body, preparing it for and maintaining the state of pregnancy	
fter the initiation of the sex organs the foetus the level rises fairly con-stently until puberty. After puberty ιe supply of LH, and therefore ιe level of testosterone, remains ɔnstant	In the foetus it initiates the develop-ment of the sex organs. At puberty it brings about development of the male secondary sexual charac-teristics and promotes the sex drive	Deficiency has the effects described above. The castrated male fails to develop secondary sexual characteristics and his body tends more towards the form of the immature female
fter about 3 months of pregnancy human females the placenta takes ver the secretion of progesterone ɔm the corpus luteum, which then ɛgenerates. It also secretes large ιantities of oestrogens	The effects of these hormones are mentioned above. It should also be noted that progesterone maintains the uterus muscles in a state of relaxation as well as causing, with oestrogen, enlargement of the glandular region of the breasts and the laying down of food stores prior to milk production	
ɛcreted in the human females, from bout the 14th to 60th days follow-ιg conception. After the 60th day ιe level declines rapidly	Gonadotrophic hormone maintains the corpus luteum until its secretory function is taken over by the placenta itself	

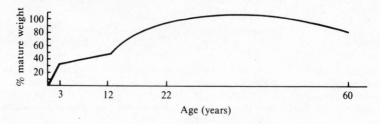

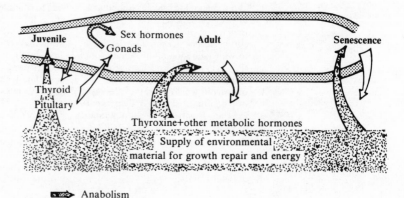

Fig. 11.4.
Hormones and the life
cycle of man. Arrows
from and to the
environment indicate
the extent of anabolic
and catabolic
processes.

changes in the anatomy and behaviour of the individual.

The last phase is that of senescence and once again this is influenced by hormonal change. In this phase, degradation of protoplasmic substances overtakes their anabolism and the rate of metabolic change and replacement is reduced (Fig. 11.4).

At each of these stages of life hormones interact to produce their effects. Thus during early development the activity of the pituitary is only made possible by the presence of thyroxine and corticosterones, which together prepare the body cells to respond to somatotrophin. Throughout the life of the individual the sex hormones and those of the pancreatic islets also encourage anabolic metabolism.

11.4 **Hormones and stress reactions in the body**

A particular type of situation which is largely coordinated by hormones is the body's reaction to stress, which may be in the form of pain, fear, heat, cold, anger, etc. The series of reactions caused by stress is shown in Fig. 11.5.

When the individual is confronted with a situation of this sort impulses from the sense organs pass to the cerebral region and thence to the hypothalamus. The hypothalamus produces CRF (corticotrophin releasing factor) which passes to the anterior pituitary and causes the release of ACTH (adrenocorticotrophic hormone) which in turn promotes secretion of hormones by the adrenal cortex. At the same time nerve impulses, initiated in the hypothalamus, travel to the sympathetic system

290

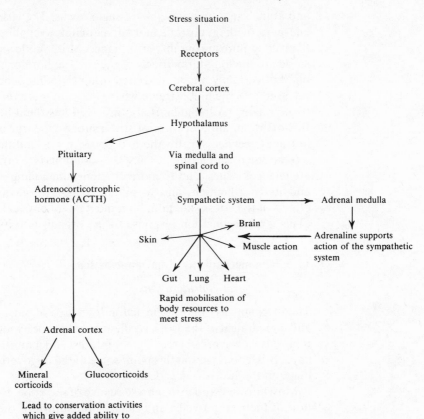

Stress situation

Receptors

Cerebral cortex

Hypothalamus

Pituitary   Via medulla and
spinal cord to

Adrenocorticotrophic
hormone (ACTH)   Sympathetic system ⟶ Adrenal medulla

Brain

Skin ←   ←   Adrenaline supports
Muscle action   action of the sympathetic
system

Gut   Lung   Heart

Rapid mobilisation of
body resources to
meet stress

Adrenal cortex

Mineral   Glucocorticoids
corticoids

Lead to conservation activities
which give added ability to
withstand stress

Fig. 11.5.
The sequence of events
in the body's reactions
to stress situations such
as danger or cold.
Activation of the
adrenal medulla
precedes that of the
cortex.

and also to the adrenal medulla, which secretes the hormones adrenaline
and noradrenaline.

Adrenaline reinforces the action of the sympathetic system and prepares the body for stress. It is known as the 'fight or flight' hormone, as it
prepares the body for one of these responses according to the nature of
the animal, and the circumstances! Thus adrenaline causes the release of
sugar from the liver, increase in heart beat and respiratory rates, decrease
in blood flow to the gut and inhibition of peristalsis, all of which increase
the amount of energy immediately available to the animal. It also causes
an increase in the release of ACTH from the pituitary. From the biochemical point of view the action of adrenaline promotes the functioning of
phosphorylating enzymes, which also lead to an increase in the amount of
energy available (see p. 296).

Noradrenaline causes blood to move out of gut muscles and areas of the
body not concerned in immediate survival of the stress situation.

While these two hormones act instantly, a slower system directed to the
same end of overcoming the stress problem also comes into play. Stimulated by ACTH from the pituitary the various hormones of the adrenal
cortex are released at an increasing rate. As indicated in Table 11.1 these
tend to conserve mineral ions essential in muscle and nerve functioning

291

and at the same time raise blood sugar levels. The latter makes this key energy-providing substrate more immediately available to tissues that are likely to require maximum performance. More subtle and less well understood are the long-term effects of the cortical hormones on wound healing, but certainly amino acids and other 'building block' molecules are released into the bloodstream where they are more readily available for tissue repair. (Adrenaline is also involved in wound-healing processes.) In starvation, another form of stress situation, cortical hormones bring into play mechanisms for the hydrolysis of fats and their respiration.

Thus the role of hormones in stress situations is to bring about immediate changes which lead to more efficient functioning of skeletal muscle and its coordination, and to initiate long-term changes whereby the effects of the stress situation, both mental and physical, can be repaired. (This is an acclimation response of the organism to its environment.)

11.5 The mechanisms of hormone action

11.5.1 *General considerations*

The mechanisms of hormone action is an area of much current research and a great deal of the work is still at the stage of hypothesis or speculation. It is, however, an area of great interest and importance and it will be appropriate to review at this point some of the discoveries that have been made in the last two decades.

Most hormones go through a sequence of secretion, followed by activation of their target cells, and finally breakdown of the hormone when their task is completed. The endocrine cell itself must be activated by a specific exciter molecule and it is supposed that each type of endocrine cell has receptor sites to which their exciter substance attaches. If we take the thyroid gland as an example, the exciter substance is the trophic hormone TSH (thyroid stimulating hormone) from the pituitary, and at the cellular level the sequence might be represented as shown in Fig. 11.6. In response to TSH the thyroid cells produce thyroxine-containing colloid that is released from the gland. Thyroxine (as indeed does any other hormone), then circulates in the bloodstream until it arrives at its particular target cells. These again will have receptor sites tailored to the shape

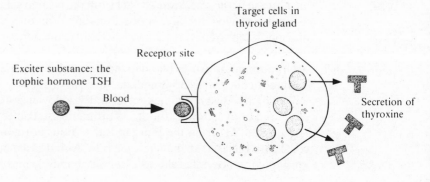

Fig. 11.6.
The action of TSH on the thyroid gland, as an example of the general mechanism of hormone secretion.

292

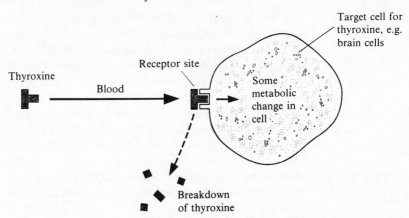

ig. 11.7.
he action of thyroxine
n its target cell, as an
xample of the
ompletion of the
ormone sequence.

of the hormone molecule. The target cell will have been programmed during its differentiation to respond to the hormone and the arrival of the hormone will trigger off a metabolic response: thus thyroxine might cause particular brain cells to differentiate. Finally the hormone, having done its job, is broken down. The completion of the typical hormone sequence can thus be shown as in Fig. 11.7.

Some hormones seem to act at the surface of their target cells triggering off a 'second messenger' which actually causes the change in the cell. This concept is discussed further on p. 296. Other hormones, such as steroid-based ones, actually enter the cell and exert a direct effect on it. It seems certain that the rapidly acting hormones such as adrenaline activate existing cell substances and pathways, while the much slower acting ones such as thyroxine promote the formation of new substances by the cell, which obviously takes much longer. It is found that many hormones are very rapidly broken down in the body; adrenaline, for example, lasts for little more than 5 minutes and insulin for about twice this time.

Finally, as already mentioned on p. 280, it should be appreciated that the majority of hormones are kept at a constant level in the blood by a negative feedback mechanism. We can now extend this notion and say that a high hormone level inhibits the secretion of its specific releasing factor from the hypothalamus and thus its own eventual secretion. This is a further example of a homeostatic mechanism in the body.

11.5.2 *Hormones acting on genes and protein synthesis mechanisms*

Cells differ from each other because they have different protein comple-ments, which in turn are determined by which of their many genes are active. Obviously all cells in the same body have the same genetic instruc-tions, once a cell has differentiated very few of these genes will actually be working. The processes of growth and secretion involve the manufacture of more of the proteins that the cell already contains, while differentiation involves the manufacture by the cell of different types of protein. Nor-mally once a cell has differentiated it cannot go back and 'unmask' genes it is no longer programmed to use.

293

The endocrine system

It will be recalled that cells make proteins by a mechanism whereby regions of their chromosomal DNA (particular gene regions in fact) pass on the information in their base-pairs, in triplets, to the particular messenger RNA (mRNA) associated with a given gene. These mRNAs move out into the cytoplasm and carry the base-pair information to the ribosomes situated on the endoplasmic reticulum. Here transfer RNAs (tRNAs) assemble amino acids to be condensed together into proteins, the exact sequence of the amino acids being determined by the 'code' of base-pair triplets on the mRNA. The general scheme is shown in Fig. 11.8. A typical cell may well contain over 1,000 different proteins that are being synthesised continuously by this process. This basic scheme must be borne in mind when the role of hormones in protein production is considered.

Hormones can affect protein synthesis in various ways. They can act on the chromosomal DNA, activating new genes and thus bringing about the synthesis of new proteins and a change in the cell as a result, they can cause the synthesis of more of a particular mRNA which will in turn increase the synthesis of a protein already being produced, and they can act at the cell membrane allowing more rapid intake of amino acids. All these modes of action, as well as some other possible effects have been claimed to exist.

One of the most elegant demonstrations of hormones activating genes comes not from vertebrates but from studies on the larva of the midge *Chironomus*. When the growth and differentiation hormone ecdysone,

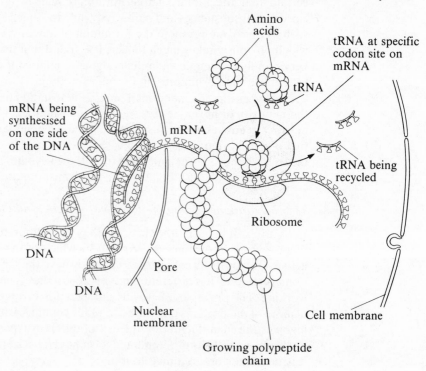

Fig. 11.8.
The general mechanism
of protein synthesis.

294

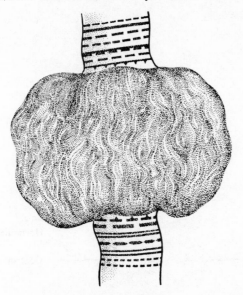

g. 11.9.
agram of a
romosome puff from
e giant chromosome
Chironomus,
duced by injection of
dysone.

itself a steroid similar to many vertebrate hormones, is administered to the larva of the midge, expansions of particular regions of the chromosomes, called puffs, begin to appear a few minutes after giving the hormone dose (Fig. 11.9). Other puffs soon follow and continue to do so over a period of 15–20 hours after the treatment. These puffs closely resemble the appearance of the chromosomes during the normal metamorphosis of the insect and there can be little doubt that the puffs represent gene areas activated for the process of differentiation of the imago. (This is also confirmed by studies using radioactively labelled mRNA, which show that in the locality of the puffs transcription is occurring.) In vertebrates the hormone thyroxine has been shown to induce DNA–RNA polymerase, an enzyme known to be necessary for transcription.

The increase in mRNA after administration of hormones has been shown many times. In certain rat liver tumour cells ACTH was found to cause the formation of the mRNA associated with the enzyme amino transferase, its rate of formation increasing some 20 times in the 5 to 10 hours after the hormone had been given. In liver cells also, somatotrophin caused increases in mRNA, tRNA and ribosomal RNA about an hour and a half after it was administered. Thyroxine and insulin have both been shown to have similar effects on appropriate target cells. The sort of results that have been obtained from numerous experiments are shown in Fig. 11.10.

It has already been mentioned that another way in which hormones stimulate protein production by target cells is by increasing the rate of uptake of amino acids across their membranes. This is one way in which the hormone insulin appears to act on muscle cells, which show increases in protein synthesis following its application.

Despite the evidence that hormones can affect protein synthesis in a

295

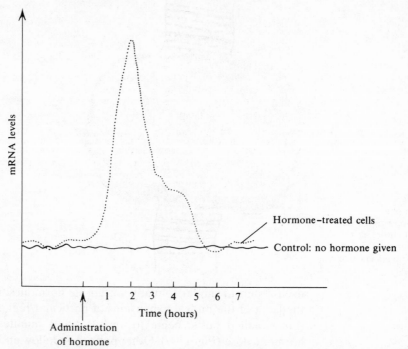

Fig. 11.10.
The effect of hormone
injection on mRNA
levels.

variety of ways there is not yet much indication of the intermediate steps,
if any exist, between the hormone and its actual effect in the cell. This
however is not true of some of the metabolic hormones, where the
existence of a 'second messenger' is definitely established.

11.5.3 *Hormones and enzymes: the 'second messenger' hypothesis*

A very important contribution to our understanding of the action of
hormones on stimulation of enzymes within the cell was made in the late
1950s by Sutherland and his co-workers. They showed that when hormones such as glucagon or adrenaline become attached at the receptor
sites of their target cells they cause activation of an enzyme, adenyl
cyclase, on the inside of the cell membrane. This enzyme causes the
formation of 3'5'-adenosine monophosphate (cyclic AMP) from ATP,
and the cyclic AMP in turn activates protein kinase. The latter enzyme is
known to be involved in the phosphorylation sequence that leads to the
formation of glucose-1-phosphate from glycogen. This sequence of cause
and effect is shown in Fig. 11.11.

In this series of events (also termed a 'cascade effect', because it
amplifies the original stimulus), the cyclic AMP has been termed the
'second messenger', as it acts as a vital intermediate between the hormone and its effect inside the cell.

Since the first demonstration of the role of cyclic AMP in the functioning of adrenaline and glucagon it has been shown to act as a second
messenger in the target cells of TSH, LH, vasopressin, ACTH and insulin.

296

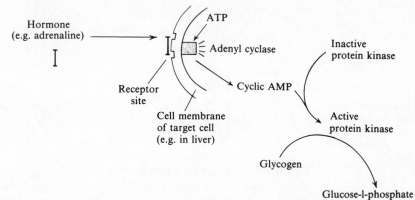

Fig. 11.11.
The sequence of
biochemical events
initiated by a hormone
such as adrenaline.

The cyclic AMP leads to activation not only of the enzymes shown in
Fig. 11.11 but also to others that are involved in muscle contraction, fat
hydrolysis, ion and water uptake, the secretion of proteins and other
functions.

While the role of cyclic AMP is clear, it may be that there are yet other
secondary messengers and cyclic GMP (guanine monophosphate) has
been suggested as being important. There is much of research on these
problems being done at the present time and although the general ways in
which hormones bring about their effects on their target cells is clearer
than before, a great deal of detail needs filling in to complete our understanding.

11.6 Examples of hormone action in non-mammalian vertebrates

The endocrine organs of mammals can, in most cases, be homologised
with those of other vertebrates, and the main endocrine controls
described operate throughout the whole of the phylum Vertebrata. On
the other hand there are some examples of hormonal coordination seen
in the lower vertebrates which are not commonly met with in mammals.
Among these are the control of colour change, migration and metamorphosis.

11.6.1 *Fishes*
Metamorphosis
Some fishes such as the eel and plaice have an immature form which is
very different from the adult. The change from such a larval stage into the
adult is called metamorphosis and is under the control of thyroid hormone.

Salinity tolerance
Salinity tolerance in fishes such as the eel which move from sea- to
freshwater or vice versa is under the control of the posterior pituitary and
adrenal cortex. As explained on p. 298 neurohypophysial extracts from
all the non-mammalian vertebrates were found to contain a hormone –

297

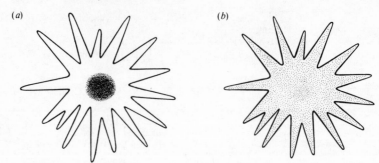

Fig. 11.12.
A melanophore.
(*a*) Pigment condensed:
colour light overall.
(*b*) Pigment dispersed:
cell dark overall.

vasotocin – that is important in water balance control. It seems probable that changes in the secretions of the posterior pituitary accompanied the change in water relations involved in colonisation of the land by verte-brates.

Colour changes

In many species of fish, colour changes are due to the secretion of a melanophore-dispersing or melanophore-condensing hormone from the posterior pituitary (Fig. 11.12). The stimulus for the release of these substances is via the eyes and brain to the pituitary. The two hormones act on the melanophores (stellate cells in the skin) and, by the spreading out or condensing of the brown pigment, produce camouflage patterns which blend the fish into its natural background. In many teleosts colour change is under nervous control. Some demersal species such as the turbot can make a remarkably accurate fit to complex external patterns (e.g. a chess board).

Changes associated with sexual dimorphism

The androgenic hormones of the male (that is, sex hormones from the gonads) are responsible for bringing about the changes in colour or other features seen at the breeding season. A well-known example is the bright colour of the male stickleback which develops during the reproductive period.

11.6.2 *Amphibians*

Metamorphosis

The change of the tadpole into the adult frog or toad is under the control of the hormone thyroxine, and for this to be secreted it is necessary that TSH is released from the anterior lobe of the pituitary. By depriving the developing tadpole of iodine, necessary for the synthesis of thyroxine, or by removing its thyroid gland, metamorphosis can be prevented. By treatment of young tadpoles with thyroxine precocious metamorphosis is brought about. (See Fig. 11.13.)

11.6 Hormone action in non-mammalian vertebrates

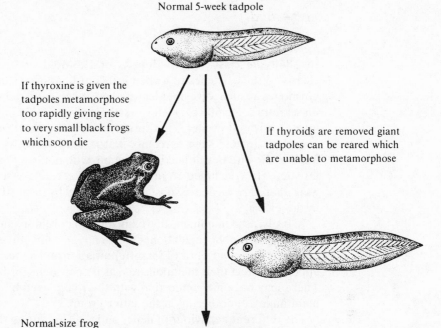

Normal 5-week tadpole

If thyroxine is given the tadpoles metamorphose too rapidly giving rise to very small black frogs which soon die

If thyroids are removed giant tadpoles can be reared which are unable to metamorphose

Normal-size frog after metamorphosis

Fig. 11.13.
The effect of thyroxine on metamorphosis of tadpoles.

Colour changes
In the amphibians colour changes are also based on the aggregation or dispersal of pigment within the melanophores of the skin, and as with fishes these movements are controlled by antagonistic pituitary hormones. Amphibians also secrete a hormone, melatonin, from their pineal gland, which causes changes in the size of the pigment cells of the skin.

11.6.3 Reptiles

Colour changes
In many lizards, such as the iguana, pituitary and adrenal hormones, as well as nervous mechanisms, control colour changes. In others, such as the chameleon, the control is exclusively nervous.

11.6.4 *Birds*

Reproduction

In 1848 the classical experiments of Berthold on cockerels pointed clearly to the existence of a chemical coordination system, even though hormones as such were not identified positively until the work of Starling on secretin some 60 years later. Berthold showed that implanting a testis into a castrated cockerel led to normal development of the secondary sexual characters such as comb, wattle and spurs, while without the implant the bird developed into a capon, with none of these male characteristics. He concluded from this that the secondary sexual characters were caused by an active substance, produced by the testis that circulated in the blood.

In birds, as in mammals, thyroxine plays a part in sexual maturation, as well as in metabolic regulation. Oddly enough, while thyroxine is important in the lower vertebrates for differentiation and maturation it does not appear to affect their metabolic rate as it does in the birds and mammals. There may be some connection with the role played by the hormone in maintaining endothermy in the latter groups.

The different behaviour of males and females during the reproductive period has been associated with gonadal hormones. An interesting subsidiary hormonal effect is the secretion of 'milk' from the pigeon's crop, which is caused by prolactin from the pituitary.

Migration

The removal of the gonads does not prevent migration taking place, so that the drive to migrate cannot be directly associated with ripening gonads. It is thought that the gonadotrophic hormones from the anterior pituitary, whose secretion is stimulated by the lengthening day, also change the physiological state of the bird. One of these changes is the deposition of fat, which acts as the 'fuel' in migratory species (it is an economical substance to store), and it may be this latter change which acts as the direct stimulus to migration.

12 Reproduction

12.1 Introduction

The life of an animal may be considered as a single chain from its conception until it has reproduced its kind. As far as biological survival is concerned the post-reproductive phase of life is unimportant for the majority of animals and, in nature, few survive long after their reproductive capacity is exhausted. In fact successful reproduction is the main objective of all individual members of a species, so during evolution we might expect this physiological process to have come under the most rigorous selection.

For reproduction to be successfully carried out a number of events must occur. In the first place the balance of hormones must shift from encouragement of growth towards the stimulation of the gonads and expression of the secondary sexual characters. There must develop both the means of reproduction – that is, gametes and sex organs – and also the drive to cooperate with another individual to ensure fertilisation. In the second place a satisfactory balance must be struck between the number of offspring needed in order for enough to survive to reproduce themselves and the number that the limited food supply of the environment can support.

The efficiency of reproduction is much increased where the parents protect and provide for the young before or after birth, and during the evolution of the vertebrates there has been a trend of increasing efficiency in the reproductive processes. Parallel adaptations are found in lower animals: for example the progressive provisioning of larvae that is seen in the Hymenoptera and other insects is a parallel of the post-natal feeding of the offspring in birds and mammals. Although there is a general tendency towards a reduction in the number of young and increasing parental care throughout vertebrate evolution there are many lower vertebrates that show convergent adaptations with birds and mammals (Table 12.1). There can be no doubt though that it was the colonisation of the land that was the major environmental change affecting the reproductive processes of vertebrates and that it was associated with a number of adaptations. Thus internal fertilisation, the cleidoic (enclosed in shell) egg, the foetal membranes – which all affect reproduction – have only exceptionally been selected for in the stable external environment of the sea.

301

TABLE 12.1. *The number of eggs laid by different vertebrates*

Animal	Eggs produced at one time
Fish	
Cod	3,000,000–7,000,000
Herring	30,000
Amphibian	
Frog	1,000–2,000
Reptile	
Adder	10–14
Bird	
Pheasant	14
Thrush	4–5
Mammal	
Dog	4
Man	1

We will now consider briefly the typical methods of reproduction within each vertebrate class, drawing attention to the special adaptations that have been made at each level.

12.2 Fishes

Most teleosts, such as cod or herring, produce a very large number of yolky eggs at a time, the exact number depending on the food supply in the environment at the time. The stimulus for ripening of the gonads is supplied by pituitary hormones which depend for their secretion on seasonal environmental changes such as day length. Nearly all fishes show antenatant (against the current) spawning migrations, so that they release eggs up-current of the feeding grounds; the larvae then drift on the current towards the latter. The majority of marine fish, an exception being the herring, thus have pelagic, or floating, eggs. Most river fish, however, make nests or secure their eggs in some way to prevent them being swept downstream.

In many marine fishes there is little sexual dimorphism and to ensure coordinated release of gametes elaborate shoaling rituals may take place. Such reproductive patterns are found in the herring and cod. In some marine fishes (such as pipefish) and many freshwater ones (such as sticklebacks) the sexes are morphologically very different in the breeding season and reproduction may involve territorial behaviour, courtship ceremonies, nest building and subsequent parental care. Both the salmon and the stickleback build a simple type of nest for their young while others such as the African catfish incubate their young within the buccal cavity. In the case of the seahorse and the related pipefish, the young are reared in a special brood pouch by the males.

Fish on the whole, though, show little parental care for their offspring and because percentage survival is low, very large numbers of offspring need to be produced to ensure that enough survive to maturity. Because of its economic importance the North Sea plaice has been studied in detail. In this species the egg has a relatively short-term supply of yolk (obviously if very large numbers of eggs are being laid the amount of nutrient available to each is strictly limited) which is only enough to feed the embryo until it hatches (Fig. 12.1). After the larval plaice hatches it has to find a regular supply of a particular diatom species in the phyto-plankton if it is to survive. The mortality rate at this time is very high and nowadays some newly hatched larvae are reared in tanks as a means of increasing survival in this hazardous planktonic stage. Once the fish are about postage-stamp size they are returned to the sea to increase the stocks. Planktonic eggs do, of course, greatly assist dispersal of the young.

The most interesting adaptations in the reproductive processes of fishes are the forms of viviparity (young born alive) that are found in

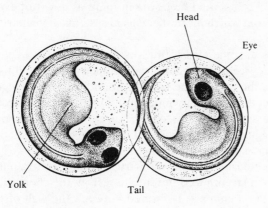

Head

Eye

Yolk

Tail

g. 12.1.
eveloping plaice eggs,
owing the embryo
h and the yolk
pply on which it
eds until it hatches.

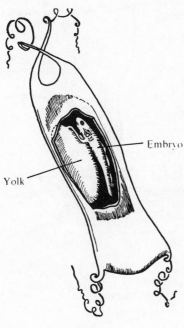

Embryo

Yolk

g. 12.2.
ogfish's egg case
ermaid's purse), cut
vay to show the
veloping embryo.

303

some teleosts but more commonly in elasmobranchs. Where viviparity occurs in teleosts, as in the guppy, the eggs always develop within the ovary, but there are large variations between species in the food supply that the mother donates to the developing embryo after fertilisation – from very little to almost 100 %.

Typically elasmobranchs have internal fertilisation, with the claspers of the male acting as an intromittent organ. Because internal fertilisation is much more reliable than the external fertilisation of, say, the plaice, where male and female gametes are just shed into the sea, this allows the fish to concentrate its reproductive efforts in a small number of eggs each with a large supply of yolk. In the common dogfish (*Scyliorhinus canicula*) fertilisation is followed by the secretion of a protective case around the eggs that is anchored by its trailing threads to objects on the sea bed. This typical development of the embryo in the egg sac, as shown in Fig. 12.2, should be borne in mind as a sort of evolutionary starting point from which more specialised systems of viviparity arose in elasmobranchs.

Thus in some elasmobranchs a condition has evolved where there is some degree of assimilation of oviduct-wall secretions from the mother. Such is the case in the smooth hound, *Mustelus vulgaris*, whose developing embryos take up mucoprotein, fat and monosaccharide sugar. The uterine secretions also contain urea, which is important in the fish's osmoregulatory mechanism. (One of the functions of the dogfish egg case is to act as a urea-containing membrane.)

Other elasmobranchs show true viviparity and a placenta (known as a yolk sac placenta to differentiate it from the chorioallantoic placenta of eutherian mammals) is formed between the wall of the uterus and the

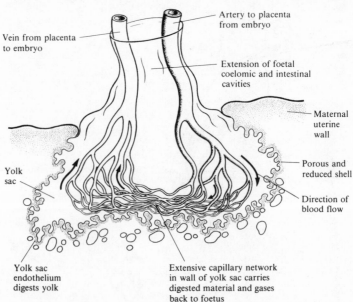

Fig. 12.3.
The york sac placenta of the elasmobranch *Mustelus laevis*.

Artery to placenta from embryo

Vein from placenta to embryo

Extension of foetal coelomic and intestinal cavities

Maternal uterine wall

Yolk sac

Porous and reduced shell

Direction of blood flow

Yolk sac endothelium digests yolk

Extensive capillary network in wall of yolk sac carries digested material and gases back to foetus

yolk sac (Fig. 12.3). Where a placenta is formed, the wall of the uterus may have large numbers of villi, and projections from the yolk sac burrow into this to make a bond of large surface area between parent and embryo. In fishes such as *Mustelus laevis*, the weight of the embryo increases several hundred per cent during development.

A transfer of substances between the female elasmobranch and its developing young is associated with a severe loss in weight of the maternal liver during gestation. This is true both of those species that form a placenta and those that secrete nutritive fluids. Loss of liver weight does not take place in mammals during gestation.

12.2.1 *Sex hormones in fishes*

Reproductive processes in fishes, as in all vertebrates, are under the dual control of gonadotrophic hormones from the pituitary, and gonadal hormones from the gonads. The former are similar in general structure throughout the class but have evolved some differences whereas the latter are identical in all vertebrates (thus testosterone will promote the secondary sexual characters in a male fish as readily as in a bird or mammal).

In all vertebrates except the lampreys, where the hypothalamus does not seem to be involved, the hypothalamus and pituitary work together in the production of the gonadotrophic hormones. Male fishes tend to have high production of spermatozoa at one season of the year and this is stimulated by a hormone analogous to the follicle stimulating hormone (FSH) that controls ovarian function in the female. Gonadal oestrogens in the females have similar effects to those seen in higher vertebrates. Prolactin (see p. 329) is found in fishes but it functions in osmoregulation and salt balance; as far as is known luteinising hormone as such is not found in fishes. Also fish gonadotrophic hormone is something of an unspecialised ancestral form. Thus the spectrum of reproductive hormones found in fish is less complex than that of mammals.

12.3 **Amphibians**

Stimulated by increasing daylight and other factors the endocrine glands of the frog which control reproduction become active during the latter part of hibernation (Fig. 12.4). Metabolic changes occur and the quantities of stored fat decrease while the gonads enlarge and ripen. Most frogs have to return to water to reproduce and they may migrate several miles to find their original spawning ground. The croaking of the male frog and the enlarged belly of the female are two of the stimuli that lead to mating, which involves the male grasping the female with his special nuptial pads. Fertilisation of the eggs takes place externally and the sperm must penetrate the egg rapidly before its coating of albumen swells. In the male the sperm passes down from the testes via the anterior part of the kidney and the Wolffian ducts. Both sexes have a cloaca via which genital

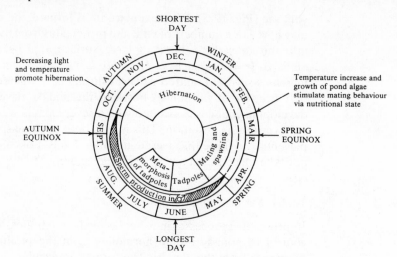

Fig. 12.4.
Annual cycle of activity
in the common frog
Rana temporaria.

as well as excretory products are passed to the exterior (Fig. 12.5). All
these features, as well as the need to return to water, indicate the primitive nature of amphibian reproduction.

Once laid, the eggs have the limited protection of their albumen
coating and the fact that the black pigment they contain has a bitter taste.
The larval amphibians or tadpoles are well adapted to life in water, having
respiratory and locomotory systems as well as sensory adaptations, such
as the lateral line, which are similar to those of fishes.

The main adverse feature of the reproductive system described above
is the necessity for an aquatic environment. Some tree frogs provide this
themselves by 'sewing' or sticking leaves together to make hollows high
above the ground where rain water collects and into which they can lay
eggs. While lack of parental care is typical of the majority of amphibians
there are a good number of exceptions, one of the most remarkable of
which is the female midwife toad, which has a number of holes in her back
into which she pushes the eggs after fertilisation. In these pockets the
tadpoles develop.

12.3.1 *Sex hormones in amphibians*

Many Amphibia have prominent sexual dimorphism at the time of breeding and the nuptial pads of frogs, the breeding crests of newts and the
vocal sacs of certain toads can all be induced to develop in males by
application of testosterone.

In female amphibians there is still a single gonadotrophin, which is
secreted in response to environmental influences (temperature, light,
nutritional state) working via the hypothalamus–pituitary complex. This
gonadotrophin stimulates the development and maturation of follicle
cells in the ovary and causes it to secrete oestrogens. These in turn cause
the manufacture of certain phospholipid proteins in the liver which will

306

(a)

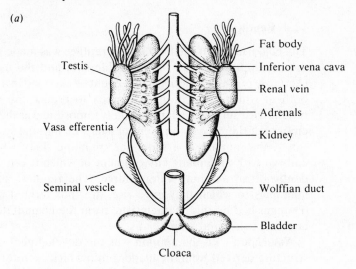

Fat body
Testis
Inferior vena cava
Renal vein
Adrenals
Vasa efferentia
Kidney

Seminal vesicle
Wolffian duct

Bladder

Cloaca

(b)

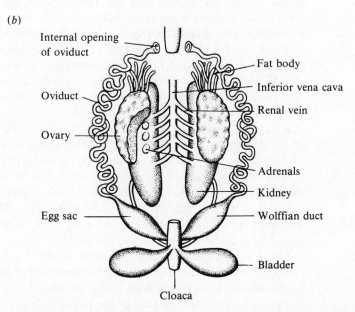

Internal opening
of oviduct
Fat body
Inferior vena cava
Oviduct
Renal vein
Ovary
Adrenals
Kidney
Egg sac
Wolffian duct

Bladder

Cloaca

Fig. 12.5.
The urinogenital
systems of (a) a male
and (b) a female frog.

later be incorporated into the eggs in the form of yolk. After ovulation corpora lutea are formed which later secrete progesterone, as in mammals, though its role in amphibians is very different: progesterone activates the oviduct to produce the albumen which will eventually cover the eggs.

While reptiles and other higher vertebrates produce sperm during the pre-nuptial season (in temperate zones – the early spring), most male amphibians produce sperm in the post-nuptial phase, that is during the summer. The sperm remains viable until the nuptial period the following spring.

307

12.4 **Reptiles**

The colonisation of the land by the reptiles was made possible partly by the development of an impermeable skin and the use of nitrogenous excretory product (uric acid) that could be excreted in a semi-solid form – both of which adaptations did much to reduce the amount of water required for survival on land. Equally important was the formation by the embryo during its development of a new membrane system, the amnion. Inside the amnion is a fluid, the amniotic fluid, which provides the embryo with an aquatic environment in which it can develop independently of an external water source. The reptiles, and the birds and mammals to which they gave rise, are thus termed amniotes and are distinguished by this characteristic from the anamniote (i.e. without an amnion) fishes and amphibians.

Associated with the amnion was the development of the allantois, a structure derived from the bladder into which excretory material can be passed during the growth and differentiation of the embryo. Internal fertilisation became a necessity because of the secretion of a leathery (sometimes lime-encrusted) shell around the embryo, allantois, amnion and yolk sac. Thus was produced the pattern of development of the young that remains unchanged in the birds and that, with modification of the allantois and outer surface of the amnion (to form the chorion), eventually allowed the development of the chorioallantoic placenta of the eutherian mammals.

Within the reptile egg the embryo develops using the yolk as a source of proteins and respiratory substrates and the albumen as a source of water. The allantois grows out to touch the chorion and provides the site for the uptake of oxygen and the release of carbon dioxide through the porous shell. It also acts as a 'kidney of accumulation', storing nitrogenous waste in the form of uric acid, allantoic acid and urates. These are all fairly insoluble substances so there is no danger of them leaking out and poisoning metabolic processes in the embryo.

The development of a shell to counteract the force of gravity on the soft materials of the egg necessitated the provision of an egg-tooth (e.g. in lizards and snakes), or a special horny modification on the head (e.g. in turtles and crocodiles) to enable the young to break out of the egg.

The parental care given after birth is very limited, although the eggs of reptiles need warmth for their development and this is often provided by the mother (as in the case of snakes, which coil their bodies around their egg clutch). Turtles and crocodiles bury their eggs to achieve a warm, uniform and protected environment.

A very few reptiles, mainly lizards living in colder habitats, are viviparous and in a few cases there is an exchange of metabolites between mother and offspring.

12.4.1 *Sex hormones in reptiles*

Sexual dimorphism is not marked in reptiles but many male lizards have erectile crests around their necks and the growth and display of these secondary sex characters is under the influence of testosterone. Unlike male amphibians the development of sperm in the males immediately precedes the nuptial period in the majority of species. In both sexes the ambient temperature appears to be the cardinal environmental factor that, via the hypothalamus, stimulates the release of gonadal hormones. The reptiles have only one gonadotrophin, except for one species, the snapping turtle, where separate pituitary gonadotrophins corresponding to the FSH and LH of mammals were found to be present.

Progesterone, originating from the corpus luteum, appears in the female reptilian hormone circulation over the reproductive period and following ovulation. One of its roles appears to be to inhibit further gonadotrophins.

12.5 **Birds**

As in the reptiles the extra-embryonic membranes provide, in the form of the amnion and its associated amniotic fluid, a uniform medium within which development can take place (Fig. 12.6). The allantois grows out from the hind gut until it comes to lie immediately next to the chorion and close to the porous egg shell. Via the allantois respiratory exchanges occur and into it waste nitrogenous products are transported.

As distinct from reptiles all birds have hard-shelled eggs and their embryos all have the sharp egg-tooth needed to cut their way through the

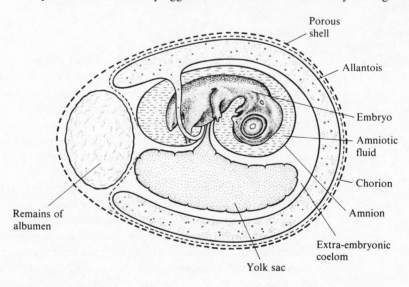

ig. 12.6.
implified diagram of
he relationship of the
xtra-embryonic
embranes to the shell
nd embryo in a bird's
gg that is about
alfway towards
atching.

Reproduction

Winter

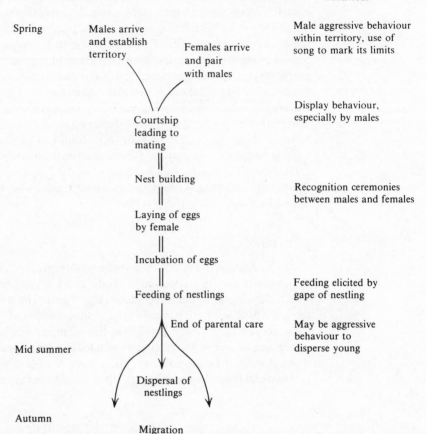

Fig. 12.7.
Scheme to show how
many bird pairs
cooperate over a
period of months for
reproductive purposes.

shell on hatching. Birds have a much wider range of colour and shape to their eggs than do reptiles, those species with open nests (e.g. plover) usually having camouflaged eggs while those with concealed or domed nests (e.g. owl and long-tailed tit) have white eggs. There is a general tendency in the more evolutionarily advanced species to reduce the number of eggs laid (thus the ostrich has many eggs while the fulmar has only one), but on the whole birds lay far fewer eggs than do reptiles. It has been shown that, within the limits for a given species, the number of eggs laid by a bird is related to the number of offspring that it can successfully provision. Variations exist within a given species according to the date of nesting, the latitude and the environmental conditions.

Parental care is well developed over the whole reproductive period and the male and female tend to cooperate in the making of the nest, incubation of the eggs and subsequent care and feeding of the young (Fig. 12.7). Nest sites are chosen to give the maximum protection from predators and are insulated with soft materials such as down and moss. The actual position of the nest in regard to those of other members of the species is also important. Land birds have quite extensive territories which they defend and which have the effect of dispersing a given species efficiently in an area. Sea birds have their nests placed close together for protection against predators but still regard the small area around their nests as territory.

Because birds are endothermic it is essential that they maintain their eggs at a constant temperature. The construction of the nest assists in this and the female (and sometimes the male) develops an increased blood supply to the skin as well as losing feathers from the breast. The female also develops the drive to incubate, which is very strong and can be clearly seen in the behaviour of a 'broody' hen.

After hatching the young are usually cared for by both parents, the gape of the fledgling's beak acting as the releaser to the parents' feeding responses. The young of nidicolous species are slow to develop and remain for a long time in the nest, which will tend to be built away from predators. Nidifugous young develop rapidly and soon leave the nest, which will be sited on the ground. Most passerines come into the first category while game birds such as duck and pheasant fall into the latter.

12.5.1 Sex hormones in birds

The onset of reproduction

Photoperiodism, which is the influence of light on reproductive activity, is widespread in birds of temperate regions. In such species increasing amounts of light in the spring coupled with rising temperatures act, via receptors such as the eyes and pineal gland, on the hypothalamus to cause releasing factors for pituitary hormones to be secreted. The gonadotrophic hormones from the anterior pituitary then lead to maturation of the gametes and secretion of gonadal hormones from the testes or ovaries. These latter hormones act on the central nervous system, lowering the threshold of response to specific external stimuli and thus triggering off the various instinctive behaviour patterns involved in the reproductive activity of the individual species.

While light and to a lesser extent temperature may act as the main environmental factors associated with the onset of breeding the 'fine control' is via numerous more subtle stimuli, which may include display by a mate, group activity and song, amount of food available, space and the precise nature of the local weather. The failure of some birds to breed successfully in captivity almost certainly indicates that normal interaction of such stimuli is lacking or imperfectly balanced.

Sexual dimorphism and hormone levels

The role of the sexes is often complementary in the successful rearing of the young, and patterns of behaviour are sustained over considerable periods of time. Behavioural interactions between mates in their court-

311

ship and throughout incubation of the eggs serve to maintain, through psychological stimuli, the level of secretion of sex hormones, which in turn leads to parental care of the offspring when they hatch.

Birds often show marked sexual dimorphism between male and female, which must certainly play an important part in the initial selection of mates. This dimorphism can affect size, type and colour of plumage, nature of the song, colour of the beak, size of the comb and other attributes as well as the behaviour pattern exhibited. On the whole the more neutral colours belong to the female and the brighter ones seem induced in the male body by the high levels of testosterone and other androgens. This is true for species such as the mallard duck and the domestic fowl among many others, and in the case of the latter the male chick can be induced by injections of androgen to develop a comb and to crow by the time it is seven days old.

The gonadotrophic hormones

Birds, like mammals, have two gonadotrophic hormones. From the anterior pituitary of the female bird FSH is secreted and this causes the maturation of a number of follicles in the ovaries. No corpora lutea are formed in birds (unlike all other amniotes) but progesterone is produced by the developing ova and their associated theca cells and this appears to act on the hypothalamus to bring about secretion of LH releasing factor. LH from the anterior pituitary then causes ovulation to occur, as it does in mammals (see p. 322). The whole sequence is thus very complicated and involves the following stages:

$$FSH \rightarrow ova \rightarrow \text{pro-gesterone} \rightarrow \text{hypo-thalamus} \rightarrow \text{ant. pituitary} \rightarrow LH \rightarrow ovulation$$

About the time of ovulation prolactin, also from the anterior pituitary, is secreted and may bring about broody and nesting behaviour in some species such as the domestic fowl. In the case of pigeons this is the hormone which causes the release of the 'milk' from the crop which will be used to feed the nestlings.

Posterior pituitary hormones are released after ovulation and cause the rhythmical contractions of the oviduct that lead to the laying of the eggs. One such hormone is oxytocin, so its effect here is very comparable to that in the mammal.

In the male bird FSH from the pituitary brings about development of the tubules of the testis and spermatogenesis, while ICSH (identical to LH in females) stimulates the Leydig or interstitial cells of the testis to produce testosterone.

The gonadal hormones

Oestrogens from the ovaries inhibit the gonadotrophic hormones (as they do in mammals) and by this means cycles of ovulation are produced. Oestrogens also bring about development of the secondary sexual characters and reproductive behaviour of the female bird. Table 12.2 shows the

TABLE 12.2. *Effects of oestrogen on some blood constituents of hens*

Blood constituent	Control bird	Hen given 125 mg oestrogen
Calcium	11.4	137 (mg/100 ml blood)
Phosphate	5.4	22 (mg/100 ml blood)
Lipid	–	Great increase
Protein	–	Considerable increase

effects of oestrogen on some blood constituents of hens. It can be assumed that these increases in minerals are related to increased activity of the shell-producing glands and that the nutrient increases are concerned with the provisioning of the yolky eggs.

In the male bird testosterone and other androgens bring about development of the secondary sexual characters and the sexual behaviour as described above.

12.6 **Mammals**

12.6.1 *The monotremes (or Prototheria)*

The monotremes are a group of very primitive egg-laying mammals of which there are only two living types: the spiny anteaters and the duck-billed platypuses. They have a large egg which after fertilisation receives nutritive acidic glycoprotein uterine secretions. The egg is shelled and after laying is kept in a pouch that forms in the mid-line of the abdomen. Many modified sebaceous glands open into this pouch and after the egg hatches these provide the baby monotreme with a secretion resembling milk. The embryo monotreme is interesting in that it has an egg-tooth (similar to that of birds) with which it breaks its way out of the shell. At the time of hatching the hind end is still in a very undifferentiated state and the young animal remains in the mother's pouch for some time. This pouch enlarges over the reproductive period and is more like a shallow groove than the much deeper and more specialised pouch of marsupials.

12.6.2 *The marsupials (or Metatheria)*

The young marsupial is also born in a very undifferentiated state. In the largest marsupials, such as kangaroos, it may weigh nearly a gram but in the smaller ones it is as light as 0.01 grams. The young is attached to the uterus by a simple type of yolk sac placenta which, unlike the true placenta of the eutherian mammals, does not produce any hormones. After birth it crawls up the mother's abdomen and finds its way to the marsupium or pouch. Once here it fixes to a teat and continues its growth and development for a long period. Thus in the kangaroo the time spent in the uterus is only 33 days while the period in the pouch is usually in excess of 200 days.

The female marsupial has two vaginae and in connection with this the males of some species have a bifurcated penis which allows deposition of sperm in both vaginae. The stimulus of the suckling of the young on the teat provides a feedback system via the hypothalamus and endocrine organs of the mother that prevents further ovulation. In some species the mother will have one free feeding offspring, one at the teat and yet another in suspended animation inside her body.

While marsupials have obviously been successful mammals in their time, the young is much more subject to desiccation, disease and damage in the mother's pouch than is the eutherian embryo inside its mother's uterus for the whole of its early life. This is one of the reasons why marsupials do not, on the whole, compete successfully with the eutherian mammals to which they gave rise.

12.6.3 *The Eutheria*

These are the most successful of the three groups of mammals and they are characterised by the possession of a chorioallantoic placenta and the internal development of the young. Parental care of the offspring is also very highly developed in this group and their reproductive efficiency is one of the reasons for their biological success.

Sexual dimorphism is found in eutherian mammals to a greater or lesser degree and in many species the roles of the male and female parents in the provisioning and protection of the young are complementary.

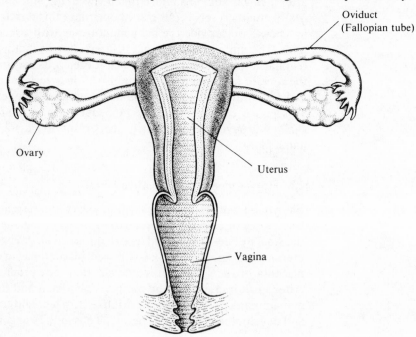

Fig. 12.8.
Human female
reproductive organs
(opening of urethra not
shown).

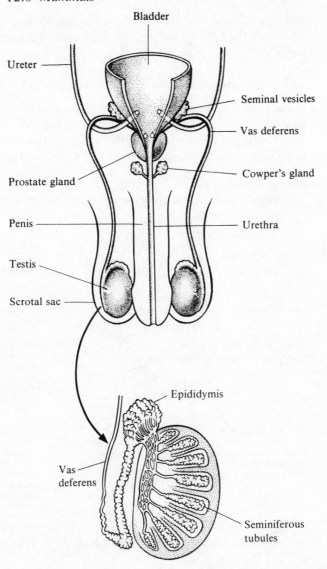

Bladder

Ureter

Seminal vesicles

Vas deferens

Prostate gland

Cowper's gland

Penis

Urethra

Testis

Scrotal sac

Epididymis

Vas
deferens

Seminiferous
tubules

Fig. 12.9.
Human male
reproductive organs.

Many species show some form of social organisation in which the
behaviour of the group may also increase the chances of survival in
newborn individuals.

The basic anatomy of the reproductive organs of the higher mammals
follows a fairly standard and well-known arrangement and it is inappro-
priate to the aims of this text to cover this in detail. However for reference
in the following discussion of function the basic female and male reproductive systems are shown in Figs. 12.8 and 12.9. The species represented
is man.

Areas of special interest are the development and structure of the
gametes and the process of fertilisation and subsequent development.
Also, much is now known of the roles of hormones in the sexual cycles

Reproduction

and in gestation and parturition and in the feeding of the newly born. These are the topics which will be dealt with in the following sections of this chapter.

12.6.4 *Gametogenesis: the development of the sex cells*

In the gonads of the mature mammal the process of gametogenesis occurs whereby diploid cells produce by meiosis haploid daughter cells – the spermatozoa of the male and the ova or eggs of the female. In the male the process is termed spermatogenesis and in the female oogenesis.

Spermatogenesis

The mammalian testis consists of many coiled seminiferous tubules

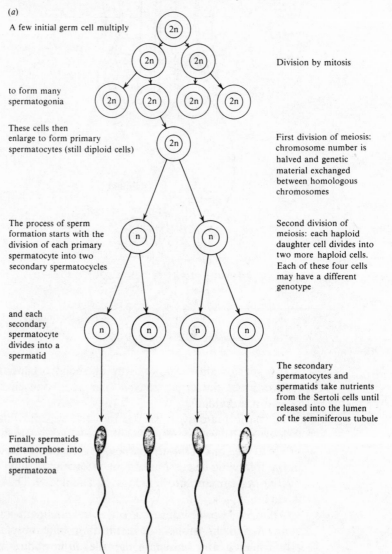

(a)

A few initial germ cell multiply

Division by mitosis

to form many spermatogonia

These cells then enlarge to form primary spermatocytes (still diploid cells)

First division of meiosis: chromosome number is halved and genetic material exchanged between homologous chromosomes

The process of sperm formation starts with the division of each primary spermatocyte into two secondary spermatocycles

Second division of meiosis: each haploid daughter cell divides into two more haploid cells. Each of these four cells may have a different genotype

and each secondary spermatocyte divides into a spermatid

The secondary spermatocytes and spermatids take nutrients from the Sertoli cells until released into the lumen of the seminiferous tubule

Finally spermatids metamorphose into functional spermatozoa

Fig. 12.10.
Spermatogenesis.
(a) The cell divisions involved (2n = diploid; n = haploid).

316

(*b*)

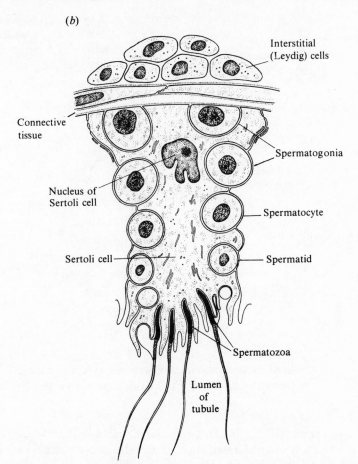

Interstitial (Leydig) cells

Connective tissue

Spermatogonia

Nucleus of Sertoli cell

Spermatocyte

Sertoli cell

Spermatid

Spermatozoa

Lumen of tubule

Fig. 12.10 (*contd*). (*b*) a transverse section through part of the testis to show the arrangement of the cells.

whose length in man may exceed 200 metres. The cells that line the tubules are termed spermatogonia and these proliferate by mitosis to make large numbers of similar diploid cells (Fig. 12.10*a*). At some stage the spermatogonia are changed, by hormonal stimuli, into primary spermatocytes with large central nuclei. These then start to undergo meiosis.

The first meiotic division produces two haploid secondary spermatocytes which have smaller and more compact nuclei. These in turn complete the meiotic process to produce four haploid spermatids, which are partially embedded in the large Sertoli cells that are between the banks of developing spermatogonia (Fig. 12.10*b*). Finally, while still attached to the Sertoli cells, the spermatids develop into mature spermatozoa.

The whole process is under control of the hormone FSH (see p. 284) and in the case of man continues steadily from the time of puberty until old age. In other mammals, such as insectivores and the orders of herbivorous and carnivorous mammals, the testes are only active seasonally, at a time that coincides with the oestrous cycles of the females.

It is generally supposed that in many mammals viable sperm are stored at somewhat below the normal body temperature and for this reason the

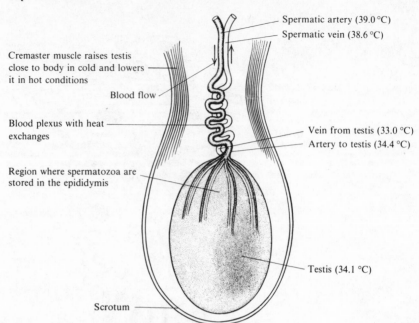

Spermatic artery (39.0 °C)

Spermatic vein (38.6 °C)

Cremaster muscle raises testis close to body in cold and lowers it in hot conditions

Blood flow

Blood plexus with heat exchanges

Vein from testis (33.0 °C)

Artery to testis (34.4 °C)

Region where spermatozoa are stored in the epididymis

Testis (34.1 °C)

Scrotum

Fig. 12.11.
A counter-current heat exchange mechanism keeps the temperature of the human testis some 5 °C below the core temperature of the body.

testis, and the epididymis where the sperm are stored, descend outside the body cavity. Such is the case in man (see Fig. 12.11).

Oogenesis

The activity of the ovary is cyclic, unlike that of the testis, and the pattern of activity and ovulation, the oestrous cycle, varies from one species to another. Most female mammals show no reproductive behaviour during their anoestrous period when the ovary is quiescent.

All female mammals with the exception of human females have a breeding season, and if they show a single cycle of ovulation during that period they are termed monoestrous. Such is the case for many carnivores, including dogs and cats. If, on the other hand, the female has many cycles of ovulation during the breeding season then the species is polyoestrous. This is the case in horses and other herd animals. In fact females of domesticated mammals tend to change their rhythms from monoestrous to polyoestrous, showing that environmental factors can produce phenotypic change. While the significance of the various cycles is not very clear it is certain that the overall result of the breeding pattern is to allow the female to produce her offspring at a time when they are most likely to survive.

In the case of the human female there is a more-or-less regular 28-day cycle of ovulation which starts with the ripening of a single (or very exceptionally two or more) Graafian follicles (Fig. 12.12). Although the human female may possess a couple of million immature follicles in her ovaries at birth the number is very substantially reduced by the time of puberty, and most of these surviving follicles will never develop but will degenerate in time.

Within each follicle the process of meiosis has started but becomes suspended at the stage of the first prophase. Maturation of the ovum only occurs at ovulation, when the primary oocyte first completes meiosis I to produce a haploid secondary oocyte and a first polar body and then this secondary oocyte completes meiosis II to produce a mature (haploid) ovum, or egg, and a second polar body. In some species, such as the rabbit, the maturation division is only completed after fertilisation.

Cells within the ripening Graafian follicle break down to provide some of the follicle liquor and some cells become granular (granulosa cells) and form the basis of the cumulus oophorus that surrounds the egg. Meanwhile the egg itself produces the zona pellucida, homologous to the vitelline membrane in lower vertebrates. Within this zona pellucida is the true egg cell, with its protoplasmic membrane rich in microvilli and a space, the perivitelline space, between this membrane and the zona pellucida.

Meanwhile in the ruptured follicle granulosa cells, under the influence of luteinising hormone (see 12.6.6), enlarge and become secretory, producing the hormone progesterone. If fertilisation has not taken place the corpus luteum (which is the name given to the modified follicle) degenerates and the oestrous cycle starts again with the ripening of another primary follicle.

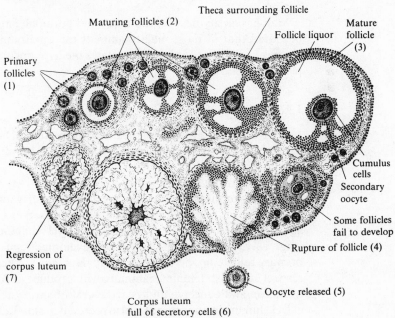

Fig. 12.12. Simplified diagram of the human ovary, showing the stages (1 to 7) in the maturation of a follicle, ovulation, and the formation and regression of the corpus luteum.

319

12.6.5 *Sex hormones in the male*

The sex hormones can be divided into those which act on the gonads, which are termed gonadotrophic hormones, and those produced by the gonads themselves, termed gonadal hormones. The pattern of secretion and action is relatively straightforward in the male.

As a result of its genetic programming the hypothalamus will, as puberty approaches, produce increasing amounts of releasing factor for the FSH (follicle stimulating hormone) that is produced by the anterior pituitary. This gonadotrophic hormone acts on the seminiferous tubules of the testis directly and causes the proliferation of the spermatozoa. At the same time ICSH (interstitial cell stimulating hormone), also from the anterior pituitary, acts on the interstitial cells in the testis, which thereafter secrete the male sex hormone testosterone.

Testosterone is a gonadal hormone that besides acting synergistically with FSH to stimulate spermatogenesis also promotes the development of the secondary sexual characters and the sex drive of the male. Further effects are on the reproductive organs themselves, which in the case of the prostate gland and Cowper's gland it stimulates into growth and secretion.

The levels of the various hormones are partly controlled by the negative feedback effect that testosterone has on the secretion of the pituitary gonadtrophic hormones. When the level of testosterone rises it inhibits FSH and ICSH secretion so the testosterone level falls off; when it does so the gonadotrophic hormones begin to increase in the bloodstream once again.

Man has an atypical reproductive and hormone pattern, and it should be noted that in most male mammals the environmental influence of increasing light in the spring stimulates the hypothalamus–pituitary complex into action.

12.6.6 *Sex hormones and cycles in the female*

In contrast to the situation in the male, the pattern of hormonal control in the female is very complex and involves more hormones.

Normal menstrual cycle control

In the human the menstrual cycle starts at puberty and normally occurs regularly every 28 days until the menopause somewhere in late middle age. The cycle starts with the secretion of FSH from the anterior pituitary (Fig. 12.13), which stimulates the development of (usually) a single primary follicle in one or other ovary. Humans are again atypical in this respect as most female mammals will produce a number of eggs at ovulation and conceive and bear several offspring at a time.

LH (luteinising hormone) is also secreted in small amounts at this time and works together with the FSH in causing the follicle to ripen. (LH is the equivalent of ICSH in the male.)

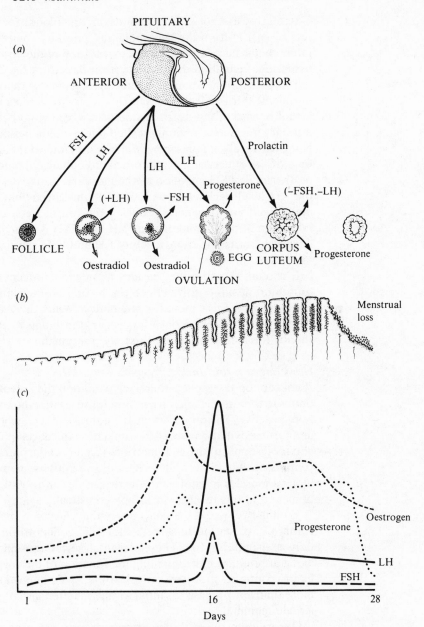

(a)

PITUITARY

ANTERIOR   POSTERIOR

FSH

LH

LH   LH

LH

Prolactin

(+LH)   −FSH   Progesterone   (−FSH,−LH)

FOLLICLE

Oestradiol   Oestradiol   EGG   CORPUS   Progesterone
LUTEUM

OVULATION

(b)

Menstrual
loss

(c)

Fig. 12.13.
The human oestrous
cycle: (*a*) the hormones
involved; (*b*) the
changes in the lining of
the uterus; (*c*)
hormone levels in the
blood.

Oestrogen

Progesterone

LH

FSH

1   16   28

Days

Meanwhile the ripening follicle starts to secrete the female sex hormone oestradiol, which is one of a class of hormones collectively termed oestrogens. Oestradiol stimulates the development of the mammary glands and causes the development of the muscle and endothelial layers of the uterus, the latter becoming glandular and well vascularised with extensive crypts. The lining of the vagina becomes cornified and the mammary glands enlarge (this process is barely detectable in human females).

321

Rising oestradiol concentrations in the bloodstream cause further release of LH from the hypothalamus–pituitary complex; it is this hormone that is necessary for actual maturation of the Graafian follicle and ovulation. Ovulation usually occurs on day 15 of the 28-day cycle.

LH then brings about the luteinisation of the ruptured follicle and during the third week of the cycle this begins to secrete progesterone. The actual release of the progesterone is now known to be due to a further pituitary hormone sometimes called luteotrophin but now recognised as being identical with prolactin (which is also important in milk secretion). One of the effects of progesterone is to enhance the negative effects of oestradiol on FSH secretion by the pituitary. As the oestradiol level falls the LH level declines as well, so that the corpus luteum is no longer maintained and degenerates during the fourth week of the cycle. The inhibition on the gonadotrophic LH and FSH is thus removed and so a new cycle starts, the rising levels of LH and FSH causing another follicle to mature.

It should be noted that besides its negative effects on the pituitary, progesterone has positive effects on the uterus and mammary glands in a way which prepares them for pregnancy. When the level declines, as it does when conception does not occur, the lining of the uterus is lost (hence the menstrual blood loss) and the mammary glands recede.

Hormone controls during pregnancy and birth

If conception takes place following ovulation the anterior pituitary continues to secrete LH and this, in turn, ensures that the corpus luteum will remain active. The progesterone it continues to secrete will suppress FSH and further ovulation as well as stimulating the development of the blood and food supply to the endometrium (lining) of the uterus. Prolactin also suppresses ovulation, while progesterone inhibits any tendencies of the uterine muscles to contract. (Oestrogen in contrast stimulates contractions and in this respect works antagonistically against progesterone.)

Within two weeks of conception the chorionic membranes of the embryo have started to secrete chorionic gonadotrophins, whose functions are similar to LH. These placental hormones replace LH from the anterior pituitary in maintaining the function of the corpus luteum. This in turn continues to secrete progesterone, an all-important hormone that, more than any other, is essential for the maintenance of pregnancy in the female mammal.

On or about the 60th day after conception chorionic gonadotrophin secretion from the placenta rapidly declines, as a result of which the corpus luteum in the ovary also degenerates and ceases to make progesterone. However as this activity has been taken over by the placenta the progesterone levels continue to rise until very shortly before birth. The placenta also makes copious quantities of oestrogens, which seem important for the increase in the rate of growth of the foetus which is evident from week 20 onwards.

The process of birth, or parturition, is due to a number of factors that

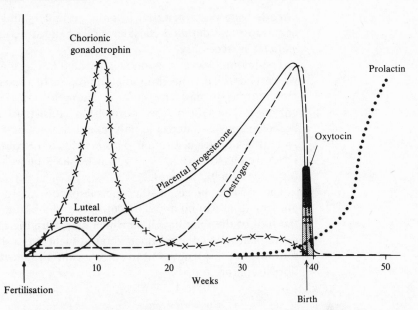

Fig. 12.14.
The changes in blood
hormone levels in the
human female during
pregnancy, birth and
lactation.

work in conjunction. Firstly it now seems certain that the foetus itself initiates the production of a prostaglandin hormone (termed PGF) from the endometrium which passes across the placenta into the mother's blood and acts either directly on the uterus or on the mother's hormone secretions to stimulate parturition. (While the details of this are not yet clear it does seem eminently reasonable that the initial stimulus should come from the full-term foetus rather than from the mother who carries it.) Secondly there is the production of spurts of oxytocin from the posterior pituitary of both mother and foetus which directly stimulate contractions of the uterus; at the same time the level of progesterone, which inhibits such contractions, declines dramatically. Thirdly there is a sharp decline in the level of oestrogen which, it will be remembered, has been working together with progesterone to maintain pregnancy. Yet a further substance, termed 'relaxin', has also been implicated in the complex process of birth; it is thought to cause relaxation of the pubic bone symphyses through which the foetus must pass.

As a result of a combination of all these and, possibly, other events the uterus muscles begin to contract with increasing frequency and strength and as a result of these contractions the foetus is finally expelled from the body of its mother.

The major changes in hormone levels during gestation and birth in the human female are shown in Fig. 12.14.

The gametes
We must now return from a consideration of the hormones involved in gametogenesis and the maintenance of pregnancy to a discussion of the processes of fertilisation and then the formation of the embryo and its extra-embryonic membranes. Spermatogenesis and oogenesis have

323

already been described, but in order to understand how fertilisation is achieved some detailed consideration of the structure of the individual gametes is necessary.

The mature spermatozoon is illustrated in Fig. 12.15. The acrosome, which is derived from the Golgi apparatus of the spermatid during its metamorphosis into the mature spermatozoon, contains proteolytic enzymes. The rest of the sperm head consists of a nucleus which is homogeneous and dense in nature and consists of some 40 % of DNA and 60 % of nuclear protein. This nucleus is, of course, haploid and the DNA weighs some 3×10^{-12} g, which is half the weight of the DNA in a (diploid) somatic nucleus.

The middle piece of the spermatozoon has a number of spiral mitochondria wound about a centriole, and a fibrillar structure which is the root of the long flagellum which makes up the tail. This tail, whose movements allow the spermatozoon to swim, has a sheath over its fibrils for most of their length but towards its end the sheath is absent and the fibrils separate out.

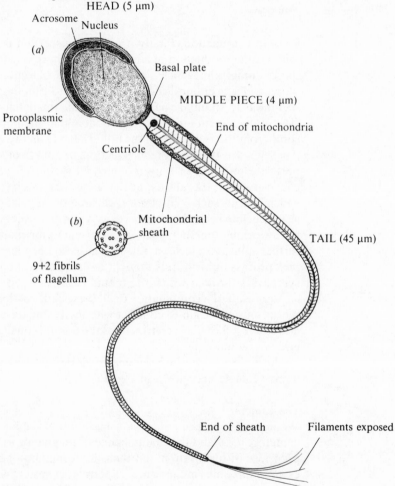

Fig. 12.15. Diagrams, based on electron microscope studies, of a mature human spermatozoon. (*a*) Whole spermatozoon; (*b*) transverse section through the middle piece.

324

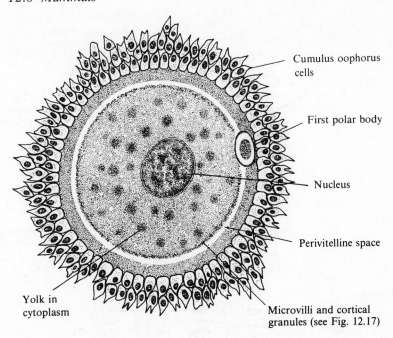

Cumulus oophorus
cells

First polar body

Nucleus

Perivitelline space

Yolk in
cytoplasm

Microvilli and cortical
granules (see Fig. 12.17)

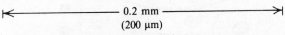

0.2 mm
(200 µm)

Fig. 12.16.
The ultrastructure of
the mammalian egg.

The egg has already been described briefly on p. 319. Its ultrastructure
is shown in Fig. 12.16. At ovulation it is still in the metaphase of the
second meiotic division, which is not completed until fertilisation occurs.
Numerous microvilli line its protoplasmic boundary and outside the
pervitelline space is the zona pellucida which is a firm lipid and glycoprotein complex. Outside this again is the cumulus oophorus, which consists
of granulosa cella held together by the connective properties of
hyaluronic acid.

Fertilisation

Fertilisation normally occurs in the oviduct (or Fallopian tube), whence
the egg has been propelled by the action of the cilia that line these ducts
and by peristalsis. In man some 4 millilitres of semen are deposited in the
vagina at copulation and this volume will contain at least 300×10^6
spermatozoa as well as the fructose, bicarbonate and phosphate buffers,
proteins and salts that make up the whole composition of the semen.
(Why such large amounts of spermatozoa are required is not at all clear. It
was previously thought that only by having so many could a sufficient
concentration of hyaluronidase be produced (see below), but recent work
has shown that even a single spermatozoon has sufficient of this enzyme
to penetrate to the surface of the egg. It may be that such numbers are
needed to ensure that at least a few sperm will end up in the vicinity of the
egg, though contractions of the uterus do aid this process. The large

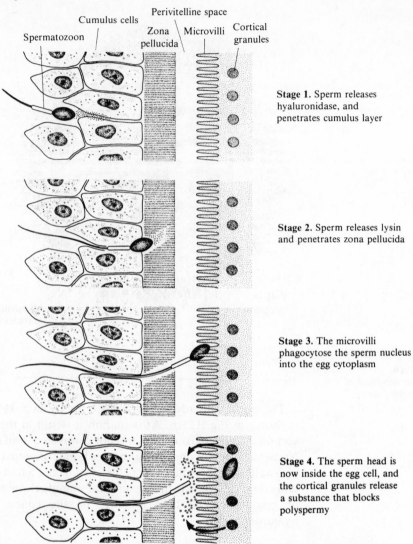

Stage 1. Sperm releases
hyaluronidase, and
penetrates cumulus layer

Stage 2. Sperm releases lysin
and penetrates zona pellucida

Stage 3. The microvilli
phagocytose the sperm nucleus
into the egg cytoplasm

Stage 4. The sperm head is
now inside the egg cell, and
the cortical granules release
a substance that blocks
polyspermy

Fig. 12.17.
The events involved in
the entry of a sperm
into the egg.

numbers do seem necessary, because semen that contains less than 60×10^6 spermatozoa per millilitre is quite infertile.)

During their last stage in the male reproductive tracts the sperm become 'capacitated' by chemicals added to the semen from glands. Capacitation changes the nature of the surface of the spermatozoa and without it the sperm is infertile.

To accomplish fertilisation the sperm acrosome needs to produce two enzymes (Fig. 12.17). The first is hyaluronidase, which is able to hydrolyse the hyaluronic acid that forms the connective material of the cumulus oophorus and allows the sperm to push its way to the surface of the egg. At this stage a second enzyme, lysin, is released from the acrosome which digests a path for the sperm through the tough zona pellucida and allows the sperm head to be pushed, by the lashings of the tail, through this zona and into the perivitelline space. After about half an hour it will have been engulfed by the microvilli on the surface of the egg and actually incorporated into the egg cytoplasm (Fig. 12.17). Within 14 hours the first cleavage of the fertilised egg takes place and within a week the developing blastula has become safely implanted in the endometrium of the uterus.

Prior to fertilisation there are some 10,000 spermatozoa in the vicinity of the egg, so the possibility exists for the egg being fertilised by more than one sperm. Polyspermy (fertilisation by more than one sperm) is prevented by the physical nature of the zona pellucida and by a substance released into the perivitelline space by the cortical granules in the egg as the act of fertilisation is taking place.

(As this edition is in preparation news has come of the first successful case of a human egg being taken from the mother's ovary, fertilised outside her body by the father's sperm, and then deliberately implanted back into the maternal uterus. This technique, which has taken many years to perfect, brings the chance of parenthood to couples who for one reason or another could not conceive by natural means.)

Implantation and the development of the placenta
Soon after fertilisation the zygote begins to divide and within 24 hours a morula of many cells has been formed. All this time it has been moving down the oviduct to the uterus, where it will become implanted. Some time between days 4 and 7 the human blastula becomes attached to the

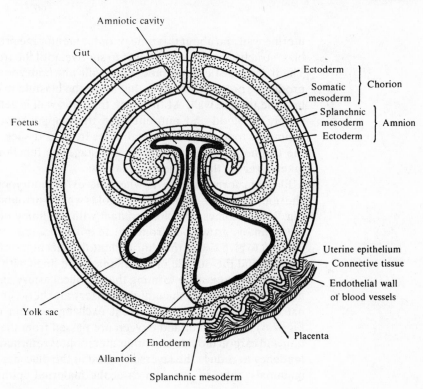

Fig. 12.18.
The formation of the chorioallantoic placenta in eutherian mammals.

327

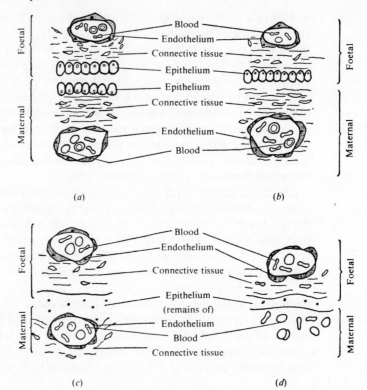

Fig. 12.19.
Types of placenta.
(*a*) Epitheliochorial
(horse);
(*b*) syndesmochorial
(cow);
(*c*) endotheliochorial
(dog);
(*d*) haemochorial
(man).

uterine wall and about this time its outer membrane produces the trophoblast structure which will allow it to survive until the true placenta forms.
The trophoblastic membrane forms villi which invade the endometrium,
producing protease enzymes that allow the blastula to digest its way well
into the uterine wall. At the same time this wall is getting an increasing
supply of blood and nutrients and the trophoblastic villi are able to
assimilate food and other metabolites from its tissues. However, though
the trophoblast is important at this stage, its functions are soon to be
taken over by the developing placenta.

Placenta formation starts when the extra-embryonic membrane, the
chorion (the outer of the amniotic folds which surround the embryo – see
Fig. 12.18) comes into close contact with the lining of the uterus. Later
the allantois grows out from the foetal endoderm and fuses with the
chorion to give rise to the chorioallantoic placenta. Both the inside of the
chorion and the outside of the allantois are lined with mesoderm and in
this blood vessels arise forming the umbilical artery and vein. The foetal
heart drives blood up the umbilical artery and between foetal and maternal circulations a counter-current exchange system may develop (see
Fig. 7.6); food, water and oxygen are passed from the maternal circulation and carbon dioxide and other metabolites returned. There has been a
tendency to reduce the layers involved in the placenta in many orders of
mammals; thus, in our own case, the maternal epithelium, connective

tissue and endothelium are not present in the placenta and the foetal tissues project directly into the blood sinuses (Fig. 12.19). In such placentas the uptake of ions and presumably other substances is greatly speeded up and in this respect they are more efficient than those in which some of the layers become lost during development. There is no reason to suppose that species with a many-layered placenta are necessarily the most primitive, as the reduction of both maternal and foetal layers has taken place by parallel evolution in a number of diverging mammalian orders.

Milk secretion and parental care in the mammals
One of the features of the whole reproductive pattern of the mammals is the high level of parental care involved. This care goes a long way to ensure the survival of the young over the first, very vulnerable, stage of their lives and thus allows the production of many fewer offspring than are found in less advanced vertebrates.

A characteristic of the parental care exhibited by the female mammal is her ability to produce milk. The way in which this happens and the actual constituents of the milk are more sophisticated in the eutherians than in the monotremes and marsupials.

Reference to Fig. 12.14 shows that prior to and following birth the hormone prolactin is secreted from the anterior pituitary. Although the mammary glands have been enlarging under the influence of progesterone and oestrogens during gestation the milk is not actually secreted until prolactin is present. (This hormone is inhibited by progesterone and it will be recalled that progesterone levels decline dramatically at the time of birth.) The suckling of the newborn young produces, via a sensory input, stimulation of the hypothalamus–pituitary complex and further amounts of oxytocin are liberated. This hormone, that has already caused contractions of the uterus, is now implicated in contraction of the muscles of the glands which produce the milk. While the mother is actively feeding her young there is a reflex inhibition of LH by prolactin and so further ovulation is suppressed. As she stops feeding the prolactin level drops and the cycle of ovulation recommences.

Although milk must represent a balanced source of nutrition for the young mammal its actual chemical composition is quite variable and depends on the environment and evolutionary position of a given species. Some of the main sorts of variation are indicated in Table 12.3.

TABLE 12.3. *The composition of milk in various species of mammal*

	Water	Fat	Protein	Lactose	Minerals
Opossum (marsupial)	86.1	4.7	4.0	4.5	0.7
Human	88	3	1.2	6.5	0.2
Cow	69	3.5	3.4	4.8	0.7
Fin whale	42.7	42	12.6	1.3	1.4
Polar bear	57.4	31	10	0.4	1.3

Milk, especially that first secreted after birth of the young (the colostrum), contains antibodies and immunoglobulins. In some mammals such as deer, where the placenta does not allow the transfer of antibodies from the mother, the milk is the offspring's only source of antibodies, but in most orders of mammals transfer of antibodies and globulins occurs both *in utero* and via the milk.

Recent investigations into the comparative composition of the milk of mammals show that it can be used as a taxonomic characteristic which substantiates the classification derived from other data. It seems that marsupials, terrestrial carnivores and the artiodactyls (hoofed mammals with an even number of toes) have the primitive composition of about even amounts of the three main classes of food substance. Primates, such as ourselves, and the perissodactyls (hoofed mammals with odd numbers of toes), have milk proportionately high in sugar, while aquatic carnivores and the cetaceans (whales and porpoises) have milk with a very high fat content. As a general rule the more recently evolved mammals have more complex protein constituents in their milk than present-day survivors of earlier stocks.

12.6.7 *Other aspects of mammalian reproduction*

Besides the material selected for presentation in this chapter some further discussion relevant to mammalian reproduction is found elsewhere in the book. The changes involved in the circulation of the foetus immediately following birth are discussed on p. 129, the counter-current exchange system of the placenta on p. 79 and the role of the mother's antibodies in providing passive immunity for the newborn mammal on p. 158.

Index

Numbers in *italic* type indicate references that include illustrations